Vertebrate Photoreception

The Rank Prize Funds
Opto-electronics Biennial Symposia

1976 Very High Resolution Spectroscopy. Edited by R. A. Smith

1977 Vertebrate Photoreception. Edited by H. B. Barlow and P. Fatt

Vertebrate Photoreception

edited by

H. B. BARLOW

Physiological Laboratory
University of Cambridge

and

P. FATT

Department of Biophysics
University College London

1977

ACADEMIC PRESS

London New York San Francisco

A Subsidiary of Harcourt Brace Jovanovich, Publishers

ACADEMIC PRESS INC. (LONDON) LTD.
24/28 Oval Road
London NW1

United States Edition published by
ACADEMIC PRESS INC.
111 Fifth Avenue
New York, New York 10003

Library of Congress Catalog Card Number: 77-74373
ISBN: 0-12-078950-7

Printed in Great Britain by William Clowes & Sons Limited
London, Beccles and Colchester

List of Participants

ALPERN, Professor M., *Professor of Physiological Optics, 5044 Kresge II, University of Michigan Medical Centre, Ann Arbor, Michigan 48109, U.S.A.*

*ARDEN, Professor G. B., *Professor of Neurophysiology, Department of Visual Science, Institute of Ophthalmology, Judd Street, London WC1H 9QS.*

ASHMORE, Dr. J. F., *Associate Research Assistant, Department of Biophysics, University College London, Gower Street, London WC1E 6BT.*

†BARLOW, Professor H. B., F.R.S., *Royal Society Research Professor, Physiological Laboratory, University of Cambridge, Cambridge CB2 3EG.*

*BAYLOR, Dr. D. A., *Associate Professor of Neurobiology, Stanford University School of Medicine, Stanford, California 94305, USA.*

BEURLE, Professor R. L., *Professor of Electronic Engineering, University of Nottingham, University Park, Nottingham NG7 2RD.*

*BLAUROCK, Dr. A. E., *Research Chemist, California Institute of Technology, Pasadena, California 91125, U.S.A.*

BOWMAKER, Dr. J. K., *Research Fellow, MRC Vision Unit, University of Sussex, Falmer, Brighton BN1 9QY.*

BOYCOTT, Professor B. B., *Medical Research Council Scientist, MRC Cell Biophysics Unit, King's College, 26 Drury Lane, London WC2B 5RL.*

†BRADLEY, Professor D. J., F.R.S., *Professor of Optics, Imperial College of Science & Technology, London SW7.*

*BROWN, Professor J. E., *Professor of Physiology, State University of New York, Stony Brook, New York 11794, U.S.A.*

BROWN, Professor K. T., *Professor of Physiology, University of California, San Francisco, California 94143, U.S.A.*

*BYZOV, Professor A. L., *Head of Laboratory, Institute for Problems of Information Transmission, Academy of Sciences USSR, Aviamotornaja 8a, Moscow, U.S.S.R.*

CAVAGGIONI, Dr. A., *Assistant Professor, Instituto di Fisiologia Umana, Universita di Parma, Via Gramsci 14, C.A.P. 43100, Parma, Italy.*

CAVONIUS, Dr. C. R., *Chief Scientific Officer, Laboratorium voor Medische Fysica, Universiteit van Amsterdam, Herengracht 196, The Netherlands.*

CHABRE, Dr. M., *Maître de Recherches, C.N.R.S., Centre d'Etudes Nucleaires de Grenoble, Avenue des Martyrs, 38 Grenoble, France.*

*COLES, Dr. J. A., *Département de Physiologie de l'Université, Ecole de Médecine, 20 Rue de l'Ecole-de-Médecine, 1211 Geneve 4, Switzerland.*

*COPENHAGEN, Dr. D. R., *Assistant Research Physiologist, Department of Ophthalmology, University of California, San Francisco, California 94143, U.S.A.*

COX, Dr. A. F. J., *Principal Scientific Officer, Signals Research and Development Establishment, Ministry of Defence, Christchurch, Hants.*

*DAEMEN, Dr. F. J. M., *Associate Professor of Biochemistry, University of Nijmegen, The Netherlands.*

DAVY, Mr. J. R., *Director, Barr & Stroud Limited, Caxton Street, Anniesland, Glasgow G13 1HZ.*

DETWILER, Dr. P. B., *Research Fellow, Physiological Laboratory, University of Cambridge, Cambridge CB2 3EG.*

DONNER, Professor K. O., *Professor of Zoology, University of Helsinki, Arkadiankatu 7, Helsinki, Finland.*

ERNST, Dr. W., *Lecturer, University of London, Department of Visual Science, Institute of Ophthalmology, Judd Street, London WC1H 9QS.*

*FAIN, Dr. G. L., *Assistant Professor, Department of Ophthalmology, Jules Stein Eye Institute, University of California, Los Angeles, California 90024, U.S.A.*

*FALK, Dr. G., *Lecturer, Department of Biophysics, University College London, Gower Street, London WC1E 6BT.*

FARVIS, Professor W. E. J., *Head of Department of Electrical Engineering, University of Edinburgh, King's Buildings, Mayfield Road, Edinburgh EH9 3JL.*

*FATT, Professor P., F.R.S., *Department of Biophysics, University College London, Gower Street, London WC1E 6BT.*

FELLGETT, Professor P. B., *Department of Engineering and Cybernetics, University of Reading, 3 Earley Gate, Whiteknights, Reading RG6 2AL.*

*FETTIPLACE, Dr. R., *Research Assistant, Physiological Laboratory, University of Cambridge, Cambridge CB2 3EG.*

FORD, Dr. M. A., *Technical Director, Perkin-Elmer Ltd., Beaconsfield, Bucks HP9 1QA.*

‡FUORTES, Dr. M. G. F., *Head of Laboratory of Neurophysiology, National Institutes of Health, Bethesda, Maryland 20014, U.S.A.*

GERSCHENFELD, Professor H. M., *Research Professor, C.N.R.S., Laboratoire de Neurobiologie, Ecole Normale Supérieure, 46 Rue d'Ulm, 75230 Paris.*

*HAGINS, Dr. W. A., *Laboratory of Chemical Physics, National Institutes of Health, Bethesda, Maryland 20014, U.S.A.*

HALLETT, Professor P. E., *Professor of Physiology, University of Toronto, Toronto, Canada M5S 1AB.*

HAMDORF, Professor K., *Head of Vision Research Laboratory, Ruhr-Universitat Bochum, Postfach 2148, 463 Bochum, West Germany.*

HEMILA, Dr. S. O., *Department of Zoology, University of Helsinki, Arkadiankatu 7, Helsinki, Finland.*

HILLMAN, Professor P., *Associate Professor of Biophysics, Neurobiology Unit, Hebrew University, Jerusalem, Israel.*

HODGKIN, Sir Alan, K.B.E., F.R.S., *J. H. Plummer Professor of Biophysics, University of Cambridge, Cambridge CB2 3EG.*

*HUBBELL, Dr. W. L., *Associate Professor, Department of Chemistry, University of California, Berkeley, California 94720, U.S.A.*

†JONES, Dr. F.E., M.B.E., F.R.S., *Director of Unitech Limited, Reading, Berks.*

JONES, Dr. R. Clark, *Research Fellow in Physics, Polaroid Corporation, 750 Main Street, Cambridge, Massachusetts 02139, U.S.A.*

KANEKO, Dr. A., *Associate Professor in Physiology, Keio University School of Medicine, Shinanomachi, Shinjuku-ku, Tokyo 160, Japan.*

KEMP, Dr. C. M., *Senior Lecturer, University of London, Department of Visual Science, Institute of Ophthalmology, Judd Street, London WC1H 9QS.*

KING-SMITH, Dr. P. E., *Lecturer, University of Manchester Institute of Science and Technology, P.O. Box 88, Manchester M60 1QD.*

KIRSCHFELD, Professor K., *Director, Max-Planck-Institut fur Biologische Kybernetik, Spemannstrasse 38, 74 Tubingen, Germany.*

*KROPF, Professor A., *Professor of Chemistry, Amherst College, Amherst, Massachusetts 01002, U.S.A.*

KUIPER, Professor J. W., *Laboratorium voor Algemene Natuurkunde, Rijksuniversiteit, Westersingel 34, Groningen, The Netherlands.*

*LAMB, Dr. T. D., *ICI Research Fellow, Physiological Laboratory, University of Cambridge, Cambridge CB2 3EG.*

LAND, Dr. E. H., *Chairman of the Board and Director of Research, Polaroid Corporation, Cambridge, Massachusetts 02139, U.S.A.*

LAND, Dr. M. F., *Lecturer, School of Biological Sciences, University of Sussex, Falmer, Brighton, Sussex BN1 9QG.*

*LASANSKY, Dr. A., *Head of Section on Cell Biology, Laboratory of Neurophysiology, National Institutes of Health, Bethesda, Maryland 20014, U.S.A.*

LAWSON, Mr. W. D., *Research Scientist, Royal Signals and Radar Establishment, St. Andrews Road, Great Malvern, Worcs. WR14 3PS.*

LENNIE, Dr. P., *Lecturer, Physiological Laboratory, University of Cambridge, Cambridge CB2 3EG.*

LIEBMAN, Professor P. A., *Department of Anatomy, University of Pennsylvania, Philadelphia 19174, U.S.A.*

LISMAN, Professor J., *Department of Biology, Brandeis University, Waltham, Massachusetts 02154, U.S.A.*

LYTHGOE, Dr. J. N., *Research Fellow, MRC Vision Unit, University of Sussex, Falmer, Brighton BN1 9QY.*

MACFARLANE, Sir George, C.B., *Red Tiles, Orchard Way, Esher, Surrey.*

MACLEOD, Dr. D. I. A., *Assistant Professor of Psychology, University of California at San Diego, La Jolla, California 92037, U.S.A.*

MACNICHOL, Dr. E. F., *Director, Laboratory of Sensory Physiology, Marine Biological Laboratory, Woods Hole, Massachusetts 02543, U.S.A..*

MAFFEI, Professor L., *Investigator of C.N.R., Laboratorio di Neurofisiologia del C.N.R., Via S. Zeno 51, 56100 Pisa, Italy.*

MARCHIAFAVA, Dr. P. L., *Research Fellow at C.N.R., Laboratorio di Neurofisiologia del C.N.R., Via S. Zeno 51, 56100 Pisa, Italy.*

MASON, Mr. W. T., *Research Student, Physiological Laboratory, University of Cambridge, Cambridge CB2 3EG.*

MCCANN, Mr. J. J., *Senior Scientist, Polaroid Corporation, Cambridge, Massachusetts 02139, U.S.A.*

MEECH, Dr. R. W., *A.R.C. Principal Scientific Officer, Department of Zoology, University of Cambridge, Cambridge CB2 3EJ.*

MINKE, Dr. B., *Lecturer, Department of Physiology, Hebrew University, Hadassah Medical School, Jerusalem, Israel.*

MOLLON, Dr. J. D., *University Demonstrator, Psychological Laboratory, University of Cambridge, Cambridge CB2 3EB.*

MORTEN, Mr. F. D., *Advanced Development Manager, Electro-Optics Group, Mullard Limited, Millbrook Industrial Estate, Southampton, Hants. SO9 7BH.*

MOSS, Dr. T. S., *Head of Radio & Navigation Department, Procurement Executive, Ministry of Defence, Royal Aircraft Establishment, Farnborough, Hants.*

MUNTA, Professor W. R. A., *Professor of Biology, University of Stirling, Stirling FK9 4LA, Scotland.*

MURAKAMI, Professor M., *Professor of Physiology, Keio University School of Medicine, Shinanomachi, Shinjuku-ku, Tokyo 160, Japan.*

NICKEL, Dr. E. E., *Department of Biology, University of Konstanz, D-775 Konstanz, West Germany.*

NUDELMAN, Professor S., *Professor of Radiology and Optical Sciences, University of Arizona, Tucson, Arizona 85724, U.S.A.*

O'BRYAN, Dr. P. M., *Department of Physiology, Boston University Medical Centre, 80 East Concord Street, Boston, Massachusetts 02118, U.S.A.*

*OWEN, Dr. W. G., *Assistant Research Physiologist, Department of Anatomical Sciences, State University of New York, Stony Brook, New York.*

PIKE, Dr. E. R., *Deputy Chief Scientific Officer, Royal Signals and Radar Establishment, St. Andrews Road, Great Malvern, Worcs.*

*PINTO, Dr. L. H., *Associate Professor, Department of Biological Sciences, Purdue University, West Lafayette, Indiana 47907, U.S.A.*

POLDEN, Dr. P. G., *Research Assistant, Psychological Laboratory, University of Cambridge, Cambridge CB2 3EB.*

REUTER, Dr. T. E., *Research Fellow, Department of Physiological Zoology, University of Helsinki, Arkadiankatu 7, Helsinki, Finland.*

ROBSON, Dr. J. G., *University Lecturer in Physiology, Physiological Laboratory, University of Cambridge, Cambridge CB2 3EG.*

*ROSE, Professor A., *Distinguished Fairchild Scholar, California Institute of Technology, Pasadena, California 91125, U.S.A.*

RUDDOCK, Dr. K. H., *Lecturer, Applied Optics Section, Imperial College of Science & Technology, London SW7.*

*RUSHTON, Professor W. A. H., F.R.S., *Professor of Visual Physiology, Trinity College, Cambridge CB2 1TQ.*

SAIBIL, Miss H. R., *Department of Biophysics, University of London King's College, 26–29 Drury Lane, London WC2B 5RL.*

†SCHAGEN, Dr. P., O.B.E., *Head of Applied Physics Division, Mullard Research Laboratories, Redhill, Surrey RH1 5RF.*

SCHOLES, Dr. J., *Medical Research Council Scientist, MRC Cell Biophysics Unit, University of London King's College, 26–29 Drury Lane, London WC2B 5RL.*

*SCHWARTZ, Dr. E. A., *Research Fellow in Neurobiology, Harvard Medical School, Boston, Massachusetts 02115, U.S.A.*

‡SIMON, Dr. E. J., *NIH Fellow, Physiological Laboratory, University of Cambridge, Cambridge CB2 3EG.*

STIEVE, Professor H., *Head of Institute of Neurobiology, Nuclear Research Centre, P.O.B. 1913, D-517 Julich 1, West Germany.*

STILES, Dr. W. S., O.B.E., F.R.S., *89 Richmond Hill Court, Richmond, Surrey.*

TORRE, Dr. V., *Laboratorio di Neurofisiologia del C.N.R., Via S. Zeno 51, 56100 Pisa, Italy.*

*TOYODA, Professor J., *Professor of Physiology, St. Marianna University School of Medicine, Takatsu-ku, Kawasaki, Japan 213.*

*TRIFONOV, Dr. YU. A., *Scientific Fellow, Institute for Problems of Information Transmission, Academy of Sciences of USSR, Moscow, U.S.S.R.*

TSACOPOULOS, Dr. M., *Laboratory of Experimental Research, Hospital Cantonal, Clinique Universitaire d'Ophtalmologie, 1211 Geneve 4, Switzerland.*

VAN MEETEREN, Dr. A., *Institute for Perception, Kampweg 5, Soesterberg, The Netherlands.*

VERINGA, Dr. F., *Reader in Physiology, Physiologisch Laboratorium der Rijksuniversiteit van Groningen, Bloemsingel 10, Groningen, The Netherlands.*

VOS, Dr. J. J., *Head of Vision Branch, Institute for Perception, Kampweg 5, Soesterberg, The Netherlands.*

*WERBLIN, Professor F. S., *Professor, Graduate Group in Neurobiology, University of California, Berkeley, California 94720, U.S.A.*

WOODHEAD, Mr. A. W., *Deputy Divisional Head, Applied Physics Division, Mullard Research Laboratories, Redhill, Surrey RH1 5RF.*

YEANDLE, Dr. S., *Physicist, Naval Medical Research Institute, Bethesda, Maryland 20014, U.S.A.*

*YOSHIKAMI, Dr. S., *National Institutes of Health, Bethesda, Maryland 20014, U.S.A.*

---

* Contributors to this volume

† Members of the Advisory Committee on Opto-electronics

‡ deceased

The Lord Rank, J.P., LL.D., 1888–1972

Foreword

by

F. E. Jones M.B.E., F.R.S

Lord Rank, who died in 1972, was an outstanding industrial personality in this country. He built up his family's flour milling and bread business into one of the major British food manufacturers called Ranks Hovis McDougall. He also built up The Rank Organisation, starting in films and now incorporating a wide range of activities, including radio and television, scientific instruments, precision engineering and leisure activities, as well as having a close relationship with the Xerox Corporation of the United States.

Lord Rank was much concerned with the quality of life and provided for many charities during his lifetime. He endowed The Rank Prize Fund for Opto-electronics with a million-and-a-quarter pounds, and it is the Committee's responsibility to see how best the funds can be used to further the subject of opto-electronics.

In September 1974 we organised an International Symposium, along similar lines to this one, on the subject of "Very High Resolution Spectroscopy", some of you here also came to that Symposium. You may wonder how the Opto-electronics Committee decided to cover the field of physiology and the eye. In fact, it was not by chance, for when the Fund came into being Lord Rank specifically agreed to include in the terms of reference "the science of opto-electronics and *closely related phenomena*". The latter was added so that we could include the eye, and hence Professor Horace Barlow's membership of the Committee. Indeed, it was he, and his colleague Dr. (now Professor) Paul Fatt, who took on the responsibility for the programme of this Symposium. To both of them we all owe a debt of gratitude.

Preface

The papers in this volume were presented at a symposium sponsored by the Rank Prize Funds and held at the Royal Society on 2–3 September 1976. The Trustees of the Funds, Mr. James Hadley—the Executive Director, and his unflagging secretary, Miss Barbara Gookey, are to be thanked for making the meeting possible and helping it to run smoothly and enjoyably.

The title given to the symposium was the rather general one of "Photoreception". However, the subject covered was more limited than this might suggest. In order to keep the subject matter within reasonable bounds the decision was taken to confine the invited papers to vision among vertebrates. This was of course partly motivated by the special interest that attaches to processes operating in that group of animals to which humans belong. In addition enough is known about vision in different vertebrates such as carp, frog, turtle and cat to infer that the mechanisms are quite uniform in this group, so that much of what is learned about the receptors of one species is likely to apply to all vertebrates. In contrast, arthropods and molluscs, many of which also have highly developed vision, differ from vertebrates in essential features.

Another problem was to decide how far centrally from the visual pigment molecule (the photoreceptor on the narrowest view) the subject matter should extend. A long chain of events follows when light enters the pupil of the eye before the generation of a conscious visual sensation, and we wanted to concentrate on those links which are most closely related to the opto-electronics of physical light detectors. For this reason the papers presented here are mainly concerned with the events that happen between the absorption of quanta of light by photosensitive pigments (rhodopsin, porphyropsin, and cone pigments) and the occurrence of electrical changes in the photoreceptor cells (the rods and cones). Synaptic mechanisms linking receptors to the nerve cells of the retina (bipolar cells, horizontal cells, amacrines and ganglion cells) were briefly touched on, as were matters of visual perception, because experiments on human sensory performance provided the best means of analysing photoreception, until it became possible to record the activity of single cells in the retina.

An area of investigation which has been virtually omitted is the biochemistry of the photoreceptor cell, in particular enzymatic processes. The reason for this is that the subject has been treated at other recent symposia such as that published in *Experimental Eye Research*, Vol. 17, No. 6 and Vol. 18, Nos. 1 and 3.

The chapters in this volume are published in the sequence in which the papers were presented at the symposium, and they will, we hope, provide an orderly succession of ideas. Two subjects occupied a large share of the time and attention of the participants. One of these is the question of the existence and nature of the intracellular transmitter postulated to be released by light in the rod outer segment and to mediate changes in ionic permeability at the receptor cell's surface membrane. The other topic is the inherent noisiness of receptors and transmission: what are the sources of the noise, and what effects does it have on the detection of light-induced signals? For the sake of those who may be unfamiliar with these questions we shall devote a few paragraphs in this Preface to explaining their background.

Intracellular transmitter. The need to postulate this sprang from several experimental findings. In some types of receptor cells (the rods of all classes of vertebrates and the cones of mammals) very nearly all of the visual pigment is contained in the membranes of flattened sacs (discs) which are stacked in the interior of the outer segment. The membrane of the individual disc forms a closed surface, i.e. it is not continuous with the membranes of adjacent sacs or with the surface membrane. About 100 msec after light is absorbed by pigment in these disc membranes the internal negativity of the cell increases (a hyperpolarization) to an extent dependent upon the amount of light absorbed. The immediate cause of this voltage change is thought to be a reduction of the permeability of the cell membrane to sodium ions, for in darkness there is a steady inward flow of sodium ions through the membrane of the outer segment. This dark current results from a metabolically mediated outward flow of sodium taking place in the receptor inner segment.

With this information the obvious suggestion was made that light absorption by the visual pigment creates a pore in the disc membrane which permits the escape from the intradiscal space of some substance accumulated there by metabolic activity operating independently of light. This substance would then diffuse over to the surface membrane of the outer segment where it would act to close the channels that allow the inward passage of sodium ions in the dark. By analogy with other situations where an intracellular transmitter operates, in particular in the initiation of contraction of muscle and in the release of synaptic transmitter from pre-synaptic nerve terminals, it was suggested that calcium ions may serve in this role. Experimental findings which show a capability of vesicles derived from the discs of rod outer segments to accumulate calcium and others which show an increase in ionic permeability of the rod disc membrane when rhodopsin is bleached, provide strong support for the involvement of calcium ions at an early stage of the intracellular transmission process. The action of calcium has been further examined by measuring the electrical behaviour of the surface membrane of the photoreceptor cell under conditions in which the calcium ion activity of the cytoplasmic space of the outer segment is altered by substances that combine with calcium ions, or by the introduction of calcium ions themselves.

Whilst the evidence for the involvement of calcium as an intracellular transmitter is strong, there is a possibility that calcium does not act directly on the surface membrane but rather that it activates some enzyme or a chain of enzymes which, through the utilization of stored chemical energy, ultimately produces a chemical change at the surface membrane. One reason for suspecting such an indirect action of calcium is the rather long delay in the onset of the light-evoked change in electrical properties of the surface membrane. This is too great to be accounted for by diffusion of intracellular transmitter and has indeed been modelled by a sequence of chemical reactions.

Retinal noise. The second major topic—that of random fluctuation in visual signal transmission—had its origin in the realization that the visual system of the human observer is so sensitive that at very low light levels it must be influenced by unpredictable fluctuations in the number of photons absorbed in the retina. With the establishment experimentally that only about 10 photons have to be absorbed in widely separated rod photoreceptors in order for a visual sensation to be evoked, it was promising to postulate the existence of a "dark light". This is meant to represent randomly occurring events which produce central effects indistinguishable from quantal absorbtions. A photic stimulus must exceed fluctuations of dark light to be acknowledged at a central region of the nervous system, where a criterion of reliability of signal detection is applied. This interpretation was supported by the results of experiments on the threshold for the perception of flashes superimposed on a steady background.

One possible source of "dark light" is the spontaneous, thermally induced, chemical conversion of rhodopsin molecules along a path similar to that produced by the absorption of a photon. This is entirely reasonable since the level of "dark light" indicated by measurements of the threshold of vision would require these spontaneous conversions to occur with a first-order rate constant of only about 10^{-9} sec^{-1} (corresponding to one such event occurring every 10 sec in each human rod containing 10^8 molecules of rhodopsin). Another possible source of fluctuation, still in the outer segment, is in the opening and closing of individual channels in the surface membrane. The amplitude of fluctuations in surface membrane potential of the photoreception cells will depend on electrical characteristics of the cells, including the extent to which they are coupled to each other by electrical junctions.

Farther from the site of absorption of photons, an important source of fluctuation is expected to occur in the release of synaptic transmitter from specialized regions of the photoreceptor cell where contact is made with other cells in the outer plexiform layer. From studies of numerous other kinds of synapses one expects a synaptic transmitter to be released in discrete quantities (packets), producing in the post-synaptic membrane conductance changes of roughly predictable magnitude. The rate of release of such packets would correspond approximately to an exponential function of pre-synaptic membrane potential. Quantitative considerations lead to the conclusion that fluctuation in synaptic transmitter release will be a major source of noise in the visual system at low light levels, i.e. it will account for a major portion of the "dark light". An additional conclusion is that the mechanism that apparently operates in all vertebrate photoreceptor cells, whereby the membrane potential is at its lowest level of internal negativity in the dark-adapted condition and light produces hyperpolarization, will serve to maximize signal-to-noise for the detection of weak light stimuli.

The coupling together of receptor cells by electrical junctions which do not involve the intervention of a synaptic transmitter is related to this problem of synaptic noise. Such junctions can be seen to be necessary where there is synaptic convergence of photoreceptor cells onto second order neurones, for the following reason. Coupling allows the effect of the absorption of a photon in a single receptor to be spread over all the receptor cells contacting a given second order cell. A level of total synaptic activity in the second order cell can thereby be achieved which would otherwise not be possible, for without it the transmitter-release mechanism of the receptor cell receiving the photon would be driven to cut-off (i.e. hyperpolarization would be more than sufficient to produce complete cessation of transmitter release). The discussion following several papers will indicate, however, that the benefit derived from this interconnection is still not altogether agreed to.

The background to other matters arising in the material presented at the conference can, it is hoped, be obtained from the references in the papers themselves. Papers in a collection such as this inevitably lack the authority and completeness of full experimental reports; the first author is the one who presented the paper, and readers may write to him for clarification, if required.

We think that all those concerned with the symposium would wish to record our profound sense of loss at the tragic death, shortly after the meeting, of one of the speakers, our colleague Elliott Simon.

April 1977

H. B. Barlow
P. Fatt

Contents

Page

List of Participants v

Foreword by F. E. JONES xi

Preface xiii

1. Vision: Human versus Electronic. ALBERT ROSE. . . . 1

2. The Molecular Photochemistry of Vision. A. KROPF . . 15

3. Calcium and Rod Outer Segments. F. J. M. DAEMEN, P. P. M. SCHNETKAMP, TH. HENDRIKS and S. L. BONTING . 29

4. Molecular Anatomy and Light-Dependent Processes in Photoreceptor Membranes. WAYNE HUBBELL, KWOK-KEUNG FUNG, KEELUNG HONG and YONG SHIAU CHEN 41

5. What X-ray and Neutron Diffraction Contribute to Understanding the Structure of the Disc Membrane. A. E. BLAUROCK 61

6. Photosensitivity of Electrical Conductance of the Rod Disc Membrane. P. FATT and G. FALK 77

7. Intracellular Transmission of Visual Excitation in Photoreceptors: Electrical Effects of Chelating Agents Introduced into Rods by Vesicle Fusion. W. A. HAGINS and S. YOSHIKAMI 97

8. Three Components of the Photocurrent Generated in the Receptor Layer of the Rat Retina. G. B. ARDEN . . . 141

9. Mechanism for the Generation of the Receptor Potential of Rods of *Bufo marinus*. L. H. PINTO, J. E. BROWN and J. A. COLES 159

10. Characteristics of the Electrical Coupling between Rods in the Turtle Retina. W. GEOFFREY OWEN and DAVID R. COPENHAGEN 169

11. Transmission from Photoreceptors to Ganglion Cells in the Retina of the Turtle. D. A. BAYLOR and R. FETTIPLACE . 193

12. Synaptic Interactions Mediating Bipolar Response in the Retina of the Tiger Salamander. F. S. WERBLIN 205

13. Responses of Second-order Neurons to Photic and Electric Stimulation of the Retina. J. TOYODA, M. FUJIMOTO and T. SAITO 231

14a. The Interaction in Photoreceptor Synapses Revealed in Experiments with Polarization of Horizontal Cells. YU. A. TRIFINOV and A. L. BYZOV 251

14b. The Model of Mechanism of Feedback between Horizontal Cells and Photoreceptors in Vertebrate Retina. A. L. BYZOV, K. V. GOLUBTZOV and YU. A. TRIFONOV 265

15. Synaptic Organization of Retinal Receptors. ARNALDO LASANSKY 275

16. Electrical Noise in Turtle Cones. E. J. SIMON and T. D. LAMB 291

17. The Threshold Signal of Photoreceptors. GORDON L. FAIN 305

18. Comparison of the Voltage Noise and the Response to One Photon in the Rods of the Turtle Retina. E. A. SCHWARTZ. 323

19. Retinal and Central Factors in Human Vision Limited by Noise. H. B. BARLOW 335

20. How does Your Research Explain our Inability to See? W. A. H. RUSHTON 357

Index 372

1

Vision: Human versus Electronic

ALBERT ROSE

California Institute of Technology, Pasadena, California, U.S.A.

Introduction and Historical Survey

It is now some thirty-five years since I first approached this subject[1] and almost thirty years since the appearance of my first attempt[2] to relate much of the data on visual acuity to the fundamental limitations imposed by the quantum fluctuations of the incoming light. These concepts of "photon noise" were readily adopted by electronic physicists[3] working with electronic imaging systems and by a small group of radiologists faced with the somewhat more obvious constraints of X-ray quanta.[4] At the same time, the invasion of these concepts into the biological literature on vision has been sufficiently restrained to give some point to including this paper in a 1976 Symposium on Photoreception.

In the light of the cautious acceptance of the photon-noise limitations on vision, it seems to me worthwhile to extend this introduction, even at the expense of the body of the paper, to give some historical perspective to the subject. Moreover, the body of the paper is dealt with in more detail in reference 5.

The first evidence of the quantum limitations on vision date back to the observations by astronomers in the early 1900's on the faintest visible stars.[6] The flux of photons entering the eye for these threshold point sources, multiplied by the storage time of the eye, yielded the empirical fact that some 100 photons were sufficient to give a threshold sensation. This set a lower limit of 1% on the quantum efficiency of the eye since at least one photon had to partake in the process.

In the early 1940's, three papers[7,8,9] appeared on the subject of how many of the 100 incident photons are actually used by the eye in generating a sensation—in brief, the quantum efficiency at absolute threshold. Also, two papers[1,10] appeared in which the criterion for visual detection over the whole range of light levels was proposed in terms of a "signal equal to the noise" where the noise was the shot noise of the photons used by the eye in detecting any elementary test object.

The three papers on the quantum efficiency at the absolute threshold all used the same statistical concept that the sharpness of the transition from "not-seeing" to "seeing" as a function of the number of photons in the light flash is a measure of the number of photons actually used by the eye. For example, if the eye used all 100 photons, the transition would take place mainly in the range of 90–110 photons, that is in the range of the rms fluctuation of 100 random events. If the eye used only one out of the 100, the transition would be quite broad, namely, in the range of about 50–200 photons incident or 1–2 photons on the average used. The concept of the experiment was straightforward. The results, however, ranged from 2 photons (van der Velden) to 7 photons (Hecht) to 25 photons (Brumberg *et al.*).

The concept of the experiment is also akin to measuring the quantum efficiency of a photo multiplier by comparing the signal-to-noise ratio of the current out of the multiplier to the signal-to-noise ratio of the photon current into the multiplier.* A physicist or electronic engineer would carry out such an experiment not at absolute threshold where the results can be contaminated by artifacts of system noise like sparking, ion currents and thermionic emission but rather at a high light level where the shot noise of the photons is large enough to dominate the system noise. It is possible that the artifacts of system noise in the visual system were responsible for the wide divergence of quantum efficiencies cited above.

DeVries[10] and Rose[1] each emphasized observations at light levels well above threshold and extending into the range of 1–100 ft lamberts. However, both authors made the common (common to them and others) error of taking a threshold signal-to-noise ratio of only unity. There were other shortcomings in each of the papers. Nevertheless, they both carried the seed of the concept of photon-noise limitation even at the higher light levels.

* The relation is:

$$\text{Quantum efficiency} \equiv 100 \times \text{quantum yield} = 100 \times \frac{\left(\frac{S}{N}\right)^2_{\text{out}}}{\left(\frac{S}{N}\right)^2_{\text{in}}}.$$

It was not until 1948[2] that the acuity data of the eye were analyzed in detail using a more realistic threshold signal-to-noise ratio of 5. (Note that the choice of unity rather than 5 *reduces* the computed quantum efficiency of the eye by a factor of 25.)

The 1948 paper was at first firmly rejected by the reviewer and later accepted by the editor, G. R. Harrison, of the Journal of the Optical Society of America. I mention this for two reasons. First, the reviewer who rejected the paper was, oddly enough, Selig Hecht, even though the essential content of the paper was simply an extension of his own quantum noise arguments towards light levels well above absolute threshold. Second, the statements by the reviewer overlap the comments I find currently, some 25 years later, in several major texts on vision. The following are selected quotes from the reviewer.

(a) ". . . the paper is based on measurements selected to prove a point."
(b) "The paper specifically assumes on p. 6 that the rate of absorption of light by the retina is directly proportional to the intensity and therefore that the absorption coefficient is not a function of prevailing light intensity.

This assumption is contrary to fact. For one thing the rods of the retina contain visual purple which is bleached by light; it has been known since Kuehne's day that a retina in the dark contains a high concentration of visual purple, *whereas a light-adapted retina has most of its visual purple bleached.*" (The italics in this quotation are mine.) ". . . The author's assumption of a constant absorption coefficient for the retina is therefore untenable, and with its removal one of the props of his argument collapses. This makes unnecessary the explanation of dark adaptation in terms of a 'gain control mechanism' in which adaptation is described as 'the resetting of the gain control . . .'. Who sets what and how?"
(c) "The positive error in the paper is that its basic tenet is contradicted by all the available measurements."
(d) "Only by neglecting most of the visual data and confining oneself to a small region of brightness can one draw a short line of slope $\frac{1}{2}$. In other words, the basic idea of the author depends for its acceptance on an accidental grouping of points and a neglect of the rest of the information."

I have made an attempt in these quotes to convey the major flavor of the reviewer's three pages of critique and to avoid any conscious distortion.

Reviewer's comment (b) has now only a historic interest. The major conclusion of my 1948 paper was to the effect that "the bleaching of the visual purple" must be ruled out on absolute grounds as the major source of the observed factor of 10^3–10^4 by which the threshold is lowered during dark adaptation. Stated in the opposite sense, if the absorption by the visual purple were reduced by a factor of 10^3–10^4 during light adaptation, it would be impossible to "see" what we do in fact "see" at, say, 1–10 ft lamberts. The signal-to-noise ratio of the absorbed photons would be too small by several orders of magnitude to transmit the information we apprehend in this range of brightnesses. Hence, "dark adaptation" must indeed be referred to an "automatic gain control mechanism". Since 1948, the direct observations by Rushton[11] on bleaching of the visual purple (or lack of significant bleaching) have confirmed the essential constancy of the light absorption by the retina in the range of 10^{-6}–10^{2} ft lamberts. Nevertheless, there is a lingering flavor in the literature on vision that the light adapted eye is somehow orders of magnitude less sensitive than the dark adapted eye.

The reviewer's comments (a), (c) and (d) have been reproduced here because they are reflected in the comments I find in several recent texts on vision. Two examples are:

T. N. Cornsweet, *Visual Perception*, Academic Press (1970), p. 88.

> "To summarize, then, while quantal fluctuations play a major role in visual phenomena at very low intensities, their importance diminishes rapidly as the illumination level increases, until at moderate or high levels, their effects are negligible. Thus the results of experiments like that of Cornsweet and Pinsker indicate that the *sensitivity of the visual system itself is proportionately reduced as the illumination is increased.*" (The italics are mine.)

J. L. Brown and C. G. Mueller in "Vision and Visual Perception" (C. H. Graham, Editor) John Wiley, 1965, pp. 228–229.

> "None of these quantum formulations of brightness discrimination describes the data over a large range of adapting luminances."

The tenor of Hecht's remarks and those of recent texts on vision is that the photon noise theories fail to explain *all* of the data on visual acuity and contrast discrimination and, hence, their validity is questionable—except, perhaps, near absolute threshold.

My answer to these remarks is now as it was in 1948:

> "Yes, the data to be compared with photon-noise limitations must, indeed, be selected in order to exclude those limitations on vision that are obviously not related to photon noise. These limitations are (1) the geometric limit on acuity of about 2 minutes of arc imposed by the finite size of the rods and cones, (2) the Weber–Fechner contrast discrimination limit of one or two percent which must be set by some system constraints such as system noise or the finite rate of transmission of nerve pulses, and (3) the largest retinal area over which photons can be summed and which is presumably set at about 100 minutes of arc by the pre-wiring of the rods into larger groups near the periphery of the retina." These exclusions leave the data from 3–100′ of arc, 2–100 % contrast and 10^{-6}–10^{2} ft lamberts to be compared with the limitations imposed by photon noise.*

Such a comparison[5] yields the outside limits to the variation of quantum efficiency, 10–1 %, in going from 10^{-6} to 10^{2} ft lamberts. These outside limits are based on the assumptions of a constant storage time, 0·2 s, and a constant

* Note that one can choose to try to describe the full range of visual data by a single mathematical expression as has Schnitzler.[12] The expression, however, contains the three arbitrary (empirical) parameters reflecting the three physically unrelated constraints described above.

threshold signal-to-noise ratio of 5. Recent data on the storage time[13] point to a factor of 2 reduction towards 0·1 sec in the range of 1–100 ft lamberts. It is also reasonable to expect smaller threshold signal-to-noise ratios (see p. 13 of ref. 5) in the neighborhood of 3–4 at very low scene brightnesses. These two corrections would reduce the range of quantum efficiencies from 10–1 % to about 5–2 % and would more nearly match the recent work by R. W. Engstrom[14] and by A. van Meeteren[15]. Their work, incidentally, is in the spirit of how further measurements on the quantum efficiency should be made—namely, by side by side comparison with an electronic system of known quantum efficiency.

A criticism of some of the assumptions leading to high quantum efficiencies will be found in Barlow's paper[16] in this volume. However, the remarks of Hecht and of Cornsweet both imply orders of magnitude change in quantum efficiency, and this is incorrect on any basis.

A few concluding remarks are appropriate to this historical survey. First, it was altogether natural in the 1940's to reach for a photon-noise model of visual performance. I was then engaged in developing television camera tubes to try to match the performance of the eye. The performance of our camera tubes was obviously limited by noise—usually amplifier noise. Even when the noise limitation approached that of photon noise, as it did in The Image Orthicon,[17] the performance of the camera tube only succeeded in matching that of the eye. There was almost no choice but to look for photon-noise-limited performance in the human eye.

Second, while the noise limitations are more or less obvious in electronic systems, the appearance of noise in the visual system is only marginal. Moreover, it is somewhat of a psychological barrier to believe that the information transmitted to the eye at, say, 10 ft lamberts is photon-noise-limited when 10^{14} photons/cm^2 s are incident on the eye. The number appears on the face of it too large to warrant any concern about the particle nature of light. We will attempt in the next section to clarify both these points since they present major obstacles to an easy acceptance of photon-limited visual performance.

Evidence for Photon-Noise-Limited Performance of the Eye

A detailed accounting of the photon-noise-limited performance of the eye is given in reference 5. The present dicussions select three aspects that can be simply presented and can convey the argument almost "at a glance".

Figure 1 is a series of photographs[18] of a subject designed to show the maximum information that can be transmitted by any imaging system, including the eye, subject to the following conditions: exposure time 0·2 s, diameter of lens opening 6 mm, quantum efficiency 10 % and distance of

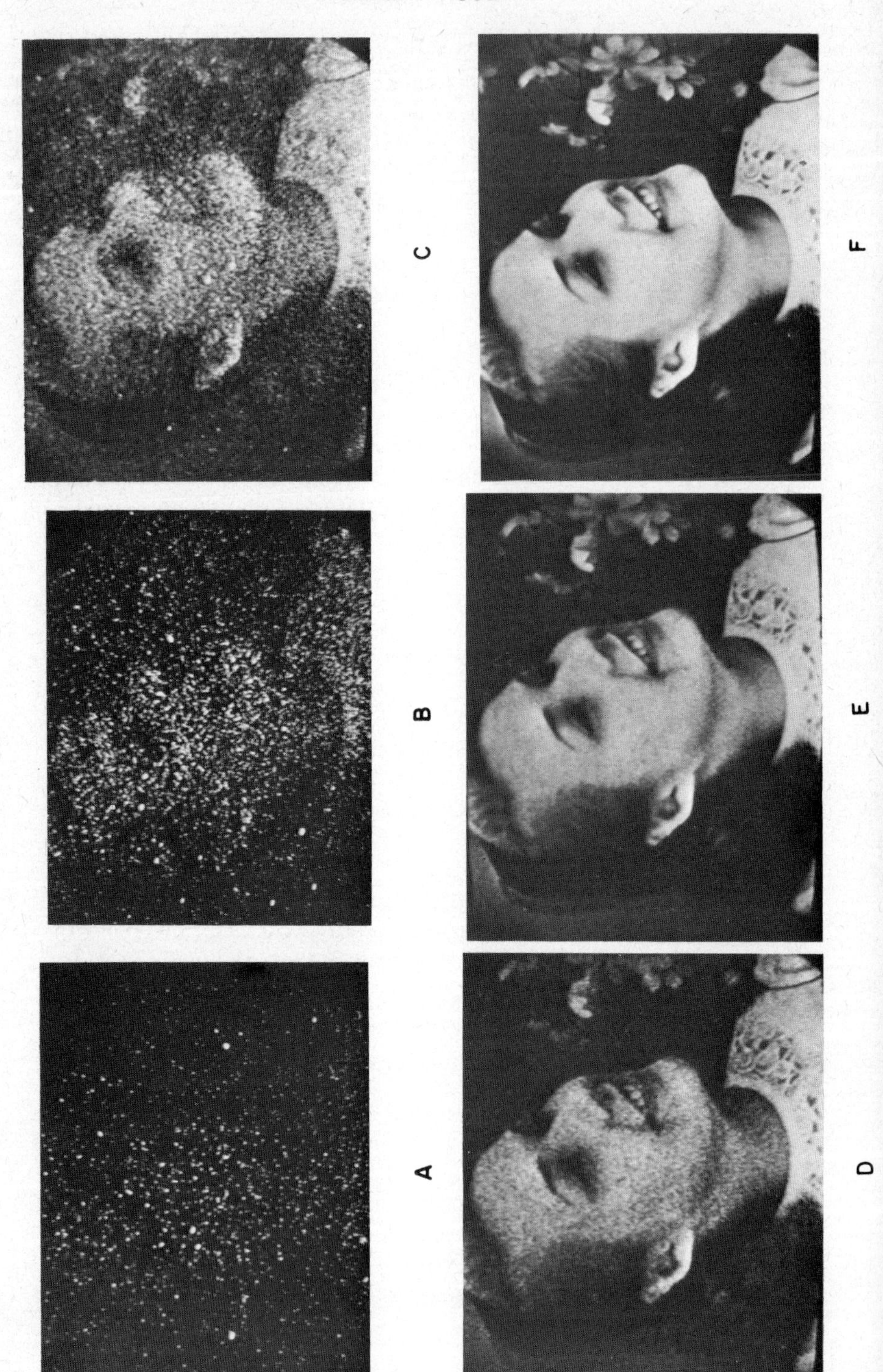
A
B
C
D
E
F

viewer from the subject 4 ft. The series covers the range of scene brightnesses from 10^{-6}–10^{-2} ft lamberts, that is, from absolute threshold to full moonlight. The total number of photons in each picture is computed as if the entire picture had the brightness of the high light areas. The series was obtained using a light spot scanner and a photomultiplier. The details are not important other than that the white dots in the first few photographs are literally traces of individual photons.

The photon-noise limitations are clearly evident in this series. Also, the reader has an opportunity to make a rough check of his own visual quantum efficiency. If what he sees does indeed match this series, his quantum efficiency in this range is about 10%. If he sees more or less, the quantum efficiency is proportionately more or less than 10%.

In reference 2, I made use of just such a comparison using a scientifically more acceptable test pattern to cross check the quantum efficiencies I had computed analytically from data in the literature.

We jump at this point to a high light brightness of 10 ft lamberts and make a spot calculation to show that the quantum efficiency of the eye is about 2%. The number of photons required to see a contrast C in a single picture element of a picture is[5]

$$N_{\text{photons}} = \frac{25}{C^2\theta} \qquad (1)$$

$C \equiv \Delta B/B$ and $0 \leq C \leq 1$ corresponds to the range of contrast from zero to 100%. θ is the quantum yield, $0 \leq \theta \leq 1$ and corresponds to the range of quantum efficiencies from zero to 100%. From the data of Blackwell,[19] 2% contrast is visible at 10 ft lamberts for a test element subtending 10 min of arc at the eye. Using a pupil opening of 0·2 cm^2 for two eyes and a storage time of 0·2 s, eqn (1) yields a quantum efficiency of 2% for green light.

Fig. 1. Series of pictures showing the dependence of picture quality on the number of photons (see text for details).

Figure	No. of photons	Equivalent scene brightness (foot-lamberts)
A	3×10^3	10^{-6}
B	$1{\cdot}2 \times 10^4$	4×10^{-6}
C	$9{\cdot}3 \times 10^4$	3×10^{-5}
D	$7{\cdot}6 \times 10^5$	$2{\cdot}5 \times 10^{-4}$
E	$3{\cdot}6 \times 10^6$	$1{\cdot}2 \times 10^{-3}$
F	$2{\cdot}8 \times 10^7$	$9{\cdot}5 \times 10^{-3}$

On these assumptions, the eye needs a quantum efficiency of at least 2% to see what Blackwell's subjects reported at 10 ft lamberts for 2% contrast in test elements subtending 10 min of arc. The eye could have a larger intrinsic quantum efficiency and not make full use of it so that its effective quantum efficiency was 2%. It is also true that test elements of other angular sizes and other contrasts may yield quantum efficiencies less than 2% owing to the invasion of constraints other than photon noise. These facts do not alter the conclusion that at 10 ft lamberts the eye must display a quantum efficiency of 2%. We remark again that the comparison of this quantum efficiency with the 5 or 10% value reported at absolute threshold, a brightness 10^7 times smaller than 10 ft lamberts, is in strong contrast with frequent statements in the literature citing a "proportionate drop in sensitivity as the scene brightness is increased".

Fig. 2(a). Original picture taken on a 2″ × 2″ slide.
Fig. 2(b). Noise-free 525 line 4·25 megacycle television reproduction of original.
Fig. 2(c). Reproduction of the original on Tri-X film in super-8 mm format. Figs. 2(a)–2(c) are reprinted from *Image Quality* by Otto Schade and published by RCA Scientific Press, Princeton, New Jersey, 1974.

(b)

(c)

The third readily available test for the photon-noise-limited performance of the eye is also applicable to the high light range of 10–100 ft lamberts. In this test, one looks at a television picture that is slightly noisy. The brightness of modern television receivers is in the neighborhood of 100 ft lamberts. The interposition of a neutral filter at the eye that attenuates the light by a factor of 10 will markedly reduce or eliminate the noise. Noise is a particularly useful test pattern since it presents an array of small area, low contrast test elements in just the range where the photon-noise-limited performance of the eye is most pronounced. This is a qualitative test in that the disappearance of the noise when viewing the television receiver through the neutral filter means that the signal-to-noise ratio of the photons at the retina is less than the signal-to-noise ratio of the television picture itself. In brief, it attests to the presence of the photon-noise-limited performance of the eye. The test can also be used to measure the quantum efficiency of the eye once the brightness and signal-to-noise ratio of the television picture itself. In brief, it attests to the presence of efficiencies computed above of 1 or 2%.

It is worth noting in passing that the recognition of textures, a highly developed pattern-recognition-faculty of the human visual system, is dependent upon the visibility of small area, low contrast elements.

Human versus Electronic Vision

The last experiment cited, using a television receiver, leads naturally into an area of considerable interest currently and in the near future.

The problem, stated briefly, is "How large a quantum efficiency must a camera have in order to record motion pictures at the ambient indoor lighting levels of about 10 ft lamberts which will be realistic and noise-free when projected at 100 ft lamberts?" This is the growing culture of "available light" photography in which pictures are recorded without the annoyance of adding supplementary lighting and are viewed in the presence of ambient light (as are television pictures) thereby avoiding the annoyance of "pulling the shades" or "turning off all the lamps".

On the face of it, one would expect a camera to have at least the sensitivity of the eye. In the present case, where the picture is projected at some ten times the brightness at which it was recorded, the quantum efficiency of the camera should significantly exceed that of the eye because the viewer has the advantage of being visually more critical at the higher brightness level. (This will be recognized as the inverse of the experiment in which television noise is eliminated by interposing a neutral filter at the eye of the observer.)

A further factor of about ten increase of the quantum efficiency of the camera relative to the eye is needed in order that the camera reproduce the realism conveyed by the large dynamic depth of focus of the eye. For a human observer, the world around him is always in sharp focus since the eye refocuses quickly and automatically to whatever is of central interest—whether near or far. The camera, in order to reproduce the effect of a world always in focus, must accomplish this statically, not dynamically, by using a lens with a large depth of focus and, therefore, smaller lens opening. The camera must make up for this loss of light by a corresponding increase in its quantum efficiency.

These two factors suggest the need for a camera whose quantum efficiency is as much as 100 times larger than that of the eye.[20] It is accordingly of central importance to know the magnitude of the quantum efficiency of the eye. If we take 1 or 2% for the visual quantum efficiency, that of the camera should approach 100%. Modern electronic cameras have good promise of meeting this goal. This is in contrast to the quantum efficiency of photographic film which is somewhat less than 1%. The inadequacy of the silver halide system for available light photography is shown in Fig. 2. The noise-free television picture of Fig. 2(b) is representative of the performance of modern television cameras under ambient light. Tri-X film was selected for Fig. 2(c) in order to have sufficient photographic speed to transmit motion pictures at scene brightnesses of a few foot lamberts. The 8 mm format was chosen in order to match the 4 mm size picture on the human retina when viewing television or motion pictures at the conventional distance of four times the vertical height of the picture.

References

1. A. Rose 1942, The relative sensitivities of television pickup tubes, photographic film and the human eye, *Proc. I.R.E.* **30**, 293–300.
2. A. Rose 1948, The sensitivity performance of the human eye on an absolute scale, *J. Opt. Am.* **38**, 196–208.
3. R. Clark Jones 1959, Quantum efficiency of human vision, *J. Opt. Soc. Am.* **49**, 645–653.
4. R. E. Sturm and R. H. Morgan 1949, Screen intensification systems and their limitations, *Am. J. Roentgenol. Radium Therapy* **62**, 617–634.
5. A. Rose 1974, "Vision: Human and Electronic", Plenum Press, N.Y.
6. P. Reeves 1918, The effect of size of stimulus and exposure-time on retinal threshold, *Astrophys. J.* **47**, 141–145.
7. S. Hecht 1942, The quantum relations of vision, *J. Opt. Soc. Am.* **32**, 42–49. S. Hecht, S. Shlaer and M. H. Pirenne 1942, Energy, quanta and vision, *J. Gen. Physiol.* **25**, 819–840.
8. E. M. Brumberg, S. I. Vavilov and Z. M. Sverdlov 1943, Visual measurements of quantum fluctuations, I. The threshold of vision as compared with the results of fluctuation measurements, *J. Phys.* (*USSR*) **7**, 1–8.
9. H. A. van der Velden 1944, Concerning the number of light quanta necessary for a light sensation in the human eye, *Physica* **11**, 179–189.
10. H. deVries 1943, The quantum character of light and its bearing upon threshold of vision, the differential sensitivity and visual acuity of the eye, *Physica* **10**, 553–564.
11. W. H. Rushton 1956, Rhodopsin density in the human rods, *J. Physiol.* **13**, 30–46.

12. A. D. Schnitzler 1973, Image detector model and parameters of the human visual system, *J. Op. Soc. Am.* **63**, 1357–1368.
13. J. J. Mezrich and A. Rose 1975, Suprathreshold estimate of temporal summation using dynamic visual noise, ARVO Meeting, Sarasota, Florida.
14. R. W. Engstrom 1974, Quantum efficiency of the eye determined by comparison with a TV camera, *J. Opt. Soc. Am.* **64**, 1706–1710.
15. A. van Meeteren 1973, "Visual Aspects of Image Intensification", Institute for Perception TNO, Soesterberg, The Netherlands.
16. H. B. Barlow 1977, Retiral and central factors in human vision limited by noise. This volume, pp. 335–355.
17. A. Rose, P. K. Weimer and H. B. Law 1946, The Image Orthicon—a sensitive television pick-up tube, *Proc. I.R.E.* **34**, 424–432.
18. A. Rose 1953, Quantum and noise limitations of the visual process, *J. Opt. Soc. Am.* **43**, 715–716.
19. H. R. Blackwell 1946, Contrast thresholds of the human eye, *J. Opt. Soc. Am.* **36**, 624–643.
20. A. Rose 1976, The challenge of electronic photography, *J. Appl. Phot. Eng.* **2**, 70–74.

Discussion

E. F. MacNichol, Jr.: The quantum efficiency of the receptors is probably much greater than approximately 10% that Hecht, Schlaer and Pirenne, and others, who have measured the absolute thresholds, have estimated. Brindley (1953), Walraven and Bowman (1960) and Walraven (1966) have made psychophysical measurements which indicate that 80–90% of the light (O.D. = 0·7–0·98) would be absorbed in the receptors.

Spectrophotometric measurements on single rods and cones made by Denton and Wyllie, Wald and Brown, Marks, Liebman, Hárosi, and others, give specific densities of from 0·007–0·0177 per micrometer (MacNichol *et al.*, 1973; Hárosi, 1975). The more recent the measurement, the better the techniques, and the larger the receptor (for technical reasons), the higher the measured specific density. Even the low figure of 0·007 measured by Dobelle *et al.* (1969) made with a human foveal cone only 0·8–1·0 μm in diameter (a very difficult measurement) would give an absorption of 52% of the light in a receptor outer segment 45 μm long. The highest figure of 0·0177 (Hárosi, 1975) would yield an absorption of 80% of the light in a receptor of the same length. Thus, there is ample evidence that photoreceptors are much better quantum catchers than was formerly supposed.

A. Rose: I agree that modern data point to an absorption by the rods and cones of 50% or more of the light incident on them and that this absorption is appreciably higher than the 10% or 20% taken from the older literature. M. Alpern and E. N. Pugh (1974) also support the 50% figure. However, if one asks what fraction of the light incident on the cornea is absorbed by the rods, the 50% figure is immediately reduced to 25% on the assumption that only half the light reaches the retina. Further, the 25% should be reduced somewhat owing to the fact that some of the light falls between neighbouring rods.

Finally, whatever value one arrives at for the absorption of light only sets an upper limit to the quantum efficiency of the eye. The latter may be still smaller since it measures only those quanta that contribute to the information content of the image. Modern photocathodes, for example, absorb about 50% of the light but have a quantum efficiency of only 10%.

It has been noted in the literature that the reason the cat's eye reflects light back through its retina is that this enhances its sensitivity by increasing the absorption of light due to the double passage. Such a mechanism is significant if the single passage gives a low absorption but the improvement becomes very small if the single passage absorption rises above 50%. Perhaps "seeing like a cat" has more to do with his short legs enabling him to put his eyes close to the object of interest than it does to his reflecting tapetum.

P. B. Fellgett: A very English illustration of quantum limitations of vision is when "bad light stopped play" (cricket of course). A photon rate that is perfectly adequate for ordinary purposes may give the batsman insufficient visual information in the short time between the delivery and the moment he must make his decision how to play the ball.

A. Rose: It is interesting that when the light level falls too low for the players, who are dependent on the quantum efficiency of the visual process, the television camera, owing in part to its higher quantum efficiency, can still transmit a good picture. Part of the advantage of the television camera must also be ascribed to its larger lens opening.

References to discussion comments of E. F. MacNichol and A. Rose

M. Alpern and E. N. Pugh 1974, *J. Physiol.* **237**, 341–370.
G. S. Brindley 1953, *J. Physiol.* **122**, 332–350.
Dobelle, Marks and MacNichol 1969, *Science* **166**, 1508–1510.
Hárosi 1975, *J. gen. Physiol.* **66**, 357–382 (Table 2, p. 364).
MacNichol, Feinberg and Hárosi 1973, *In* "Colour 73" (ed. R. W. G. Hunt), Adam Hilger, London, Table 1, p. 206.
P. L. Walraven 1966, *Nature* **206**, 311–312.

2

The Molecular Photochemistry of Vision

A. KROPF

Department of Chemistry, Amherst College, Amherst, Massachusetts, U.S.A.

Introduction

Probably the most thoroughly studied of the sensory processes is vision, and this includes the entirety of the hierarchy of visual processes, from elementary photochemistry to psychophysics. Though the dominating role of vision for our own perception of the world is undoubtedly a key factor in this continuing interest, the relatively rapid progress which has been achieved in understanding the bases for sensory transduction and neural information processing seems to have had an autocatalytic effect on research efforts.

An area which is of immense current interest and which also relates to vision is that of membrane structure and function. Membranes are ubiquitous structures in cells and are thought to be essential for such key cellular processes as cell-cell interaction, permeability and transport and excitability, to name a few. Included in such a list should be sensory transduction, especially since it is the study of this process which has placed the visual cells, especially the rod and cone cells of the vertebrate retina, at the center of interest of membrane biophysics.

Visual photochemistry may seem like a minor aspect of the overall effort to understand how a photosensitive membrane system can transduce photon capture into a reliable electrochemical signal, but this may be a result of too strong an anthropomorphic bias, particularly in the temporal realm. In another context Kamen[1] has called this bias temporal solipsism, and in Fig. 1

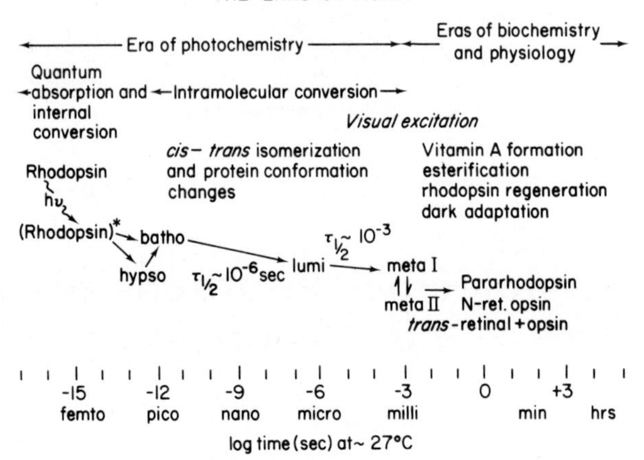

Fig. 1. The time course of rhodopsin bleaching and its relationship to molecular events and biochemical and physiological reactions. Visual excitation refers to the earliest recorded electrical response in a retina.[26] The reaction times are order of magnitude estimates and vary from species to species as well as for visual pigments in different *in vitro* or *in vivo* environments. The sequence of formation of the products of bleaching (pararhodopsin, *N*-retinylidene opsin and *trans*-retinal + opsin) is still unresolved.[27]

Kamen's viewpoint has been adopted in order to emphasize the importance of the basic physico-chemical events of visual excitation.

With Fig. 1 as a preface, it may be easier to state the problem that will be discussed below. The problem is that of constructing a molecular model for the events occurring during the "era of photochemistry" as well as quantifying the photochemical reactions and their rates. Though the analysis may appear at times to be abstract and though there are far too few data available for the formulation of a definitive molecular photochemical scheme, a model can be presented which is consistent with experimental findings and which can provide a context for continuing refinement.

Energy Analysis

The chain of molecular events constituting visual photochemistry can be abstractly conceived of as the motion of a point on an appropriate set of potential energy surfaces. The initial set of coordinates in this multidimensional space corresponds to an 11-*cis* chromoprotein ground state, and the coordinate point, specifying the position of each atom in the visual pigment, progressively moves through a series of photochemical and thermal intermediates until it terminates at those coordinates corresponding to an all- *trans* chromophore detached from its apoprotein, opsin. Though the

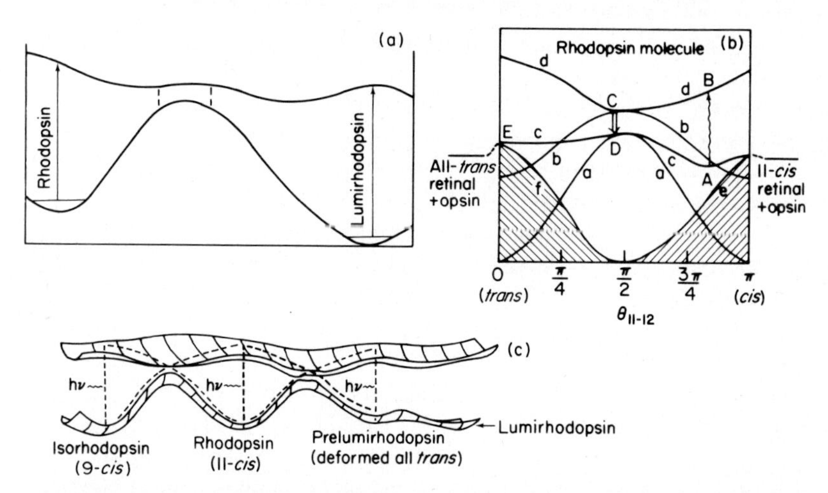

Fig. 2. Calculated potential energy curves for the photoisomerization of rhodopsin. In diagram (a) Abrahamson and Ostroy[2] show the ground state and first excited singlet state of rhodopsin and lumirhodopsin at room temperature. Presumably there is a crossing over from the excited state curve to the ground state curve in the region between the vertical dashed lines. Diagram (b), calculated by Kakitani and Kakitani,[3] shows potential energy curves for the ground and excited states of the Schiff base of retinal, curves a and b respectively, and the ground and excited states of rhodopsin, c and d respectively. Curves e and f are potentials imposed upon the Schiff base by opsin. Symbols ⇝ and ⇒ represent respectively radiative and non-radiative transitions. θ_{11-12} is the angle of rotation about the 11–12 double bond. Warshel's bicycle-pedal potential energy surface[4] is shown in diagram (c). The thick dashed line describes the initial photochemical step in the bleaching sequence, while the thin dashed lines describe other possible phototransitions. The designation prelumirhodopsin is synonymous with bathorhodopsin.

construction of an accurate set of potential surfaces is a task beyond the capabilities of present theoretical or experimental methods, there have been some limited attempts which have incorporated certain experimental observations, such as photochemical *cis* to *trans* isomerization,[2] the high quantum efficiency for visual pigment photolysis,[3] and the extremely short time necessary to reach the initial photoproduct, bathorhodopsin.[4] Figure 2 displays the potential energy diagrams just referred to.

Since a complex system such as a visual pigment presents formidable problems to the would-be draftsman of a chemical topographic map, an accurate potential energy surface, it should surprise no one that glaring instances of disagreement can be cited between predictions implicit in the potential energy surfaces shown in Fig. 2 and experiment. In the case of diagram (a) the prediction was made that pre-lumirhodopsin (i.e. bathorhodopsin) would not exist as a distinct species at room temperature, a prediction in clear disagreement with experiment.[5] The diagram shown as (b) does not explicitly provide for the appearance of the well-known intermediates batho-, lumi- and meta-rhodopsin, nor does it allow for the possibility of a photochemical *trans* to *cis* isomerization of the pigment chromophore, a process known to occur under certain conditions.[21]

The recently published energy surface calculated by Warshel[4] is undoubtedly the most sophisticated proposal of this kind which has yet been made. In proposing a novel conformational pathway for *cis-trans* isomerization, the so-called bicycle-pedal model, he has rationalized the initially surprising experimental finding that bathorhodopsin can be detected within 6 picosec. of the absorption of a photon by rhodopsin.[5] Apparently as a consequence of Warshel's model the 9-*cis* chromoprotein, isorhodopsin, should be photochemically transformed *directly* to rhodopsin, the 11-*cis* isomer. Since a sharp isosbestic point can be recorded during the low-temperature photoconversion of isorhodopsin to bathorhodopsin,[6] that prediction of Warshel's model appears to be in conflict with experimental findings. Another broad criticism can be made of the calculated energy relationships, particularly between rhodopsin and bathorhodopsin. Warshel places prelumirhodopsin (i.e. bathorhodopsin) and rhodopsin at approximately the same level of energy though it is shown below in Fig. 4 that bathorhodopsin is at least 11 kcal/mol higher in free energy than rhodopsin.

Rather than relying upon theoretical estimates for the energy states of visual pigments and their photochemical products, we can utilize the experimental data which are now available to make reasonable estimates of the relative free energies of cattle rhodopsin and its bleaching intermediates and products. Though such a free energy ordering does not correspond exactly to a multidimensional potential energy surface, especially since the configurations of the chromophore and/or the protein for each intermediate are not known with confidence, it is nevertheless instructive to examine such an empirically determined free energy diagram. The photochemical and thermal scheme on which it is based is shown in Fig. 3.

In order to fix the relative free energies of all-*trans* retinal + opsin and rhodopsin we must first consider the free energies of 11-*cis* and all-*trans*

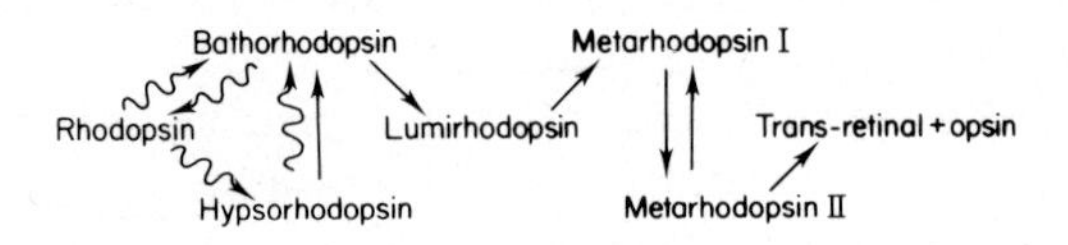

Fig. 3. Reaction sequence and intermediates in the photolysis or bleaching of rhodopsin. Reactions requiring light are indicated by $\rightsquigarrow$. Thermal, dark reactions are indicated by $\rightarrow$. The status of hypsorhodopsin in the sequence is still unclear.[28] Bathorhodopsin was formerly called pre-lumirhodopsin.[21] Several other intermediates on the pathway from metarhodopsin II to *trans*-retinal and opsin have been described[32] but their status is still uncertain.

retinal. Hubbard[7] showed that these two isomers in solution differed in free energy by 1·1 kcal/mol at 25°C. Thus the system 11-*cis* retinal + opsin is placed 1·1 kcal higher in free energy than all-*trans* retinal + opsin. 11-*cis* retinal + opsin spontaneously combine to form rhodopsin, and the reaction is reported to be complete[8] when the reactants are initially at concentrations around 10^{-5} M. The equilibrium constant,

$$K_{eq} = \frac{(\text{rhodopsin})_{eq}}{(11\text{-}cis\ \text{retinal})_{eq}(\text{opsin})_{eq}},$$

can be estimated to be as large as 10^9 M^{-1}, i.e.

$$\frac{(10^{-5}\ \text{M})}{(10^{-7}\ \text{M})(10^{-7}\ \text{M})},$$

assuming that complete reaction corresponds to at least 99 % reaction. We are now in a position to calculate the free energy difference, ΔG, between: 11-*cis* retinal (10^{-5} M) + opsin (10^{-5} M) and rhodopsin (10^{-5} M). The relationship is:

$$\Delta G = -RT \ln K_{eq} + RT \ln Q,$$

where Q is the ratio of concentrations of products to reactants at the appropriate concentrations,[9] which is 10^{-5} in this case. Thus

$$\Delta G = -RT \ln 10^9 + RT \ln 10^5 = -5{\cdot}5\ \text{kcal/mol} \quad \text{at } 25°\text{C}.$$

With this result we can now peg all-*trans* retinal + opsin at +4·5 kcal/mol relative to rhodopsin.

Having fixed the relative free energies of the initial and final products of rhodopsin photochemistry, we can work backwards to the relative free energy of each intermediate in the bleaching sequence. Metarhodopsin II spontaneously decomposes to all-*trans* retinal plus opsin at temperatures greater than 0°C.[10] We will conservatively assume, as before, that complete reaction in this system occurs at 15°C and corresponds to 99 % formation of products. Such an assumption is consistent with the limits of measurement in these solutions where we monitor concentrations by their absorbance and our usual precision is about ±1 %. Applying the formula

$$\Delta G = -RT \ln K_{eq} + RT \ln Q$$

again where

$$Q = \frac{(trans\ \text{retinal})(\text{opsin})}{(\text{metarhodopsin II})}, \quad K_{eq} = \frac{(10^{-5}\ \text{M})(10^{-5}\ \text{M})}{(10^{-7}\ \text{M})}, \quad \text{and} \quad T = 288°\text{K},$$

we obtain $\Delta G = -2{\cdot}7$ kcal/mol for the transformation; metarhodopsin II (10^{-5} M) → *trans*-retinal (10^{-5} M) + opsin (10^{-5} M).

Table 1
Free Energies of Reaction and Activation for the Synthesis and Photolysis of Cattle Rhodopsin in Solution

Reaction	T(°K)	K_{eq}/Q	$\Delta G\left(\frac{\text{kcal}}{\text{mol}}\right)$	k (sec^{-1})	$\Delta G^{\ddagger}\left(\frac{\text{kcal}}{\text{mol}}\right)$ (Calc.)	$\Delta G^{\ddagger}\left(\frac{\text{kcal}}{\text{mol}}\right)$ (Exptl.)
11-*cis* retinal + opsin → rhodopsin	300	$10^9/10^5$	−5·5	—	—	—
Hypsorhodopsin → bathorhodopsin	25	$10^2/1$	−0·2	10^{-3}	1·7	
Bathorhodopsin → lumirhodopsin	148	$10^2/1$	−1·4	10^{-3}	10·5	9·9[19]
Lumirhodopsin → meta I	240	$10^2/1$	−2·2	10^{-3}	17·1	16·5–19[32]
Meta I → meta II	273	1/1	0	—	—	12·5–37[32]
Meta II → *trans*-retinal + opsin	288	$10^{-3}/10^{-5}$	−2·7	10^{-3}	21	18·8–22·2[33]

Each of the remaining intermediate steps involves an intramolecular transformation where the number of moles of reactants and products are the same. The free energy difference, then, is independent of the concentrations of the particular pair of intermediates and only dependent upon the ratio of product to reactant and the temperature of transformation. Again considering the transformation to be complete when 99 % of the hypso-, batho- or lumi-intermediate has reacted, we have calculated and listed in Table 1 the free energy changes for each step of reaction shown in Fig. 3.

Though we have positioned the ground state of each intermediate in the bleaching sequence of cattle rhodopsin, we should be wary of the fact that the free energy changes may be more or less temperature dependent, a dependence given by the well-known thermodynamic equation $(\partial \Delta G/\partial T)_p = -\Delta S$.[9] Ideally, we should know ΔS and its temperature dependence for each transformation, in order to be able to calculate and compare the ΔG's at a single, common temperature. As the temperature dependences of the assumed equilibrium transformations have not been determined, we have no simple way of knowing the values of ΔS and/or ΔH at present, and can only assume that the ΔH term predominates in the equation $\Delta G = \Delta H - T\Delta S$ and that ΔH is only weakly dependent upon temperature. If these conditions are met, even approximately, then our free energy diagram can serve as a useful approximation to a potential energy surface at 25°C. Figure 4 is such a diagram.

In addition to estimating the free energy levels of the stable species encountered in the bleaching sequence, we can locate the maxima of the free energy barriers separating these species by utilizing the relationship between the free energy of activation, ΔG^{+}, and the specific reaction rate constant, k, derived from absolute reaction rate theory.[11] In the equation

$$\Delta G^{+} = -RT \ln\left(\frac{kh}{k_B T}\right),$$

k_B is Boltzmann's constant, h is Planck's constant, R is the universal gas constant and T is the absolute temperature. To calculate ΔG^{+}, the free energy necessary to reach the top of the energy barrier separating reactants and products, we need to estimate k in an analogous manner to our earlier estimate of K_{eq} when determining ΔG. To do this we first assume that the decay of each of metarhodopsin II, lumirhodopsin, bathorhodopsin and hypsorhodopsin are first order processes. Then we can assign a value for k, the specific rate constant, which will correspond to an average, measurable rate of decay of the intermediate in question at its reported transition temperature. A rate constant of $10^{-3}\,s^{-1}$ was chosen since it corresponded to a half-life of approximately 11 min, close to the measured time needed to effect the transformations at their reported temperatures. The values for ΔG^{+} obtained

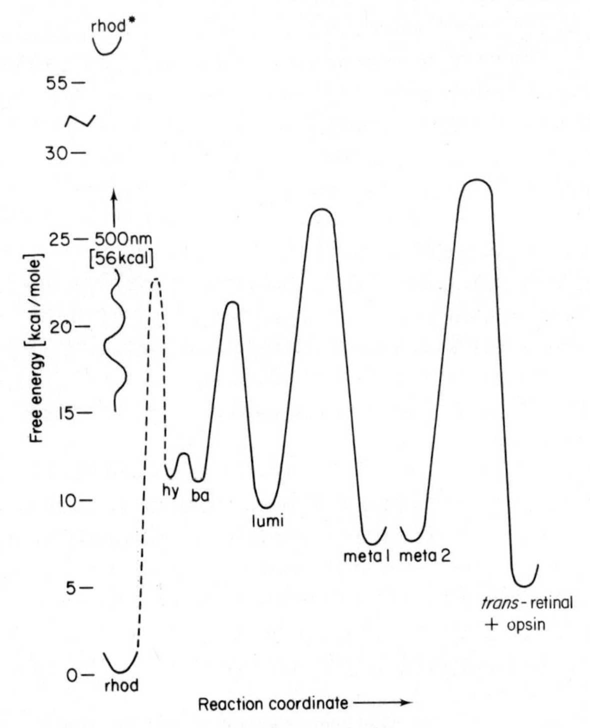

Fig. 4. Calculated free energy diagram for the reaction sequence shown in Fig. 3. The energy values are calculated from measurements made on rhodopsin solutions. The positions of the intermediates and products of the reaction are obtained from the values given in Table 1. Values for the free energy barriers are also obtained from Table 1. The barrier between rhodopsin (rhod) and hypsorhodopsin (hy) is hypothetical. The free energy minimum for excited rhodopsin (rhod*) is calculated from the λ_{max} value of the rhodopsin absorption spectrum.

in this way are listed in Table 1 where we have also listed the available experimental values of ΔG^{+}. The agreement is seen to be quite tolerable and encourages one to believe that this simple method of estimation may be useful in situations where experimental values are difficult to obtain. A cautionary note should be attached to the value of 1·7 kcal/mol calculated for the hypso- to batho-transition at 25°K. Since the validity of the absolute reaction rate expression at this low temperature is questionable, one should treat this result with qualification.

Molecular Analysis

Free energy states and potential energy surfaces allow us to calculate the rates of reaction and the amounts of each species present at equilibrium but, since thermodynamics deals with macroscopic changes, a free energy diagram such as Fig. 4 does not explicitly describe the molecular changes responsible for the observed macroscopic behavior. On the other hand, the calculated potential energy curves of Fig. 2 are based upon specific molecular transformations, all involving *cis* to *trans* isomerization of the retinylidene chromophore and, in (b) and (c), the interaction of the chromophore with its protein environment. But following the reservations described earlier, it seems prudent at this time to refer again directly to experimental data in order to try and assign conformations to the intermediates named in Fig. 3.

Most workers now agree that a stable pigment, whether natural or artificial, consists of a retinylidene chromophore bonded through a Schiff's base linkage to the ε-amino group of a particular lysine residue of opsin. Though only the 11-*cis* chromophore has been found in naturally occurring visual pigments, it seems likely that this selectivity is not due to steric factors alone. The relatively large number of geometric isomers, as shown in Fig. 5, having shapes which are different from the 11-*cis* chromophore but which are nevertheless accommodated in the chromophoric "pocket" of opsin, argues against a

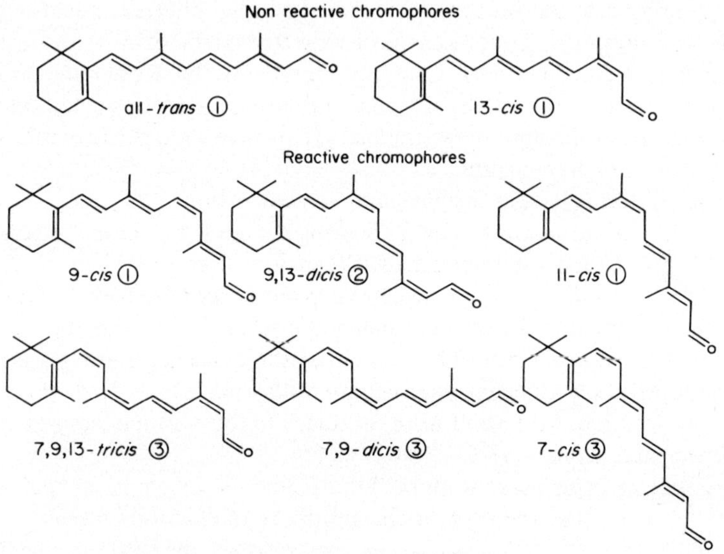

Non reactive chromophores

all-*trans* (1)

13-*cis* (1)

Reactive chromophores

9-*cis* (1)

9,13-*dicis* (2)

11-*cis* (1)

7,9,13-*tricis* (3)

7,9-*dicis* (3)

7-*cis* (3)

Fig. 5. Geometric isomers of retinal whose reactions with opsin have been determined. 11-*cis* retinal reacts with opsin to produce rhodopsin and the pigment product of 9-*cis* retinal with opsin is called isorhodopsin, or isorhodopsin I. 9,13-*dicis* retinal forms isorhodopsin II. Names have not yet been proposed for the pigments formed from isomers having a *cis* configuration at the 7–8 double bond. Compounds designated by (1) were studied by Hubbard and Wald [29], (2) was studied by Crouch *et al.*[30], while the 7-*cis* compounds, (3) were studied by De Grip *et al.*[31]

restrictive geometric requirement for the chromophore. The notion of rhodopsin as an opsin "lock" fitted with a chromophoric "key" needs to be modified. Perhaps an opsin "glove" with an oily surface grasping the chromophore would be more apt as a metaphor, although hardly the quantitative model we will need in order to accurately describe the molecular architecture of a visual pigment.

Though the first photochemical intermediate, bathorhodopsin, had not then been observed, it was argued in 1958[12] that the only thing light did to a visual pigment was to isomerize its chromophore. The discovery of bathorhodopsin[13] with its unusual spectral properties[14] and its very rapid appearance following light absorption by rhodopsin[5] has led to the suggestion that the initial photochemical step was the production of a distorted or strained *trans* chromophore[14] attached to the protein opsin which retained the configuration complementary to the *cis* chromophore. Though this view continues to command support,[15] alternative suggestions have been made, including photochemical proton transfer.[16]

Although some experimental[16] and theoretical[17] support for this scheme is reported, recent studies with pigments formed from desmethyl analogues of retinal seem to rule out proton transfer as the initial photochemical step.[18] The evidence supporting photochemical *cis* to *trans* isomerization as the initial photochemical step has been carefully marshalled by Rosenfeld and her colleagues[15] and supports the conclusion that bathorhodopsin consists of a distorted all-*trans* chromophore attached to a protein which has not relaxed to a complementary stable state.

The next intermediate in the bleaching sequence, lumirhodopsin, is usually formed in either of two ways. Cattle rhodopsin is cooled by liquid nitrogen to around −195°C and then irradiated. When it is slowly warmed, there is a transition at around −140°C[14] whereupon the bathorhodopsin component of the pigment mixture converts to lumirhodopsin, as is evidenced by a shift of λ_{max} to shorter wavelength. The second method, flash photolysis, detects lumirhodopsin as a transient intermediate following a brief flash given to a rhodopsin solution. Lumirhodopsin produced in this way has been observed at temperatures from −67°C[19] to 29°C.[5]

The spectral properties of those lumi-pigments which have been studied[20,21] suggest that the protein-chromophore interactions resemble those in the corresponding rhodopsins. As can be seen in Table 2, the lumi-intermediates differ in λ_{max} from the native visual pigment by 7 nm at most, except for the invertebrate and cusk pigments. This near coincidence of λ_{max} and the corresponding similarities of absorption band half-widths argue that the secondary interactions of chromophore and protein are similar in rhodopsin and the lumirhodopsin derived from it. The primary covalent bonding must be the same since lumi-pigments are formed before hydrolytic

Table 2
Absorption Properties of Six Rhodopsins and Their Corresponding Lumirhodopsins Measured in 2:1 Glycerol : Digitonin Sol'n at −65°C

Species	λ_{max} (nm)		Half band width* (10^{14} sec^{-1})		Ref.
	rhod.	lumirhod.	rhod.	lumirhod	
Bullfrog	508	515	1·17	1·23	(20)
Cattle	504	497	1·17	1·13	(20)
Chicken	512	513	1·13	1·09	(20)
Cusk	500	519	1·26	1·09	(20)
Monkey	500	500	1·26	1·05	(20)
Octopus	483	508	1·31	1·13	(20)
Squid	500	530	1·28	—	(20, 21)

* Half band widths are obtained, by subtraction, from the two frequencies at which the absorbance is half its maximum value.

reactions can reasonably have occurred. This brief discussion, then, leads to the conclusion that lumirhodopsin consists of an all-*trans* chromophore joined as a protonated Schiff base to a specific lysyl residue of opsin. The secondary interactions believed necessary to account for the bathochromic shift in visual pigments[22] must persist in the lumi-intermediate, since their

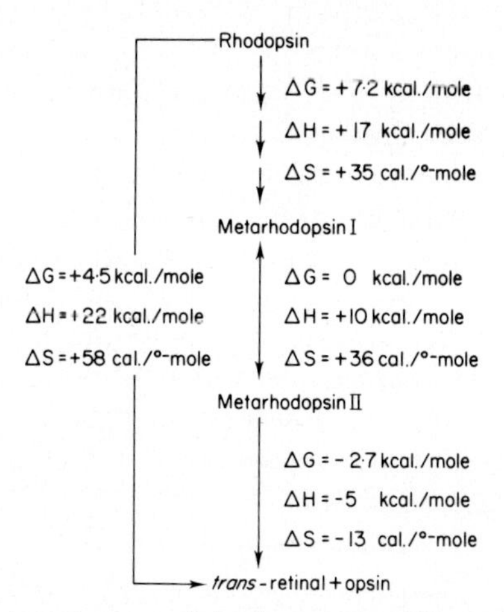

Fig. 6. Thermodynamic estimates of the free energy change (ΔG), enthalpy change (ΔH) and entropy change (ΔS) accompanying rhodopsin photolysis. ΔG values are from Table 1 and are subject to the restrictions mentioned in the text in conjunction with those values. ΔH values are from Cooper and Converse[24] and ΔS values are obtained from the formula $\Delta S = (\Delta H - \Delta G)/T$. The temperature chosen for the transformations rhodopsin to meta I and meta I to meta II is 276°K, while 300°K is chosen for the change from meta II to *trans*-retinal plus opsin.

spectral properties are so similar. If these secondary interactions occur in the vicinity of the ionone ring, as experiments with visual pigment analogues suggest,[23] then the bicycle-pedal model[4] for the isomerization of the chromophore would provide a simple rationale for the persistence of the secondary interactions in the visual pigments and their intermediates, especially lumirhodopsin.

The penultimate products of photolysis which are shown in Fig. 3 are metarhodopsin I and II. They are believed to be in tautomeric equilibrium[10] involving the proton of the Schiff base nitrogen atom as well as other protons bonded to amino acid residues. Due to the relative stability of these intermediates, it has been possible to carry out calorimetric studies of the photoproduction of metarhodopsins I and II and the subsequent hydrolysis to *trans*-retinal and opsin.[24] The enthalpy changes obtained by this direct thermal measurement are shown in Fig. 6, together with the calculated values of ΔG listed in Table 1 and values of ΔS derived from the equation:

$$\Delta S = \frac{(\Delta H - \Delta G)}{T}.$$

Having now obtained values of the entropy changes for some of the steps in bleaching the temptation is present to speculate about the disorganization and reorganization involved in proceeding from meta I to meta II and then to *trans*-retinal and opsin. Furthermore, if we were to hold fast to the view that the early steps in bleaching, up to the formation of lumirhodopsin, involve only small amounts of protein and solvent reorganization, then most of the 35 entropy unit change in the rhodopsin to meta I transition would arise in the lumi- to meta I change. Since the major positive entropy changes would then occur at the stages of metarhodopsins I and II, intermediates appearing at about the same time as electrical activity is first detected in the retina,[25] we might be tempted to link protein "unfolding" in metarhodopsin to visual excitation. Lacking more information about the nature of the entropy changes, the structures of earlier intermediates and further support for our estimates of the free energy changes shown in Table 1, it is probably best to refrain from such temptation at present.

Summary

Potential energy surfaces have been calculated which predict the pathway and time course for the photochemical reactions of visual pigments. Though it appears premature to rely on such theoretical constructs, they may contain suggestions for the mechanism of chromophore photoisomerization within the constraints of an enveloping protein. Alternatively, an empirical approach is presented which attempts to estimate the relative free energies of rhodopsin and its photochemical intermediates. This latter approach also leads to estimates of the free energy barriers separating intermediate states. Where comparisons with free energies of activation can be made, agreement is found to be tolerably good. Though an empirical free energy diagram does not lead directly to a mechanistic description, it provides for an energy ordering which rationalizes equilibrium and kinetic measurements in visual pigment photochemistry.

A brief review of mechanistic proposals suggests that *cis* to *trans* isomerization remains a better model for the initial photochemical step than does proton transfer from chromophore to protein. The subsequent steps, passing from bathorhodopsin to lumirhodopsin and then on to metarhodopsins I and II are still only vaguely understood. Describing these reactions as involving conformational changes, principally in the protein, does little to clarify the mechanism. Unfortunately, until more powerful quantitative methods can be utilized in studying these intermediates, we must content ourselves with the current vagaries.

Acknowledgment

This work was supported by the U.S. Public Health Service in the form of Research Grant EY 00201.

References

1. M. D. Kamen 1963, "Primary Processes in Photosynthesis", Academic Press, New York.
2. E. W. Abrahamson and S. E. Ostroy 1967, *Prog. Biophys. and Molec. Biol.* **17**, 179.
3. T. Kakitani and H. Kakitani 1975, *J. Phys. Soc. Jpan* **38**, 1455.
4. A. Warshel 1976, *Nature* **260**, 679.
5. G. E. Busch, M. L. Applebury, A. A. Lamola and P. M. Rentzepis 1972, *Proc. Nat. Acad. Sci. USA* **69**, 2802.
6. T. Yoshizawa, personal communication.
7. R. Hubbard 1966, *J. Biol. Chem.* **241**, 1814.
8. P. K. Brown and G. Wald 1956, *J. Biol. Chem.* **222**, 865.
9. G. W. Castellan 1964, "Physical Chemistry," Addison-Wesley, Reading, Mass.
10. R. G. Matthews, R. Hubbard, P. K. Brown and G. Wald 1963, *J. Gen. Physiol.* **47**, 215.
11. S. Glasstone, K. J. Laidler and H. Eyring 1941, "The Theory of Rate Processes", McGraw-Hill, New York.
12. R. Hubbard and A. Kropf 1958, *Proc. Nat. Acad. Sci. USA* **44**, 130.
13. T. Yoshizawa and Y. Kito 1958, *Ann. Rep. Sci. Works, Fac. Sci. Osaka Univ.* **6**, 27.
14. T. Yoshizawa and G. Wald 1963, *Nature* **197**, 1279.
15. T. Rosenfeld, B. Honig, M. Ottolenghi, J. Hurley and T. G. Ebrey 1977, *Pure App. Chem.* **49**, 341.

16. M. R. Fransen, W. C. M. M. Luyten, J. van Thuijl, J. Lugtenburg, P. A. A. Jansen, P. J. G. M. van Breugel and F. J. M. Daemen 1976, *Nature* **260**, 726.
17. K. van der Meer, J. J. C. Mulder and J. Lugtenberg 1976, *Photochem. Photobiol.* **24**, 363.
18. A. Kropf 1976, *Nature* **264**, 92.
19. K. H. Grellman, R. Livingston and D. C. Pratt 1962, *Nature* **193**, 1258.
20. R. Hubbard, P. K. Brown and A. Kropf 1959, *Nature* **183**, 442.
21. T. Yoshizawa 1972, *In* "Handbook of Sensory Physiology", vol. 7, part 1 (ed. H. J. A. Dartnall). p. 146. Springer-Verlag, Heidelberg.
22. A. Kropf and R. Hubbard 1958, *Ann. N.Y. Acad. Sci.* **74**, 266.
23. A. Kropf, to be published.
24. A. Cooper and C. A. Converse 1976, *Biochem.* **15**, 2970.
25. W. A. Hagins 1972, *Ann. Rev. Biophys. and Bioeng.* **1**, 131.
26. M. G. F. Fuortes and P. M. O'Bryan 1972, *In* Handbook of Sensory Physiology", vol. 7, part 2 (ed. M. G. F. Fuortes) p. 321. Springer-Verlag, Heidelberg.
27. T. G. Ebrey and B. Honig 1975, *Quart. Rev. Biophys.* **8**, 129.
28. T. Yoshizawa and S. Horiuchi 1973, *In* "Biochemistry and Physiology of Visual Pigments" (ed. H. Langer), p. 69, Springer-Verlag, New York.
29. R. Hubbard and G. Wald 1952, *J. Gen. Physiol.* **36**, 269.
30. R. Crouch, V. Purvin, K. Nakanishi and T. Ebrey 1975, *Proc. Nat. Acad. Sci. USA* **72**, 1538.
31. W. J. De Grip, R. S. H. Liu, V. Ramamurthy and A. Asato 1976, *Nature* **262**, 416.
32. E. W. Abrahamson 1973, *In* Biochemistry and Physiology of Visual Pigments (ed. H. Langer), p. 47, Springer-Verlag, New York.
33. E. W. Abrahamson and J. R. Wiesenfeld 1972, *In* "Handbook of Sensory Physiology", vol. 7, part 1 (ed. H. J. A. Dartnall) p. 69. Springer-Verlag, Heidelberg.

Discussion

A. E. Blaurock: In one of the figures it appears that, for all the geometric isomers of retinal which bind to opsin to form coloured complexes, the distance from the aldehyde O (oxygen) to the ring is shorter than for the two isomers (all-*trans* and 13-*cis*) which do not bind. This observation would need, of course, to be verified on CPK models. Is this the case for these models? If so, might the increase in length explain why the all-*trans* retinal does not bind, i.e. is it too long for the presumed binding pocket of the opsin?

A. Kropf: Tracings of CPK models of 7-*cis*, 7,9-*dicis*, 7,9,13-*tricis*, 11-*cis*, 9-*cis* and 9,13-*dicis* retinals bear out the suggestion that the distance between the aldehyde oxygen and the ionone ring may be the critical factor in pigment formation. When such tracings are superimposed, they resemble each other in broad outline and differ from similar tracings of the 13-*cis* and all-*trans* aldehydes, the latter two isomers showing a larger span between the oxygen and ionone ring.

3

Calcium and Rod Outer Segments

F. J. M. DAEMEN, P. P. M. SCHNETKAMP,
TH. HENDRIKS and S. L. BONTING

Department of Biochemistry, University of Nijmegen, Nijmegen, The Netherlands

Introduction

It has been established that hyperpolarization of the membrane of vertebrate rod photoreceptor cells follows photoactivation of rhodopsin molecules embedded in the rod sac membranes of the outer segment. The suggestion of Baylor and Fuortes[1] that a transmitter might propagate the signal through the intracellular space from the rod sac membrane to the plasma membrane of the outer segment, has been elaborated by Yoshikami and Hagins.[2, 3] In their model, in darkness high calcium activities are maintained inside the rod sacs and a low activity in the intracellular space, the cytoplasm. Upon illumination, rhodopsin would initiate release of calcium ions from the sacs into the intracellular space. These calcium ions would diffuse to the plasma membrane, where they would reversibly block the sodium channels, thereby causing the transient hyperpolarization which is observed experimentally.

The initial evidence for this calcium transduction hypothesis of visual excitation rests largely on electrophysiological experiments, without detailed calcium analyses. Therefore, biochemical experiments on this problem would seem to be most desirable. However, a number of attempts in this direction are not very convincing, especially since there is poor agreement between the different reports on several essential points. Most authors have chosen to study isolated frog or cattle rod outer segments for their biochemical studies. This is understandable, since the transduction takes place in the outer segment and generally accepted methods are available to isolate them in bulk. However, it is equally true that the rod outer segment is a very delicate structure, the handling of which, also in relation to its environment, is critical. Therefore, after our initial experiments on light-induced release of calcium ions in frog rods,[4] we have focused on the behavior of calcium in various cattle rod outer segment preparations. Present experience suggests that only if we acquire a clear insight into the "calcium metabolism" of rod outer segments, may optimal conditions for the assessment of a light effect be found.

In this paper we review our studies on the binding of calcium by cattle rod outer segment membranes, on the calcium content of isolated cattle rod outer segments and on accumulation of calcium in cattle rod outer segments.

Binding of Calcium by Cattle Rod Outer Segment Membranes

Passive binding of calcium to isolated cattle rod outer segment membranes has been studied by means of equilibrium dialysis.[5] Cattle rod outer segments are prepared by means of sucrose gradient centrifugation.[6] The initial, mild homogenization of the retinas is carried out in 0·16 M Tris/HCl, 1 mM EGTA (pH 7·4). The rhodopsin-containing layer of the gradient is lysed in 15 mM EDTA and twice washed with water. The final sediment, consisting of rod outer segment membranes, is then taken up in the desired medium (0·4–0·8 mg protein/ml) and immediately used for equilibrium dialysis experiments.

Samples of 5 ml are placed in dialysis bags. The closed bags are placed in test tubes containing identical buffer solution to which trace amounts of ^{45}Ca and various concentrations of ^{40}Ca are added. The stoppered tubes are attached to a vertically rotating disk (1 rev/min). Dialysis takes place for 40 h at 4°C in the dark, during which time equilibrium has been reached. Then aliquots are taken from both the inside suspension and the outside solution and counted in a liquid scintillation counter. From these counts, the total amount of radioactivity applied and the total amount of calcium per tube, the amount of calcium bound to the membranes is easily calculated. Recoveries of radioactivity were more than 95 %, while effects due to volume changes or Donnan equilibria could be excluded on the basis of control experiments.

Table 1 shows the calcium binding capacity of cattle rod outer segment membranes in various media at a fixed calcium concentration of 10^{-5} M. When the dialysis medium contains only 20 mM Tris/HCl, 6·1 nmoles calcium are bound per mg protein (0·4 mol calcium per mol rhodopsin). Addition of 100 mM NaCl decreases the calcium binding by about 75 %. Substitution of the 100 mM NaCl by 100 mM KCl or a mixture of 50 mM NaCl and 50 mM KCl causes the same decrease in the amount of bound calcium. Addition of 5 mM $MgCl_2$ to these media lowers the amount of bound calcium by a further 55 %.

Table 1
Effect of Medium Composition on Calcium Binding by Cattle Outer Segment Membranes in the Presence of 10^{-5} M Calcium

Dialysis medium		Calcium binding (nmol Ca/mg protein)	Number of experiments
20 mM Tris/HCl		$6{\cdot}1 \pm 0{\cdot}4$	12
,,	+ 100 mM NaCl	$1{\cdot}4 \pm 0{\cdot}2$	8
,,	+ 100 mM NaCl, 5 mM $MgCl_2$	$0{\cdot}6 \pm 0{\cdot}1$	4
,,	+ 50 mM NaCl, 50 mM KCl	1·2	1
,,	+ 100 mM KCl	$1{\cdot}4 \pm 0{\cdot}4$	2
,,	+ 100 mM KCl, 5 mM $MgCl_2$	$0{\cdot}6 \pm 0{\cdot}1$	3

Thus, the amount of calcium bound is strongly affected by the composition of the medium. The presence of sodium or potassium lowers calcium binding to an equal extent. The fact that even at Na or K to calcium ratios of about 10,000 there is still a significant amount of calcium bound, indicates the presence of rather specific binding sites for calcium. Addition of 5 mM $MgCl_2$ decreases

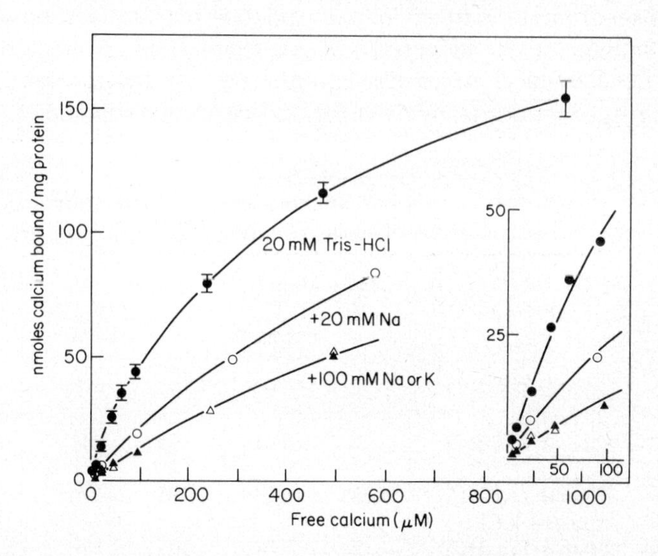

Fig. 1. Effect of calcium concentration on calcium binding by cattle rod outer segment membranes. Medium: 20 mM Tris/HCl. ● ($n = 3$): 20 mM NaCl, 20 mM Tris/HCl. ○: 100 mM NaCl, 20 mM Tris/HCl. ▲: 100 mM KCl, 20 mM Tris/HCl. △: pH 7·4. Protein concentration as determined by the Lowry method was about 0·35 mg/ml.

calcium binding to less than half at 10^{-5} M calcium (Mg/Ca ratio 500) but only by 10% at 10^{-3} M calcium (Mg/Ca ratio 5), again indicating the high specificity of the binding sites for calcium.

The effect of the free calcium concentration on the binding of calcium has been studied in four different media (Fig. 1). As expected, the binding of calcium increases at higher free calcium concentrations. Under all conditions less calcium is bound in media of higher ionic strength. Calcium binding is affected equally by the presence of sodium and potassium. In 20 mM Tris/HCl and at 10^{-3} M calcium, as much as 155 nmoles calcium are bound per mg protein (10·3 mol calcium per mol rhodopsin).

The calcium binding data of Fig. 1 have been analyzed by means of Scatchard plots. These plots are biphasic, indicating the presence of (at least) two classes of binding sites. Linearized plots are used to calculate the number of binding sites of each type and their apparent dissociation constants, assuming independence of the binding sites. The results are presented in Table 2, which indicates that the higher affinity sites can maximally accommodate 5 and the lower affinity sites 195 nmoles calcium per mg protein, or 0·3 and 13·0 mol calcium per mol rhodopsin, respectively.

The calcium binding properties of the rod outer segment membranes are not changed after prior storage overnight at either 4 or −70°C. Neither does lyophilization of the material have a significant effect. Together with the relatively small variation found in many different preparations, this suggests that our findings are typical for rod outer segment membrane fragments in the absence of anatomical structures derived from e.g. the rod sacs. This is confirmed by experiments in which the divalent cation ionophore A 23187

Table 2
Comparison of Calcium Binding Sites in Cattle Rod Outer Segment Membranes

Assay conditions	High affinity n	High affinity $K_{diss.}$ (mM)	Low affinity n	Low affinity $K_{diss.}$ (mM)
20 mM Tris/HCl	5	0·006	195	0·5
20 mM NaCl–20 mM Tris HCl	5	0·013	195	0·9
100 mM NaCl–20 mM Tris HCl	5	0·040	195	1·6
100 mM KCl 20 mM Tris HCl	5	0·040	195	1·6

Values derived from Scatchard plots assuming independency of sites. n: amount of Ca^{++} bound in nmoles per mg protein; $K_{diss.}$: dissociation constant.

(10^{-7}–10^{-5} M) is added to the equilibrium dialysis mixture. Under these conditions the amount of bound calcium is not decreased, indicating that there is no accumulation of calcium inside sacs or vesicular artifacts derived from them.

We have also examined the effect of light on the binding of calcium by rod outer segment membranes. Although the results suggest that, independent of the amount of calcium bound, illuminated membranes bind slightly less calcium than dark-kept controls, this difference does not appear to be significant.

At first glance, the binding capacities and dissociation constants, given in Table 2, might explain the relatively high calcium content of rod outer segments. If, for example, 100 mM NaCl, 1 mM $CaCl_2$ is taken as an approximation for the composition of the extracellular fluid, a calcium concentration of about 8 mM over the entire rod outer segment can be calculated, if the rod outer segment membranes were in a simple equilibrium condition with regard to the extracellular fluid. However, as will be shown later on, the extracellular calcium concentration has very little influence on the calcium concentration of isolated rod outer segments. Moreover, the presumably very low cytoplasmic calcium concentration in rod outer segments does not fit such a simple picture. These considerations strongly suggest that the high calcium content of rod outer segments requires, in addition to passive binding, a translocation of calcium into the rod sac interior against a concentration gradient.

Calcium Content of Cattle Rod Outer Segments

If calcium is to play an important role in rod outer segments, an essential parameter is, of course, the calcium content of the rod outer segments. Widely diverging numbers have so far appeared in the literature: for isolated rod outer segments from 0·25–10 mol calcium per mol rhodopsin.[4, 7, 8]

Isolation of reasonable amounts of pure cattle rod outer segments requires in our experience continuous gradient centrifugation. Necessarily then, we have to deal consecutively with the separation of the outer segments from the retina, the gradient centrifugation itself and the isolation of the band containing the rod outer segments. Each of these steps may have a profound influence on the calcium content. In fact, each manipulation or change of medium tends to cause loss of calcium. Thus, shaking off the outer segments from the retina leaves more calcium than mild homogenization in a Potter device. Isolation in media containing fair amounts of ions (160 mM Tris/HCl, 160 mM NaCl, Ringer solution) is unfavorable as compared to media containing high sucrose concentrations and low ion concentrations. A single low speed (2000 × g) centrifugation step causes up to 60 % loss of calcium in some outer segment preparations. Loss is also caused by lyophilization, sonication and handling of suspensions with micropipettes.

On the basis of these findings we have used two standard procedures, yielding what we have called either "Tris-rods" or "sucrose-rods". In the first procedure, cattle retinas are homogenized in 160 mM Tris/HCl (pH 7·4). After filtration through a wire screen the rod outer segment suspension is applied to a continuous sucrose gradient. After centrifugation the band containing rod outer segments is collected, diluted with three volumes of the isolation medium and centrifuged at 2000 × g. A variant of this procedure, in which 1 mM EGTA is added to the isolation medium, yields "calcium depleted Tris-rods".

In the second procedure we have attempted to reduce as much as possible the loss of calcium due to medium conditions and experimental handling. Cattle retinas are collected in 600 mM sucrose, 10 mM ascorbate, 20 mM Tris/HCl (pH 7·4). Rod outer segments are shaken off by agitation on a vortex mixer and filtered through a fine wire screen. A continuous gradient is made from the filtrate and an equal volume of 20 % (660 mM) sucrose, 14 % Ficoll-400. Following gradient centrifugation the rod outer segment layer is collected, and after dilution with isolation medium sedimented by centrifugation at 2000 × g.

Table 3
Endogenous Calcium Content of Various Cattle Rod Outer Segment Preparations

	"Tris-rods"	Ca-depleted "Tris-rods"	"Sucrose rods"
calcium content (mol Ca/mol rhodopsin)	2·3	0·43	4·4
S.E.	0·7	0·09	0·5
range	0·9–4·0	0·18–0·66	2·2–7·6
no. of preparations	5	5	11

The calcium content of these three types of rod outer segment preparations, as determined by atomic absorption spectroscopy, is presented in Table 3. The rather variable results again indicate that the outer segments easily lose calcium during isolation. Addition of EGTA during the final washing procedure reduces the calcium content by less than 20 %, which shows that little externally bound calcium is involved, the more so since some calcium efflux is likely to occur during this period of time. The data of Table 3 indicate that the calcium content of cattle rod outer segments *in vivo* probably amounts to more than 4–5 mol calcium per mol rhodopsin.

Calcium Incorporation in Cattle Rod Outer Segments

Incorporation of calcium in isolated rod outer segments has been studied by incubation with the radioisotope ^{45}Ca. Isolation after incubation occurs with a rapid filtration method over borosilicate glass fibre filters with a washing medium of the same composition as the incubation buffer, but without ^{45}Ca and with enough EGTA added to chelate all calcium present. The entire washing procedure takes about 5 s. Blanks consist of ^{45}Ca-free rod outer segment suspensions and radioisotope solution mixed in the washing medium and immediately subjected to the same washing procedure. Radioactivity is determined in a liquid scintillation counter. Unless otherwise indicated, the data to be presented here, refer to steady-state filling levels, i.e. the condition where influx of ^{45}Ca equals ^{45}Ca efflux. Upon incubation at 25° this level is reached after 10 to 30 min. Hence, standard incubation times of 45 min have been used. In the presence of EGTA ^{45}Ca incorporation does not significantly exceed the isotope blank level. Incorporation in the absence of EGTA extrapolates at zero time to the blank level. These control experiments indicate that incorporation of ^{45}Ca has occurred inside the rod sacs (as opposed to

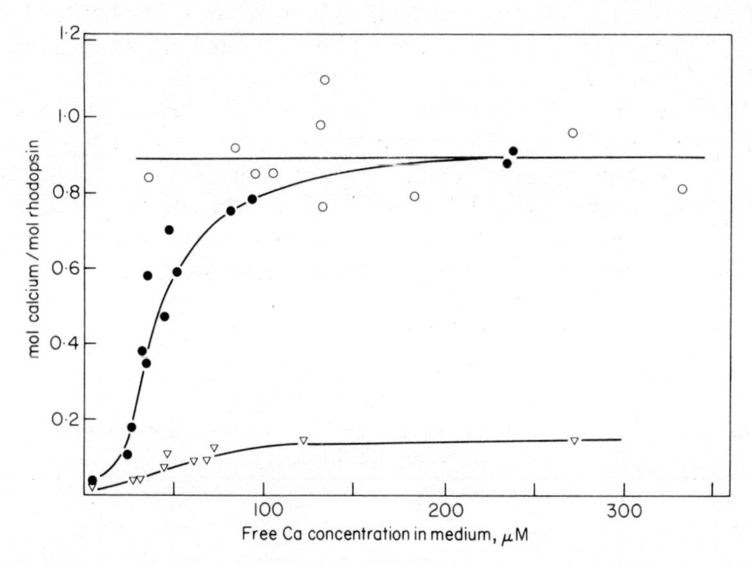

Fig. 2. Effect of external calcium on ^{45}Ca accumulation by depleted and non-depleted "Tris-rods". Medium: 100 mM KCl, 2 mM $MgCl_2$, 20 mM Tris/HCl (pH 7·4). Free external calcium concentrations are calculated by subtracting from the total amount of calcium present, endogenous and exogenous, the amount of calcium incorporated in the outer segments. The results of 3 experiments are combined. ○: non-depleted preparations, 1 mM ATP present in medium. ●: depleted preparations, 1 mM ATP present in medium. ▽: depleted preparations, no ATP present.

passive binding to the rod sac membranes). In calculating the amount of accumulated calcium the isotope is assumed to be equally distributed over endogenous and exogenous calcium in the steady state. This assumption is supported by the findings presented in Fig. 2.

So far most experiments have been carried out with "Tris-rods", and particularly with calcium-depleted preparations. Incorporation of ^{45}Ca in depleted rods, both in the presence and absence of 1 mM ATP, is not seriously influenced by the composition of the medium, except in the presence of 100 mM NaCl, when the calcium accumulation is decreased by about 70 %. As standard incubation medium 100 mM KCl, 2 mM $MgCl_2$, 20 mM Tris/HCl (pH 7·4) has been used throughout.

In Fig. 2 the effect of external calcium concentration on the calcium content is shown for both depleted and non-depleted "Tris-rods". Non- depleted rod outer segments have, in the presence of 1 mM ATP, a calcium content of 0·9 mol calcium per mol rhodopsin, which is independent of the external calcium concentration over a wide range from 40–350 µM. In the absence of ATP a similar independence is found at a somewhat lower level of 0·6 mol calcium per mol rhodopsin. Thus, in these preparations no net uptake of calcium takes place, but rather calcium/calcium exchange. Depleted rod outer segments show in the presence of 1 mM ATP a net calcium accumulation to a level equal to that of the non-depleted rods. Without ATP the depleted outer segments show a very low incorporation, even at high external calcium levels.

The effect of ATP on ^{45}Ca incorporation is rather small in non-depleted rods, but very high in calcium depleted preparations (Table 4). Replacement of ATP by β,γ-methylene-ATP, which is not hydrolyzed by ATP-ases and kinases, abolishes the stimulation of incorporation. This suggests that the stimulating effect of ATP involves enzymatic hydrolysis.

Interesting are the effects of the temperature and of ionophore A 23187 on calcium incorporation (Table 5). Low temperature only slightly reduces ^{45}Ca

Table 4
Effect of ATP and β,γ-Methylene ATP on ^{45}Ca Accumulation in Cattle Rod Outer Segments

	no ATP	ATP 1 mM		β,γ-methylene ATP 1 mM	
	%	%	n	%	n
depleted "Tris-rods"	≡100	529 ± 50	4	69 ± 2	2
non-depleted "Tris-rods"	≡100	154 ± 3	3	90 ± 2	2

Medium composition: 100 mM KCl, 2 mM $MgCl_2$, 50 µM $CaCl_2$, 20 mM Tris/HCl (pH 7·4). Results are averages for depleted "Tris-rods" prepared from the same batch of eyes, with S.E. and no. of experiments.

Table 5
Effect of Temperature and Ionophore A 23187 on ^{45}Ca Accumulation in Depleted "Tris-Rods". Medium Composition as in Table 4

Incubation conditions	^{45}Ca incorporation (%)
Standard medium:	
45 min at 25°C	≡100
45 min at 0°C	53
45 min at 25°C, then 45 min at 0°C	91
Standard medium + 1 mM ATP:	
45 min at 25°C	403
45 min at 0°C	64
45 min at 25°C, then 45 min at 0°C	395
Standard medium + 1 mM ATP + 10 μM A 23187:	
45 min at 25°C	36

incorporation in calcium depleted rods in the absence of ATP, but completely inhibits the stimulation of ^{45}Ca accumulation by ATP. The latter observations support the enzymatic nature of this process. However, once the rods are loaded with ^{45}Ca, decrease of temperature does not lead to a significant loss of calcium, which suggests that the storage of this ion does not require metabolic energy.

The presence of the divalent cation ionophore A 23187 inhibits accumulation of calcium (Table 5), whereas addition of A 23187 to a rod suspension after accumulation leads to rapid efflux of calcium. These observations suggest the existence of calcium gradients in rod outer segments.

The effect of various substances, known to influence calcium accumulation in other tissues, has been studied with calcium depleted rods. None of these substances (oligomycin, 5 mg/ml; ruthenium red, 100 μM; lanthanum chloride, 50 μM) has a clear effect, which indicates that the rod accumulation system differs from that in mitochondria, sarcoplasmic reticulum and erythrocytes. Lack of inhibition by ouabain suggests that the Na-K activated ATPase system is not directly involved.

In a further experiment the efflux of ^{45}Ca from preloaded rods has been studied. Non-depleted rods are preloaded with ^{45}Ca in a 100 mM KCl medium, and suspended in 9 vols of the 100 mM KCl medium with 0·25 mM EGTA. Whereas only a slow efflux of ^{45}Ca is seen in this medium ($t_{1/2} \simeq 20$ min), even after increasing the KCl concentration to 200 mM, a very rapid ($t_{1/2} \simeq 15$ s) and extensive efflux occurs when the medium is made 100 mM in NaCl or 1 mM in $CaCl_2$ by adding a small volume of very concentrated solution. These observations support the existence of sodium/calcium and calcium/calcium exchange systems.

Our present experience with "Tris-rods" can be summarized as follows:

1. Incorporation of ^{45}Ca in non-depleted rod outer segments involves mainly ATP-independent calcium/calcium exchange, since their calcium content is virtually independent of the external calcium concentration.
2. Incorporation of ^{45}Ca in calcium depleted rod outer segments is strongly stimulated by ATP to a level, which is comparable to that of non-depleted preparations. The ATP effect appears to involve hydrolysis of ATP.
3. Sodium specifically inhibits the incorporation of calcium and promotes its efflux, suggesting the existence of a sodium/calcium exchange system. Incorporation would then consist of translocation across the rod sac membrane by means of a calcium/sodium exchange carrier system, which requires ATP for net uptake of calcium, but not for exchange transport of calcium.

Our preliminary experiments with the "sucrose rods", which behave rather differently in a number of respects from the "Tris-rods", suggest strongly that in the "Tris-rods" the plasma membrane is not a permeability barrier, whereas in the "sucrose rods" it is. Their high endogenous calcium content (Table 3) suggests that the original outer segment structure is better preserved than in "Tris-rods". Calcium/calcium exchange occurs very rapidly, but again the total calcium content is independent of the external calcium concentration over a wide range of concentrations. Addition of 1 mM ATP to the medium does not stimulate the uptake of calcium. Sodium/calcium exchange can also be demonstrated to occur. Upon adding ionophore A 23187 the total calcium content becomes dependent on the external calcium concentration. Lanthanum appears to inhibit calcium exchange. Transfer of "sucrose rods" to a medium, high in ions, changes its properties to those described for "Tris-rods". All these observations are consistent with the idea that "sucrose rods" behave like a three-compartment system, rather than like a two-compartment system, with the bulk of the calcium stored in the rod sacs and with an intracellular space, low in calcium, surrounded by an intact plasma membrane. We feel that rod outer segments, isolated and maintained in sucrose media, approach the physiological state of rod photoreceptor cells better than the "Tris-rods" and preparations used by other investigators.

Conclusion

Our present studies on the role of calcium in rod outer segments confirm their high calcium content, which cannot be explained by mere passive binding of calcium to rod sac membranes. A translocation system for calcium appears to be operative both in the rod sac membranes and in the plasma membrane. This system may, tentatively, be described as a sodium/calcium exchange carrier.

Exchange transport of calcium (sodium/calcium as well as calcium/calcium exchange), both across the rod sac membrane and across the plasma membrane, may occur without ATP. However, net transport of calcium into the rod sacs, and possibly net extrusion of calcium across the plasma membrane, does seem to require ATP.

It is hoped that a complete picture of the calcium metabolism of rod outer segments can be derived on the basis of these observations, but many details must await further experiments.

Acknowledgment

The research in our laboratory has been supported by the Netherlands Foundation for Basic Research (ZWO) through the Netherlands Foundation for Chemical Research (SON).

References

1. D. A. Baylor and M. G. F. Fuortes 1970, Electrical responses of single cones in the retina of the turtle, *J. Physiol.* **207**, 77–92.
2. W. A. Hagins 1972, The visual process: excitatory mechanism in the primary receptor cells, *Ann. Rev. Biophys. Bioeng.* **1**, 131–158.
3. W. A. Hagins and S. Yoshikami 1975, Ionic mechanisms in excitation of photoreceptors, *Ann. N.Y. Acad. Sci.* **234**, 314–325.
4. Th. Hendriks, F. J. M. Daemen and S. L. Bonting 1974, Light-induced calcium movements in isolated frog rod outer segments, *Biochim. Biophys. Acta* **345**, 468–473.
5. Th. Hendriks, P. P. M. van Haard, F. J. M. Daemen and S. L. Bonting 1977, Calcium binding by cattle rod outer segment membranes studied by means of equilibrum dialysis, *Biochim. Biophys. Acta* **467**, 175–189.
6. W. J. de Grip, F. J. M. Daemen and S. L. Bonting 1972, Enrichment of rhodopsin in rod outer segment preparations, *Vision Res.* **12**, 1697–1707.
7. P. A. Liebman 1974, Light-dependent Ca^{++} content of rod outer segment disc membranes, *Invest. Ophthalmol.* **13**, 700–701.
8. H. H. Hess 1975, The high calcium content of retinal pigmented epithelium, *Exp. Eye Res.* **21**, 471–479.

Discussion

F. Veringa: Could Dr. Daemen give an estimate of the time needed by Ca^{++}-ion to escape from the space between the rod discs, so as to finally reach the plasma membrane? The question is asked because the large time constants encountered in electrophysiology, as well as in psychophysics, continue to lack a full explanation at the molecular level, and a process of internal transmission might be a likely candidate.

F. J. M. Daemen: Hagins (ref. 2, p. 144) argues that "a solute with a diffusion coefficient of 10^{-5} cm^2 sec^{-1} released at the center of a rat rod 1·8 μ in diameter would reach 90% of its equilibrium concentration at the plasma membrane in less than 2 msec".

4

Molecular Anatomy and Light-Dependent Processes in Photoreceptor Membranes

WAYNE HUBBELL, KWOK-KEUNG FUNG, KEELUNG HONG and YONG SHIAU CHEN

Department of Chemistry, University of California, Berkeley, California, U.S.A.

It is generally accepted that the primary function of vertebrate rhodopsin is to directly or indirectly modulate the cytoplasmic activity of an internal transmitter substance in response to light stimulation. The theoretical and experimental bases for this notion have been discussed thoroughly.[1,2] There is little doubt that photochemical events in rhodopsin are coupled to transmitter modulation via conformational transitions in the protein, and two distinct types of coupling mechanisms can be formulated. In the first mechanism, rhodopsin is imagined to function as a transmembrane channel, releasing transmitter from its storage site in the disc lumen when photon absorption produces an "open channel" configuration. A detailed hypothesis based on this mechanism and suggesting Ca^{++} as the transmitter has been described by Yoshikami and Hagins[3] and has received significant experimental support.[4–9]

In the second type of mechanism, the functional activity of rhodopsin is imagined to be confined to the disc membrane surface. For example, photon absorption may result in the formation of a rhodopsin conformational state which is capable of binding and thereby activating an endogenous enzyme system. The active enzyme would produce (or degrade) transmitter, thus achieving the desired modulation of activity. The discovery of a light-activated phosphodiesterase activity in the rod outer segment (ROS) provides an experimental basis for this mechanism.[10]

These two mechanisms involve rhodopsin in fundamentally different roles, in one case as a transmembrane channel and in the other as a direct enzyme activator. Of course it is possible for a substance released from the disc lumen by a channel mechanism to in turn activate an enzyme system, but the basic problem to which we wish to draw attention is not whether combinations of these mechanisms are possible but whether rhodopsin acts on or across membrane surfaces in the transduction process.

It is clear that the rhodopsin molecule has more than one role in the receptor function. For example, both ROS protein kinase[11] and the phosphodiesterase activation[10] have an action spectrum coincident with the rhodopsin absorption spectrum. Protein phosphorylation in the ROS is too slow to be involved in transduction[11] but cyclic nucleotide degradation may be sufficiently rapid to play a direct role in the excitation. Under any circumstance, it is quite likely that these enzymatic phenomena occur as a result of processes at the disc membrane surface. We have recently initiated efforts to uncover any transmembrane properties of rhodopsin, and the present communication describes results of some new and preliminary studies in this direction as well as results regarding conformational regulation of rhodopsin by light.

Organization of Rhodopsin in the Disc Membrane

Freeze-fracture images of disc membranes show that rhodopsin remains preferentially with the cytoplasmic membrane face during fracture.[12] Shallow etching reveals holes or pits in the opposite face, strongly suggesting that rhodopsin penetrates the thickness of the membrane in the intact structure, and is "pulled out" during fracture to create the pits. In order to confirm this conclusion we have applied lactoperoxidase catalyzed iodination and asymmetric chemical labeling methods to rhodopsin containing membranes. These experiments are based on a unique advantage offered by the use of reconstituted membrane vesicles. These vesicles are single-walled and sealed, the interior space being inaccessible to membrane impermeable substances. In experiments to be described throughout this paper, extensive use is made of reconstituted membranes prepared either from highly purified rhodopsin and phospholipids according to Hong and Hubbell[13] or from crude rhodopsin and phospholipids using a similar procedure but employing sodium cholate as the solubilizing agent. We cautiously assume that the rhodopsin properties in these reconstituted membranes are sufficiently similar to those in the native membranes to justify their use. This assumption has held up under several tests.[14, 15]

Figure 1 shows a schematic representation of the structures of the native and reconstituted disc vesicles based on freeze-fracture analysis[12] and assuming a

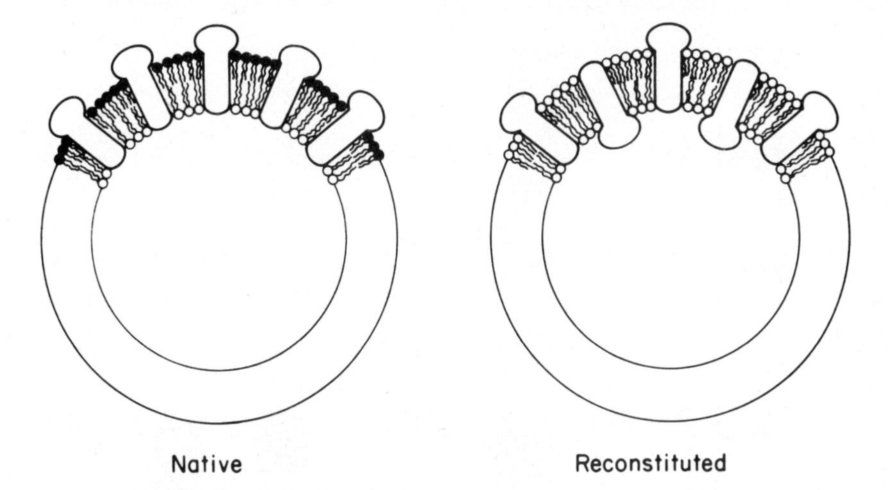

Fig. 1. Schematic representation of vesicles derived from the native and reconstituted disc membrane. The asymmetric rhodopsin molecules have a preferred orientation in the native membrane, while in the reconstituted membrane the orientation is random (or nearly so). In reality, the reconstituted vesicles are much smaller (diameter ≈ 400 Å) than those derived from the native membrane (diameter 0·2–1 μ).

transmembrane distribution of rhodopsin. The significant difference between these structures lies in the symmetry of the protein orientational distribution with respect to the central plane of the membrane; in the native structure the distribution is asymmetric, while in the reconstituted membrane it is symmetric. Thus, in the reconstituted membrane, both "ends" of the protein will be exposed at the outer membrane surface *if* the protein is transmembrane. Figure 2 shows SDS-polyacrylamide gel electrophoretograms of extracts of the native and reconstituted membrane vesicles before and after treatment with the proteolytic enzyme papain under the conditions given in the figure legend. The enzyme partially degrades all of the rhodopsin in the native membrane vesicles to produce two membrane-bound fragments of apparent molecular weights 27,000 (Rh27) and 12,000 (Rh12) daltons, a result similar to that found with thermolysin proteolysis.[17] In the reconstituted membranes, however, about 60% of the protein is degraded to produce the same fragments, with the remainder of the protein remaining intact, even after prolonged treatment. This result is interpreted to mean that the protein population with the inverted orientation (relative to the native membrane orientation) is either completely inaccessible to the protease or that no susceptible cleavage points exist on any externally exposed surface.* Although

* The ratio of the two rhodopsin populations is not expected to be 1:1 since the reconstituted membrane vesicles prepared by cholate dialysis are quite small (diameter ≃400 Å) and the ratio of the inner:outer surface areas is significantly less than unity.

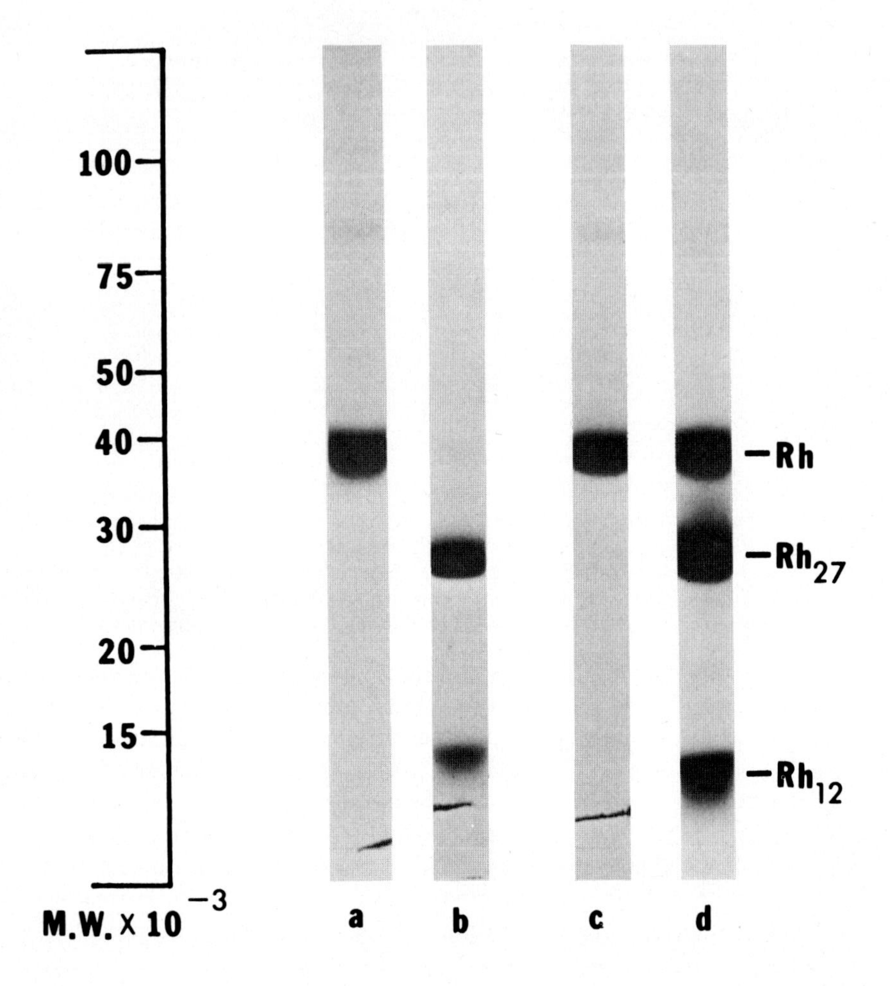

Fig. 2. SDS-polyacrylamide gel electrophoresis patterns of native and reconstituted vesicles before and after papain proteolysis. (a) native ROS membranes; (b) native ROS membranes after papain proteolysis; (c) reconstituted membranes; (d) reconstituted membranes after papain proteolysis. *Rh* marks the position of rhodopsin: Rh_{27} Rh_{12} mark the positions of the rhodopsin proteolytic fragments of molecular weight 27,000 and 12,000, respectively. Proteolysis of either ROS or reconstituted membrane vesicles (4 mg/ml rhodopsin) was carried out in 10 mM NaCl, 10 mM morpholinopropane sulfonic acid, 10 mM cysteine, 1 mM EDTA, pH 7·2, at a mole ratio of papain to rhodopsin of 1:30. After 3 h of proteolysis for ROS membranes or 8 h for the reconstituted membrane, the reaction was terminated by adding an equal volume of 100 mM iodoacetamide and incubating for 15 min. Longer periods of proteolysis produced no further changes in the gel patterns. The membranes were prepared for electrophoresis on polyacrylamide gels essentially as described by Fairbanks.[16]

this result does not allow us to conclude whether or not rhodopsin spans the membrane, it does provide a direct way to distinguish the two rhodopsin orientational populations indicated in Fig. 1, and this will be used to advantage in the experiments to be described below.

As shown in Fig. 3, rhodopsin in the native and reconstituted membrane vesicles can be enzymatically iodinated with ^{125}I using lactoperoxidase with a balanced amount of hydrogen peroxide generated by a glucose oxidase-glucose system.[18] Figure 3(b) shows the distribution of radioactivity on an SDS-polyacrylamide gel of the reconstituted membrane following papain treatment and lactoperoxidase catalyzed iodination with ^{125}I. Both orientational populations of rhodopsin are iodinated, suggesting that the molecule must span the membrane thickness. This tentative conclusion rests on the assumption that lactoperoxidase is excluded from the vesicle interior and that all iodination occurs by enzymatic catalysis. The first assumption seems justified since papain, a considerably smaller molecule than lactoperoxidase, is excluded from the vesicle interior. The second assumption is valid since no iodination occurs in the absence of lactoperoxidase when all other components of the reaction mixture are present. However, peptide mapping experiments are currently underway to experimentally demonstrate that iodination occurs exclusively at the external membrane surface.

Other experiments have been carried out following the same design but using the membrane impermeable reagent (I) instead of lactoperoxidase iodination.

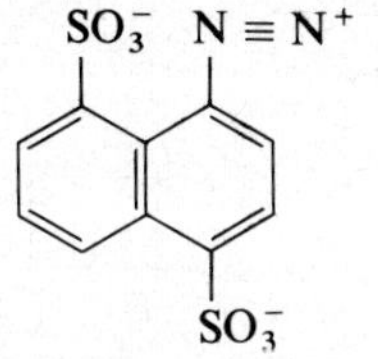

SO_3^- $N \equiv N^+$

SO_3^-

(I)

In this case, the labeled peptides are fluorescent, and it is again found that both orientational populations are labeled.

Thus, all available evidence leads to the conclusion that rhodopsin is a transmembrane protein, and the structure of the disc membrane is at least consistent with a permeability gating role for rhodopsin.

Calcium Permeability of Reconstituted Rhodopsin-Phospholipid Membranes

The possibility that rhodopsin may act as a light-dependent Ca^{++} permeability gate is an attractive one which we have chosen to investigate in compositionally defined reconstituted membranes. For the experiments to be

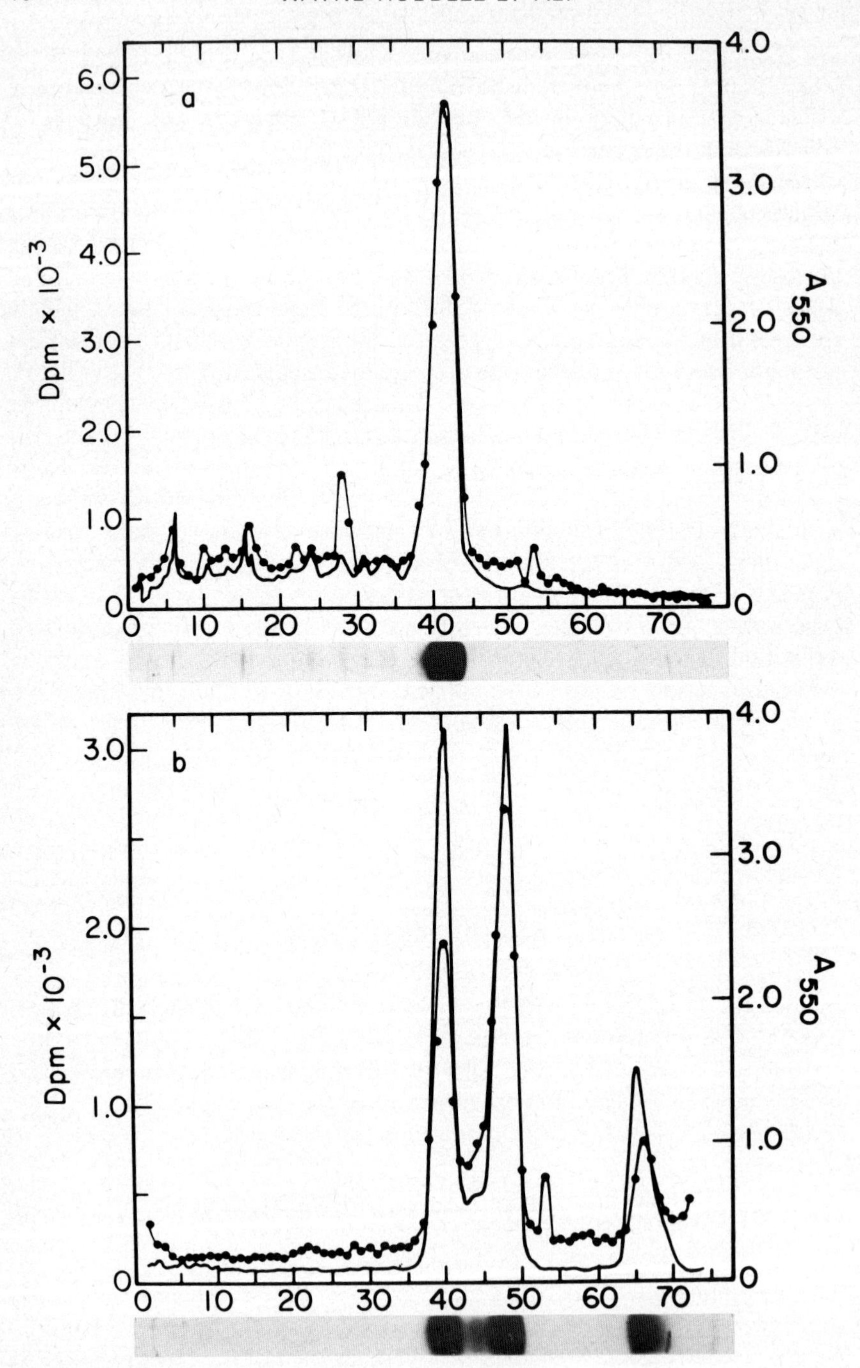
a
Dpm × 10^{-3}
A_{550}
6.0
5.0
4.0
3.0
2.0
1.0
0
10
20
30
40
50
60
70
4.0
3.0
2.0
1.0
b
Dpm × 10^{-3}
A_{550}
3.0
2.0
1.0
0
10
20
30
40
50
60
70
4.0
3.0
2.0
1.0

described here, membranes were prepared from highly purified rhodopsin and egg yolk phosphatidylcholine (PC) or mixtures of PC and egg yolk phosphatidylethanolamine (PE).[14] In all cases, the molar ratio of rhodopsin to total phospholipid in the membrane was approximately 1:100, similar to that in the native membrane.

Vesicles of reconstituted membranes containing trapped radioactive inulin, sucrose or $^{45}Ca^{++}$ may be produced by mild sonication of the membranes in a medium containing the desired substance. Figure 4 shows electron micrographs of a sonicated vesicle preparation; the average diameter of the vesicles is about 500 Å. Since the vesicles readily trap ^{14}C-sucrose or ^{3}H-inulin and appear quite impermeable to these compounds, we infer that the membranes are closed and have a definite internal volume. From the amount of trapped sucrose, the internal volume is estimated to be approximately 30 l/mol of rhodopsin. This value is the same order-of-magnitude as that estimated from the size distribution as determined by electron microscopy (assuming the vesicles to be of discoid shape as suggested by the internal volume measurements of Kornberg and McConnell[20] on sonicated vesicles of pure PC). According to this internal volume, vesicles prepared by sonication in the presence of 5 mM Ca^{++} should trap 0·12 moles of Ca^{++} per mole of rhodopsin. Experimentally a somewhat larger amount is found, possibly due to a population of bound as well as free internal Ca^{++}.

The efflux of $^{45}Ca^{++}$ from the vesicles has been studied by two different methods, both of which yield similar results.

In one method, the vesicles are passed onto a small glass column containing Sepharose 4B with concanavalin A covalently bound to the surface. Since rhodopsin binds to concanavalin A,[21] the vesicles are strongly adsorbed to the column bed. The desired eluant is then passed through the column and fractions collected as a function of time. The radioactivity profile of the

---

Fig. 3. SDS-polyacrylamide gel electrophoresis patterns of native and reconstituted vesicles labeled with ^{125}I. (a) native membrane vesicles: ●——●, radioactivity; ——, absorbance at 550 nm. (b) reconstituted membrane vesicles after papain proteolysis: ●——●, radioactivity; —— absorbance at 550 nm. The electrophoresis pattern of reconstituted membranes before proteolysis is identical with (a), except for a lower background staining and radioactivity due to the absence of minor proteins normally present in the native membranes. Papain proteolysis was carried out as described in the legend to Fig. 4. Iodination was carried out with 2 mg/ml rhodopsin in 0·1 M phosphate buffer, pH 7·2, containing 2 μM NaI, 300 μCi ^{125}I, 35 milliunits lactoperoxidase, 35 milliunits glucose oxidase and 10 mM glucose. After 30 min the reaction was terminated by addition of $Na_2S_2O_3$ to a final concentration of 0.1 mM. The membranes were washed twice with 0.1 mM $Na_2S_2O_3$, 0·1 M phosphate buffer and twice with 1 mM NaI, 0·1 M phosphate buffer to remove all unreacted ^{125}I before electrophoresis according to Fairbanks.[16] Gels were stained with Coomassie Blue and scanned at 550 nm. For radioactivity profiles, gels were sliced at 1 mm intervals and counted.

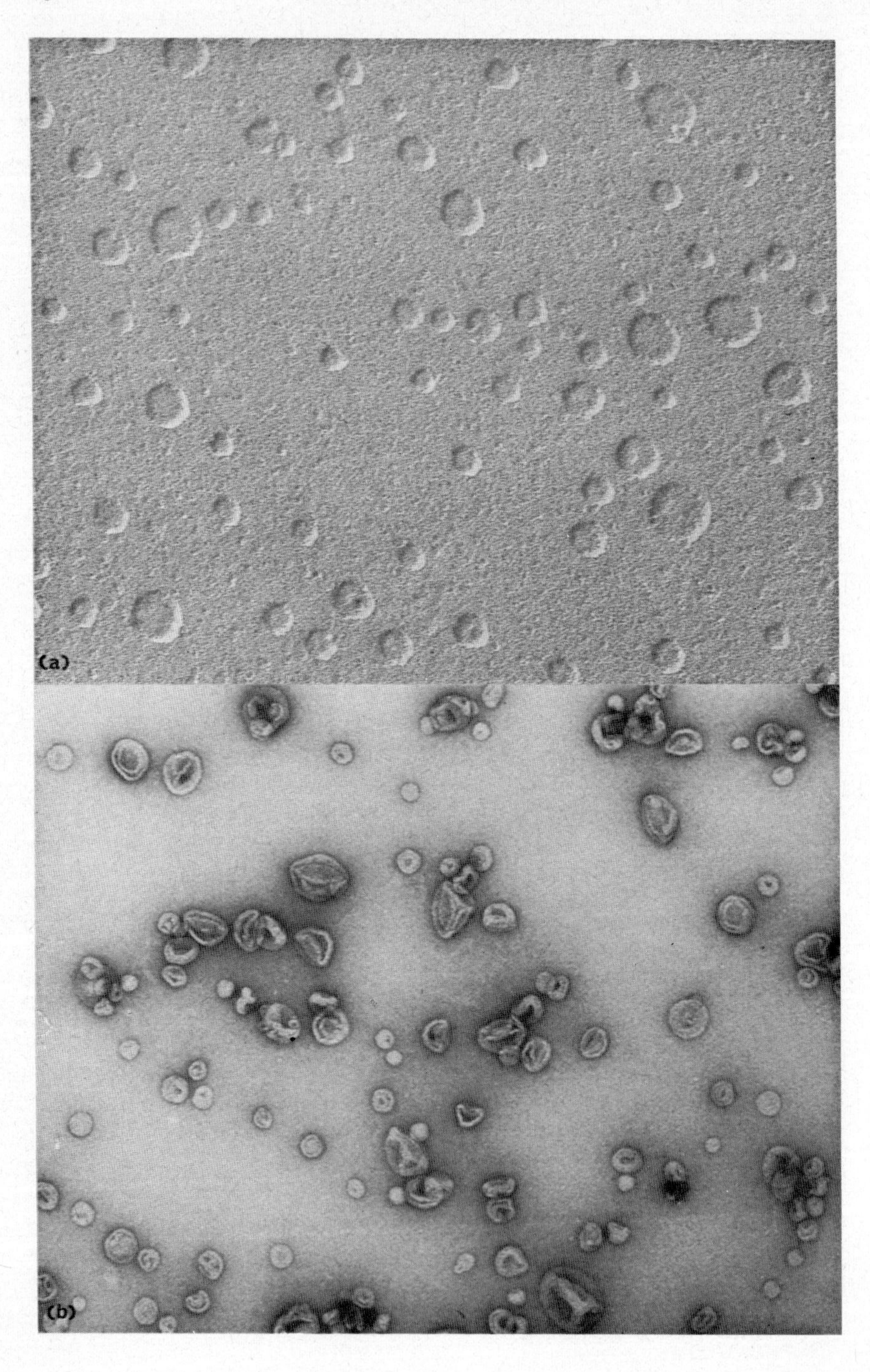
(a)
(b)

fractions describes the time-dependent isotope efflux from the adsorbed vesicles. In the second method, loaded vesicles are passed through a gel filtration column equilibrated with the desired medium and capable of resolving free Ca^{++} or small molecules from the vesicles. After separation of the vesicles from excess isotope (remaining from the loading procedure), but before elution of the vesicles themselves, a light flash is delivered. The elution profile of radioactivity after the light flash as compared to a control run entirely in the dark then gives relative rates of escape of isotope from the vesicles.

Figure 5 shows the results of an experiment using the first method when $^{45}Ca^{++}$ containing vesicles (prepared from PC and purified rhodopsin) are adsorbed to the column bed. The dotted line in the figure is the radioactivity elution profile obtained when the ionophore X537A in an EGTA buffer (see legend) is applied to the column in the dark. The rapid decrease in radioactivity of the early fractions (before ionophore addition) is due to washout of excess $^{45}Ca^{++}$ from the loading procedure. In the absence of ionophore, this washout curve simply declines to background count levels and remains essentially constant in time. The obvious ionophore-induced $^{45}Ca^{++}$ release indicates that a pool of unbound Ca^{++} is contained within the vesicle interior. This conclusion is supported by the fact that the amount of ionophore-induced $^{45}Ca^{++}$ release (0.14–0.19 mol Ca^{++}/mol rhodopsin) is in tolerable agreement with that predicted on the basis of vesicle internal volume as estimated by sucrose entrapment. The ionophore A23187 gives similar results, but the release kinetics are considerably more rapid.

Following the ionophore treatment, some additional $^{45}Ca^{++}$ may be released from the column by application of detergent solution containing α-methyl glucoside, a competitive inhibitor for rhodopsin binding to concanavalin A. This behavior again suggests the existence of a bound $^{45}Ca^{++}$ population.

The solid curve in Fig. 5 shows the radioactivity elution profile following a light flash which bleaches approximately 80% of the rhodopsin. The amount of $^{45}Ca^{++}$ released by the light flash (0·10–0·12 mol Ca^{++}/mol of rhodopsin) is similar to that released by the ionophores. Furthermore, application of the

---

Fig. 4. Electron micrographs of sonicated rhodopsin-phospholipid vesicles. Vesicles were supported on a polylysine coated[19] carbon surface and (a) shadowed with platinum carbon, ×78,600; (b) negative stained with uranyl acetate, ×78,600. The membranes were prepared from purified rhodopsin and egg phosphatidylcholine, 1:100 mole ratio, respectively, according to Hong and Hubbell.[13] Vesicles were prepared by sonication for 10 min (Heat Systems Ultrasonics Model 185 fitted with the microtip) at 4 degrees in a saline buffer containing: NaCl, 120 mM; KCl, 3 mM; $MgCl_2$, 3 mM; $CaCl_2$, 5 mM; morpholinopropane sulfonic acid, 10 mM, pH 6·6.

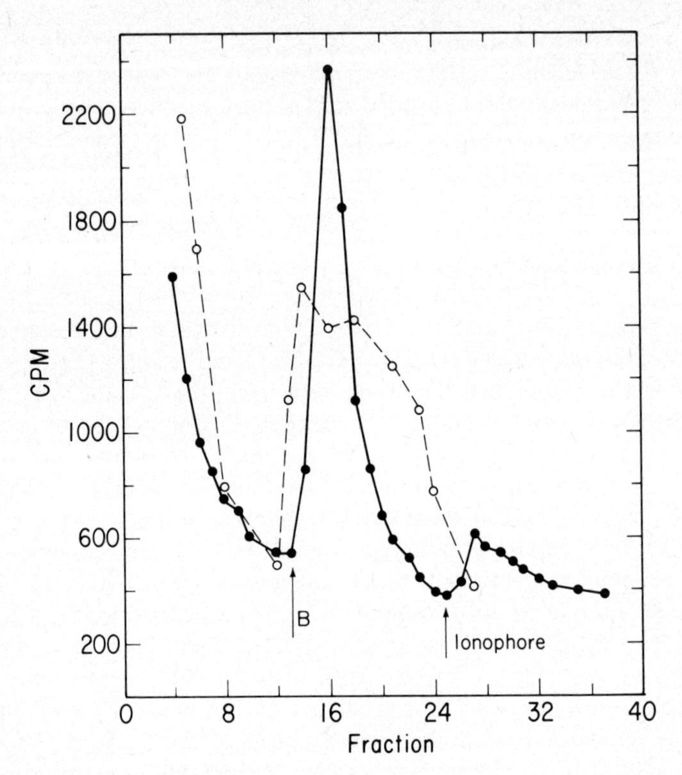

Fig. 5. Light-induced calcium release from reconstituted membrane vesicles. Vesicles (150 μl of a suspension containing *circa* 16 nanomoles of rhodopsin) preloaded with 5 mM radioactive $CaCl_2$ (2,000 cpm/nanomole Ca^{++}) in saline buffer (see legend, Fig. 4) were adsorbed on the bed surface of a concanavalin A-sepharose column (total volume 0·5 ml). Typically, 6–12 nanomoles of protein were retained on the column. Saline buffer without $CaCl_2$ and supplemented with 2 mM EGTA was passed through the column and fractions (200 μl) collected as a function of time and assayed for radioactivity. Each fraction represents 90 sec of effluent. The *filled circles* show the effect of light on the $^{45}Ca^{++}$ content of the effluent; the initial decrease is due to the washout of excess, external $^{45}Ca^{++}$. A light flash was delivered at the point marked "B". Following the light-induced release, X537A (20 μM) was applied at the time shown. The *open circles* show the effect of X537A (20 μM) applied to the column in the dark at fraction 12. A light flash following the ionophore produces no additional release.

ionophore following a light flash (see Fig. 5) or vice versa produces only a very small additional release.

Release profiles of the type shown in Fig. 5 have also been obtained for reconstituted vesicles prepared from mixtures of PC and PE and purified rhodopsin. In these membranes, the light-induced efflux is considerably slower than for pure PC membranes, and we have found significant enhancement in the presence of 2 μM valinomycin when the medium contains KCl. Valinomycin alone in the dark produced no $^{45}Ca^{++}$ release. This effect might be expected if the permeability increase were highly selective for Ca^{++}, since a transmembrane potential would then arise during ion efflux. Once the

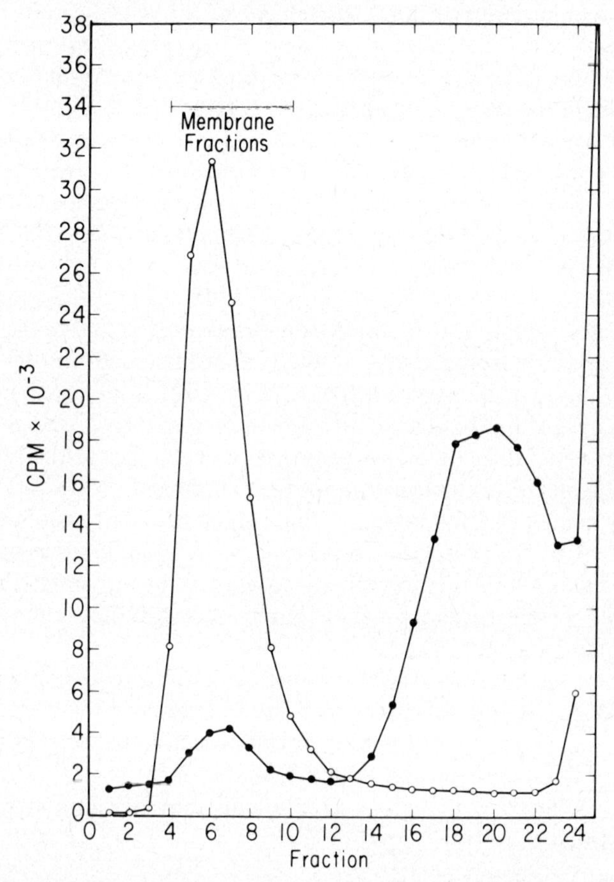

Fig. 6. Light-induced calcium release from reconstituted membrane vesicles. Vesicles (150 μl of a suspension containing 38 nanomoles of rhodopsin) preloaded with 40 mM radioactive $CaCl_2$ (4,000 cpm/nanomole Ca^{++}) in saline buffer (see legend, Fig. 4), were applied to a column (16 ml volume) of BioGel A 0·5 M (BioRad Laboratories) equilibrated with the saline buffer without $CaCl_2$ and supplemented with 2 mM EGTA. Fractions of 300 μl were collected each 90 seconds and assayed for radioactivity. ○: show the radioactivity elution profile in the dark. The vesicles elute in fractions 4–12; the steep front of radioactivity beginning at fraction 23 is due to the excess $^{45}Ca^{++}$ from the loading procedure. ●: show the radioactivity elution profile following a light flash delivered when the vesicles have migrated approximately half-way down the column. At this point, the vesicles have separated from the excess $^{45}Ca^{++}$. Again, the vesicles themselves elute between fractions 4–12.

membrane capacitor were charged at the equilibrium potential, efflux would be limited by the leakage rates of other ions. Valinomycin in the presence of K^+ would effectively short-circuit the membrane capacitance and reduce the transmembrane potential.

In order to eliminate possible involvement of the concanavalin A in the processes being studied, similar experiments were carried out using the gel filtration method described above. Figure 6 (filled circles) shows $^{45}Ca^{++}$ elution profile in the dark using an agarose gel filtration column. The initial peak of radioactivity coincides with vesicle elution at the void volume, and is followed by the intense front of excess isotope from the loading procedure. The radioactivity peak associated with the vesicle fraction is relatively symmetric, indicating that the trapped Ca^{++} is not rapidly "leaking" from the vesicles as they proceed down the column. The open circles in Fig. 6 show the elution profile obtained when the rhodopsin is bleached on the column after the vesicles have completely separated from the excess $^{45}Ca^{++}$ in the medium. Now, relatively little isotope elutes at the void volume with the vesicles, but later appears in a relatively symmetrical peak which could only have been produced by a rapid light-induced $^{45}Ca^{++}$ efflux from the vesicles. Studies are now in progress to utilize the information contained in the elution line-shapes to place limits on the release kinetics. In the experiment shown in Fig. 6, the vesicles were loaded with 40 mM Ca^{++}, but similar profiles are obtained when 5 mM Ca^{++} of the same specific activity is used, although the number of counts per fraction are correspondingly reduced. This illustrates the important point that the release phenomenon is independent of the amount of Ca^{++} contained.

Similar experiments have been conducted with vesicles containing internalized ^3H-inulin, and we find no light-induced permeability changes to this compound, suggesting that bleaching does not simply cause rupture of the vesicles or a highly non-specific permeability increase.

The conclusions from the above experiments may be summarized by the following points:

(1) Rhodopsin-phospholipid vesicles contain a well-defined interior volume.
(2) The vesicles can be loaded with $^{45}Ca^{++}$, and the ionophores X537A and A23187 release an amount of isotope consistent with that expected to occupy the interior volume.
(3) Illumination of loaded vesicles releases an amount of $^{45}Ca^{++}$ nearly equal to that released by the ionophores.
(4) Illumination and the ionophores release the same isotope pool.
(5) Illumination does not release trapped ^3H-inulin under the same conditions used to observe $^{45}Ca^{++}$ release.

To conclude that rhodopsin regulates transmembrane permeability, it must be unequivocally demonstrated that the $^{45}Ca^{++}$ released by light originates from an unbound pool in the vesicle interior, and is not the result of vesicle rupture. The evidence that this is the case is contained in conclusion (2) and (5) above along with the assumption that here, as in other systems, the ionophores induce transmembrane movements of free ions. Under the conditions of our experiments, the kinetics of X537A release are quite slow, requiring 20–30 min for complete release of the available isotope pool. On the other hand, little release is seen in the same time period in the absence of ionophore. We have no means of predicting *a priori* what the kinetics of ionophore transport should be in our system, since a thorough kinetic analysis of this ionophore has not yet been reported. However, in other experiments using A23187 we find a much more rapid release. The possibility that the ionophore directly removes bound $^{45}Ca^{++}$ from the external surface by complexation is remote, since the membranes are continuously bathed in a solution containing EGTA, a compound with much greater affinity for Ca^{++}.

An important aspect of the Ca^{++} channel hypothesis is that it provides signal amplification; one photon activates one channel which can release hundreds to thousands of ions in the requisite time. In the present experiments, it is not possible to obtain the movement of hundreds or thousands of ions per bleached rhodopsin molecule under any condition, simply because the internal volume of the vesicles is very small. As pointed out above, the trapped volume measurements imply that at 5 mM Ca^{++}, there can be only *circa* 0·14 mol Ca^{++}/mol rhodopsin. No fundamental significance should be attached to this number, since it is an experimental variable depending on vesicle size and the Ca^{++} concentration used during the loading procedure. This is illustrated by the data in Fig. 6, where more than 1 mol Ca^{++}/mol rhodopsin is trapped and released when the vesicles are loaded at 40 mM Ca^{++}. An important extension of these experiments will be to perform partial bleaching. It would be predicted that if bleached rhodopsin acts as a long-lived channel, the number of $^{45}Ca^{++}$ released per rhodopsin would increase as the per cent of bleached rhodopsin decreases. These experiments are difficult to carry out on the column system used here because extreme light scattering problems result in highly non-uniform bleaching at low bleach levels. Modifications of this method to allow such experiments are currently under investigation.

Conformational Transitions and Light-Dependent Chemistry of Rhodopsin

The results of the preceding experiments suggest that rhodopsin is a transmembrane protein, and that the permeability of membranes containing rhodopsin as the only protein can be changed by light. Anticipating that these tentative conclusions will withstand further test, we are now looking for conformational transitions in the protein which are expected to mediate the photochemical and permeability events. In this final section we present direct physical evidence for the existence of relatively rapid, reversible conformational changes of rhodopsin which occur following photon absorption in the native membrane. The evidence is derived from EPR spectra of rhodopsin selectively spin-labeled on particular sulfhydryl (SH) groups. Rhodopsin has a total of six free SH groups,[22] two of which react in the dark adapted native membrane with the sulfhydryl reagent 4,4′-dithiopyridine (4-PDS), and these will be referred to as S1 and S2. A third group, S3, is found to react with certain spin label reagents, as will be discussed below.

The spin label reagents used in the experiments are the organomercury compounds (II) and (III):

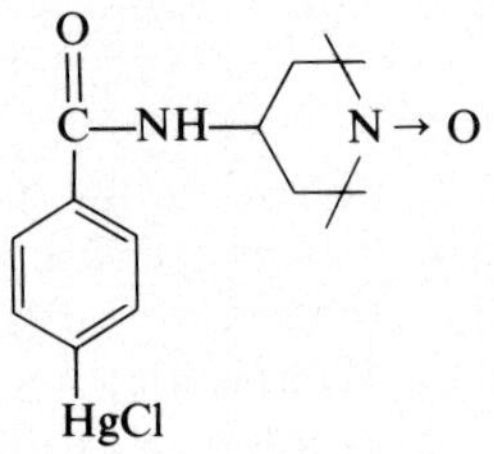

O ‖ C—NH— N→O

HgCl

(II)

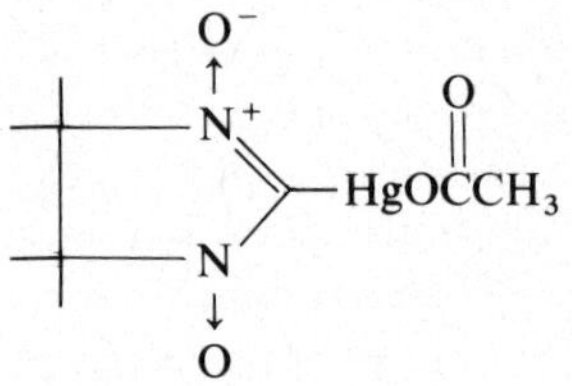

O^- ↑ N^+ —$HgOCCH_3$ (O ‖) N ↓ O

(III)

Both compounds have a high degree of specificity for SH groups. If S1 and S2 are blocked by reaction with 4-PDS in the dark, (II) reacts selectively with S3 to give the EPR spectrum shown by the dashed curve in Fig. 7. Since 96 % of the bovine ROS sulfhydryls are attributable to rhodopsin,[23] there is no question that the spectrum originates from labeled rhodopsin. The solid curve in Fig. 7 shows that bleaching the entire complement of rhodopsin produces a substantial decrease in mobility of the label reflecting a change in the local organization of the protein. The detailed EPR line shape and magnitude of the bleaching effect vary from sample to sample and the change shown in Fig. 7 is one of the largest we have observed, but was obtained in many different experiments. In other preparations, the bleaching change was only a fraction of the magnitude shown here but was always present. We have no clue as to the origin of the variability, but two observations suggest that this conformational change may have functional significance in excitation. First, the EPR spectral change is reversed when the bleached protein is regenerated from 11-*cis* retinal in the dark. Second, the transition is complete within about 200 msec at 20° following a 6 μsec bleaching flash from a dye laser.* The magnitude of the

* See Addendum, p. 58.

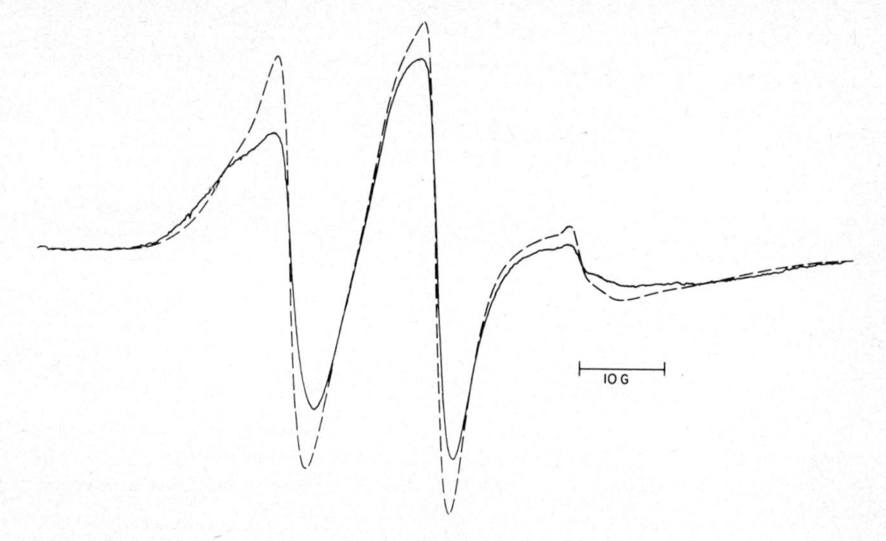

Fig. 7. EPR spectra of bovine ROS membranes spin labeled with (II). (------) dark adapted; (———) bleached.

spectral change depends on the amount of rhodopsin bleached, but we do not know if the relationship is linear.

If dark adapted ROS membranes are treated with (II) without prior blocking with 4-PDS, only one of the two dark reactive sulfhydryls (S1 or S2) is labeled. A conformational transition with bleaching is also detected at this position in the molecule, but is considerably smaller than that in the vicinity of S3.

Label (III) is a nitronyl nitroxide[24] which also selectively labels a single group which we believe to be S3 when S1 and S2 are blocked with 4-PDS. The EPR spectrum of the labeled membranes in the dark (Fig. 8, solid line) shows the radical to be strongly immobilized as indicated by the 72 G separation of the nitrogen hyperfine extrema (the maximum separation obtained in a rigid glass is 75 G). The response of the spectrum to a brief light flash which bleaches a large portion of the rhodopsin corroborates the result obtained with spin label (II) on S3, that is, the spectrum becomes even more immobilized with the maximum hyperfine splitting increasing by about 2 G (not shown). Interestingly, the response of the system to *continuous* light is quite different and results in the steady decline of the strongly immobilized spectrum and a concommitant increase in intensity of a sharp, 10 line pattern as shown in Fig. 8, dotted line.* The kinetics of this light-dependent process have not yet been studied, but the reaction is complete in approximately

* See Addendum, p. 58.

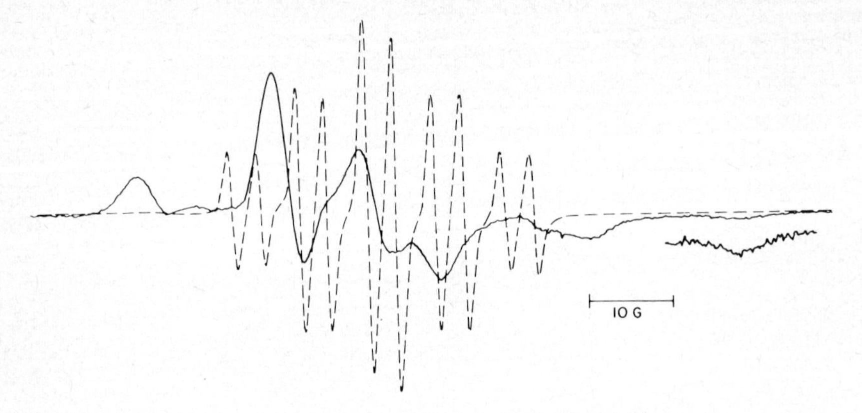

Fig. 8. EPR spectra of bovine ROS membranes spin labeled with (III). (———) dark adapted. A portion of the high field region was recorded at increased spectrometer gain to show the broad transition. (– – – – – – – – –) bleached with continuous irradiation (see text).

20 min under intensities similar to room light. Unlike the strongly immobilized radical in the dark membranes, the radical giving rise to the 10 line spectrum can be removed from the membranes by washing, and is readily identified as the nitronyl nitroxide (IV),

O^- ↑ N^+ ⟩—H ; N ↓ O (IV)

The elimination most likely proceeds according to the following net reaction,

O^- ↑ N^+ ⟩—Hg—S—(rhodopsin)—SH ; N ↓ O $\xrightarrow{h\nu}$ O^- ↑ N^+ ⟩—H + S—Hg—S—(rhodopsin) ; N ↓ O

Before interpreting this response, it is helpful to consider the result of one additional experiment. When purified ROS membranes are incubated in the dark with 4-PDS, S1 and S2 are labeled, as mentioned above. After bleaching all of the rhodopsin, S1 and S2 are still the only groups labeled with 4-PDS, *provided the reaction is carried out in the dark* after the bleach. Even though no additional groups are reactive, the reaction velocities of 4-PDS with both S1 and S2 are increased, indicating an altered molecular environment in the vicinity of the sulfhydryls. A similar observation has been made for rhodopsin in digitonin solution.[25] With continuous illumination, however, the situation is quite different. Under steady room light or light passed by a 475 nm cut-off filter ($\lambda \geq 475$ nm), additional SH groups are found to react with 4-PDS. The number of additional groups depends on the composition of the medium, but always lies between 1 and 3. These reactions are not unique to 4-PDS, and similar results are found with ^{14}C-N-ethylmaleimide. In addition, the reduction of the simple nitroxide TEMPO (V)

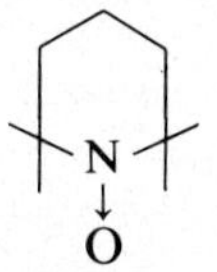

(V)

by suspensions of ROS also appears to be enhanced under continuous illumination.[26]

The elimination of the nitronyl nitroxide, the apparent increase in number of reactive sulfhydryls and the reduction of (V) with steady illumination are believed to be manifestations of the same process, for which we have considered two alternative explanations. First, the reaction may involve primary photochemical processes in addition to isomerization, perhaps similar to sensitized photochemical oxidations of proteins with aromatic dyes.[27] For example, the excited polyene chromophore may abstract a hydrogen atom from a nearby SH group, thus producing the hyper-reactive thiyl radical. Detailed mechanisms will not be discussed here, but suffice it to say that the generation and subsequent reactions of this radical could explain the continuous light-dependent reactions discussed above. An interesting feature of these reactions is that they do not seem to consume retinal, and if this mechanism obtains, the polyene must be involved in a catalytic role. Since the 4-PDS reactions can be driven with light of $\lambda \geq 475$ nm,* it is likely that only protein-associated retinal is being excited, and this model would require the close proximity of at least one SH and the bound chromophore (this SH group cannot be one of those normally exposed on the protein surface in the dark membranes, since both of these have been protected with 4-PDS in all experiments discussed here).

The second possible mechanism involves chromophore isomerization as the only photoprocess. When rhodopsin absorbs a single photon, isomerization of retinal occurs and initiates a sequence of protein transformations which proceeds thermally without the further involvement of light. Several intermediates along this dark reaction pathway have been identified by their characteristic absorption spectra, but there may be chemically distinct isochromic forms.[28] All of the thermal intermediates are light sensitive by

* The wavelength dependence of the nitronyl nitroxide elimination and the TEMPO reduction have not yet been investigated.

virtue of photoisomerization of retinal, and absorption of a photon by some of them results in the regeneration of rhodopsin, which can again be bleached by the absorption of yet another photon. Thus, under continuous irradiation, several cycles of continuous bleaching and regeneration are possible. Each cyclic pathway involves some of the normal thermal intermediates, and, in principle, a finite steady-state concentration of all dark intermediates could exist under constant irradiation. Now if it is imagined that a short-lived intermediate exists, and that this intermediate has a conformation with a closely apposed pair of available SH groups, then the reactions described earlier can be readily understood, since only under continuous irradiation would the short-lived intermediate be constantly available for any of the sulfhydryl reactions to reach completion. The 4-PDS reactions would be explained by the increased SH availability in the conformational state of the intermediate; the nitronyl nitroxide elimination and the TEMPO reduction would be explained by the proximity of the two SH groups in the assumed intermediate. Further studies are underway to distinguish between possible mechanisms, but under any condition, the elimination reaction of the nitronyl nitroxide requires the close approach of two SH groups.

Finally, it must be remembered that the information presented here concerning light-induced conformational transitions and reactions was derived from rhodopsin labeled in various ways. The extent to which the properties discussed are unique to the modified protein can only be assessed as more information becomes available on the chemistry and function of the protein. However, it should be clear that the rhodopsin sulfhydryl groups, if not directly involved in the function, are at least located in "conformationally active" regions of the molecule.

Addendum

Since the submission of this manuscript, further experiments have revealed that in several preparations the elimination of the nitronyl nitroxide (III) does not require continuous illumination, but proceeds in the dark after initiation by light. In the instances that appear to require continuous illumination, it may be that light simply accelerates the elimination, and if sufficient time were allowed, the reaction may proceed in the dark after the initial bleach.

Further experiments on the kinetics of the structural transition detected by label (II) have revealed that the transition has a half-life of about 3 ms at 20°C.

References

1. R. A. Cone 1972, *In* "Biochemistry and Physiology of Visual Pigments" (ed. H. Langer), pp. 275–282. Springer-Verlag, Berlin.

2. W. A. Hagins 1972, *Ann. Rev. Biophy. Bioeng.* **1**, 131–158.
3. S. Yoshikami and W. A. Hagins 1972, *In* "Biochemistry and Physiology of Visual Pigments" (ed. H. Langer), pp. 246–256. Springer-Verlag, Berlin.
4. W. A. Hagins and S. Yoshikami 1974, *Exp. Eye Res.* **18**, 299–305.
5. M. Meller, N. Virmaux and P. Mandel 1975, *Nature* **256**, 68–70.
6. P. Liebman 1974, *Investigative Opthal.* **13**, 700–701.
7. Th. Hendriks, F. J. M. Daemen and S. L. Bonting 1974, *Biochim. Biophys. Acta* **345**, 468–473.
8. N. T. Mason, R. S. Fager and E. W. Abrahamson 1974, *Nature* **247**, 562–563.
9. E. Z. Szuts and R. A. Cone 1974, *Federation Proceedings* **33**, 1471.
10. M. W. Bitensky, N. Miki, J. Keirns, M. Keirns, J. Baraban, J. Freeman, M. Wheeler, J. Lacy and F. Marcus 1975, *In* "Advances in Cyclic Nucleotide Research" Vol. 5 (eds. G. Drummond, P. Greengard and G. Robinson), pp. 213–240. Raven Press, New York.
11. D. Bownds, J. Dawes, J. Miller and M. Stahlman 1972, *Nature New Biol.* **237**, 125–127.
12. Y. S. Chen and W. L. Hubbell 1973, *Exp. Eye Res.* **17**, 517–532.
13. K. Hong and W. L. Hubbell 1972, *Proc. Nat. Acad. Sci. USA*, **69**, 2617–2621.
14. K. Hong and W. L. Hubbell 1973, *Biochemistry* **12**, 4517–4523.
15. M. Applebury, D. Zuckerman, A. Lamola and T. Jovin 1974, *Biochemistry* **13**, 3448.
16. G. Fairbanks, T. L. Steck and D. F. H. Wallach 1971, *Biochemistry* **10**, 2606.
17. J. Pober and L. Stryer 1975, *J. Mol. Biol.* **95**, 477–481.
18. A. L. Hubbard and Z. A. Cohn 1972, *J. Cell Biol.* **55**, 390–405.
19. K. Fisher 1976, *Proc. Nat. Acad. Sci.* **73**, 173–177.
20. R. Kornberg and H. M. McConnell 1971, *Biochemistry* **10**, 1111.
21. A. Steinemann and L. Stryer 1973, *Biochemistry* **12**, 1499.
22. W. J. De Grip, G. L. M. Van De Laar, F. J. M. Daemen and S. L. Bonting 1973, *Biochim. Biophys. Acta* **325**, 315–322.
23. W. J. De Grip, S. L. Bonting and F. J. M. Daemen 1975, *Biochim. Biophys. Acta* **396**, 104–115.
24. J. Osiecki and E. F. Ullman 1968, *J. Am. Chem. Soc.* **90**, 1078–1079.
25. E. A. Kimble and S. E. Ostroy 1974, *Biochim. Biophys. Acta* **325**, 323–331.
26. M. Delmelle and M. Pontus 1975, *Vision Res.* **15**, 145–147.
27. J. Spikes and M. MacKnight 1970, *In* "Photochemistry of Macromolecules" (ed. R. Reinish), pp. 67–83. Plenum Press, New York–London.
28. T. P. Williams 1970, *Vision Res.* **15**, 69–72.

5

What X-ray and Neutron Diffraction Contribute to Understanding the Structure of the Disc Membrane

A. E. BLAUROCK

Department of Chemistry, California Institute of Technology, Pasadena, California, U.S.A.

Introduction

This paper on the diffraction studies of disc membrane is divided into three sections. The first section will deal with the primary result, which is a map of the electron density. It will be shown that there is general agreement amongst X-ray diffractionists on the electron-density profile of the disc membrane, with but one dissenting view. The situation with respect to the in-plane diffraction from the disc membrane has been less harmonious, but I believe that we are coming to a consensus. The second section will deal with the interpretation of the profile and of the in-plane data. A number of different structures have been put forward, which are all based on the bilayer structure. I will describe these structures and indicate which of them seems the most probable—but in the end the choice of one of them must be the reader's. Finally, I will describe briefly some X-ray and neutron diffraction experiments on the effects of bleaching the rhodopsin.

It should be understood that the division into sections is artificial. Thus a lipid bilayer-like structure generally was assumed in order to compute a profile for the membrane. Nonetheless, this division of the subject has the advantage of showing the broad agreement on the profile itself, as distinct from the substantial disagreement as to the interpretation of the profile. In this way I want to help the reader sort through the confusing multiplicity of structures which have been proposed.

First some remarks about working methods.

It is widely accepted that the structure of a molecule can be deduced from diffraction experiments, provided that it can be crystallized. One standard approach is to label the molecule with heavy atoms. In this case the analysis assumes the structure of the heavy atom. Because we feel very confident about this structure, the results of the diffraction experiment are accepted almost without question.

Alas, this confidence cannot be transferred to the low-angle work. It is important to understand that there is no standard method of deducing the electron-density map from low-angle data. The heavy-atom method, for example, is out of the question: neither the physical resolution, nor the dimensionality of the data, nor the degree of crystallinity are adequate to resolve single atoms. We must accept that the analysis of a low-angle diffraction problem generally makes important assumptions at the beginning, and these assumptions often are not as certain as the structure of an atom. These assumptions will be based on other data such as electron microscopy, interference microscopy and, probably most important, on the available chemical data. What we seek is a structure to account for the X-ray diffraction data, and it must be consistent with other relevant observations.

Those of us doing X-ray work have been slow to indicate the weaknesses in our findings. Often it has not been emphasized that a certain interpretation is not unique, and other possibilities have been ignored without comment. There has at times been some carelessness in collecting data, or in knowing precisely to what the data relate, and erroneous structures have been put forward as a result.

Lest I lower unduly the esteem held for X-ray work, I will point out that seven years ago there were some independent reports that the molecular weight of rhodopsin is below 30,000. Is there anyone who cares to defend this value today? What precisely is the proportion of protein in the disc membrane? Early reports set it at 60%—is that still the consensus? I suspect not. I have found that electron microscopists are apt to overemphasize the asymmetry of membrane structures when these are examined by means of freeze-fracture. For reasons such as these, I ask that the X-ray work should not be ignored. I think the X-ray work will be found to be useful if its strengths and weaknesses are understood.

The Membrane Profiles and In-plane Diffraction

Figure 1 shows the electron-density profiles calculated by four groups of workers. (a) is a set of profiles derived by Wilkins and Blaurock.[1] We first put forward a profile by assuming a bilayer membrane and by referring to electron micrographs showing the two membranes in the disc. We then proved the

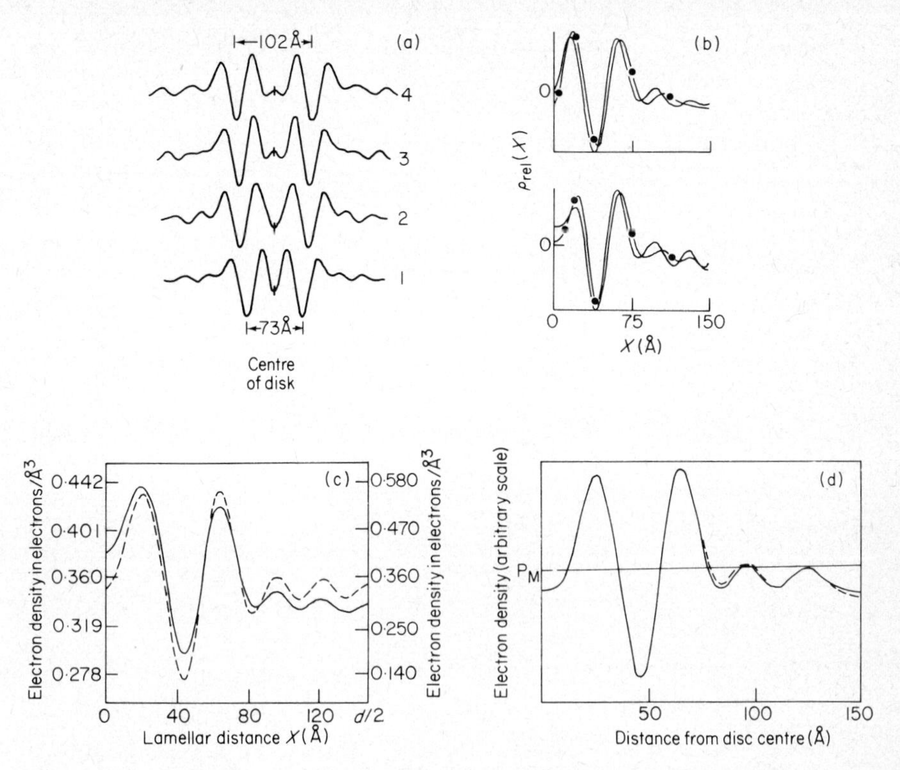

Fig. 1. Electron-density profiles of the disc membrane. The curves are from (a) Blaurock and Wilkins[1]; (b) Corless[2]; (c) Worthington[3]; and (d) Chabre.[4]

correctness of the bilayer profile by doing osmotic swelling and shrinking experiments. Each profile in the series shows the two membranes making up the disc, with the cytoplasmic fluid to either side of the disc. Both the space between discs and the space within the disc can be seen to vary in width, but a constant membrane profile is seen. I note that this has been the most successful working method in low-angle studies although it is not without pitfalls.

The profiles derived by Corless[2] are shown in (b). In this case the phasing was similarly checked by osmotic shrinking of the disc. The two curves in each pair show the shift in position of a constant membrane profile. The profile reported by Worthington[3] is shown in (c), and (d) is the one calculated by Chabre and Cavaggioni.[4]

In all these cases the profiles are similar because the bilayer structure was assumed at the beginning of the respective analyses. The two peaks in each single-membrane profile are identified with the electron-dense headgroups of the lipids. The trough at the center of each membrane is identified with the lipid fatty chains in the core of the membrane since only these are less electron dense than the cytoplasm. The rather large ripples in the cytoplasm are thought to be artifacts, and recent work by Schwartz *et al.*[5] supports this conclusion.

In contrast to the lipids in the membrane, the protein does not stand out. This is not surprising since the possible range of electron density in a protein is limited: all of the twenty amino-acid residues are considerably more dense

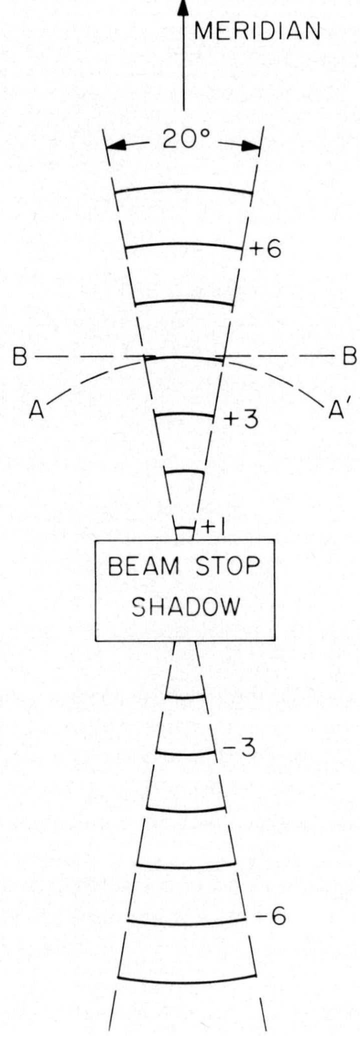

Fig. 2. Schematic diffraction pattern to show the ways of measuring the arc length of the disc-stacking reflections. A–A′: Blaurock and Wilkins (unpublished). B–B′: Worthington.[3]

than water.[6] However, the membrane protein makes itself felt: the area in the central trough is much smaller, in proportion to the area under the two peaks, than for a bilayer of the membrane lipids.[6] Thus the protein raises the density of the central trough. The lipid headgroups and the protein are both electron dense, and protein inserted into the layers of headgroups will not be so easily detected.

You will appreciate that by examining only these profiles we cannot say very much about how the protein is distributed in the membrane. One observation is obvious, however: the membrane profile is close to symmetric. Insofar as an asymmetric distribution of protein in the bilayer would force an asymmetric distribution of the lipids, I conclude that the protein is arranged in a fairly symmetric way in the disc membrane. Model calculations support this conclusion, as I will show in the second part of my paper.

The profile calculated by Worthington[3] is somewhat different from the others. While the discrepancy is not large it does lead to a different structure for the membrane. I am therefore obliged to comment on the discrepancy, which is due to the different correction factor used.

In general, the diffracted intensity from a specimen is subject to a correction which depends on the nature of the specimen and on the kind of diffraction camera used. If all of the outer segments in a retina specimen were parallel and if they contained uniformly stacked discs, then the Bragg reflections would all be spots the size of the X-ray beam. In fact, using a diffraction camera in which the X-rays are point-focused to a spot roughly 0·1 mm in diameter, we have always observed arcing. The arcing indicates to us that the outer segments are not strictly parallel to one another. We examined the arcing more closely, as follows.

A densitometer tracing was made by rotating the diffraction pattern about its center and recording the variation of density along an arc at constant radius from the center. I used a rotating film table made at King's College especially for this purpose. Figure 2 shows a schematic diffraction pattern; the tracing was along the path A–A′. I measured the density at intervals along some of the stronger reflections, for which the data will be the most accurate. Figure 3 shows the density recorded every 2°. The angular widths are the same for orders 3 and 6 to within the error of measurement: $18° \pm 1°$. Order one, which was visible on one side only of the diffraction pattern, is wider by 2°. The greater angular width is attributed to the finite size of the X-ray beam, which must be convoluted with the natural arc length to account for the arc observed.

The observed arc length of order one is 0·15 mm. Assuming two one-dimensional Gaussians as models for the natural arc length and for the finite size of the beam, the beam would need to have been 0·06–0·07 mm wide (full width at half maximum). This is an acceptable value. The correction we made

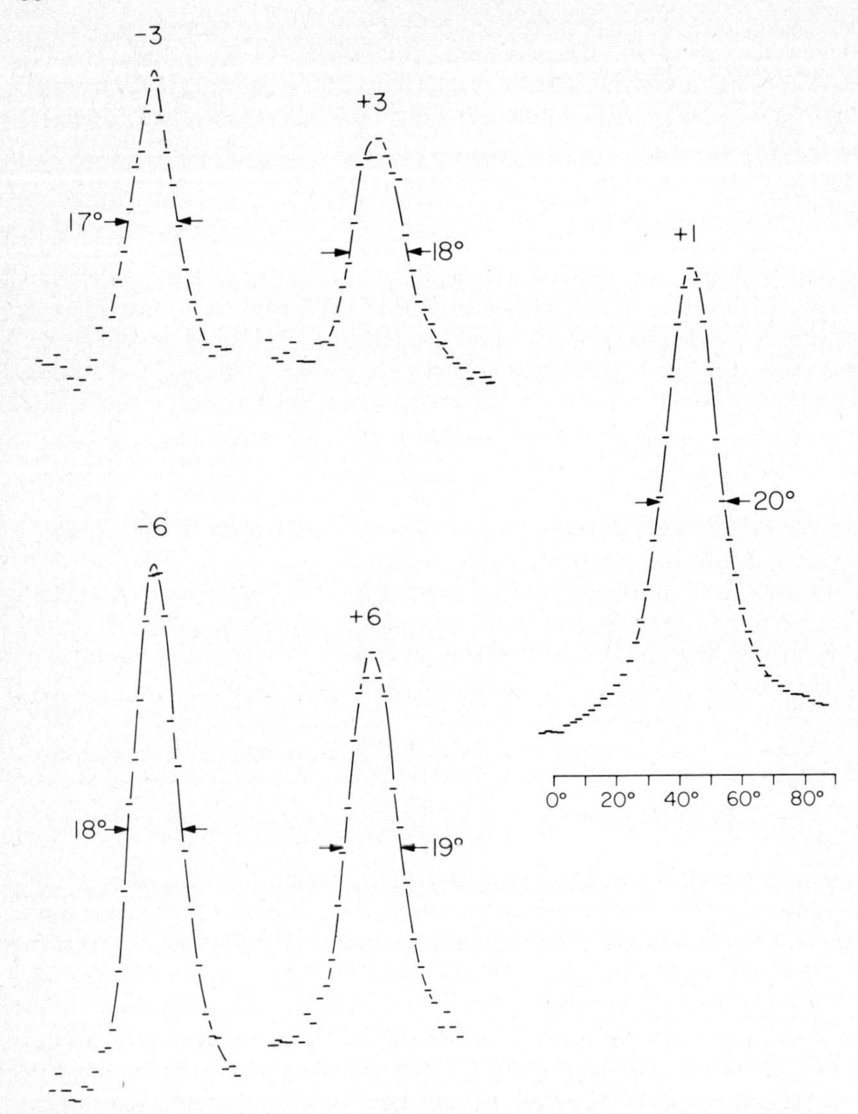

Fig. 3. Angular density profiles of some of the disc-stacking reflections from the excised retina (Blaurock and Wilkins, unpublished data). The density was measured every 2° (see text) along the path A–A′ in Fig. 2. The plus and minus signs refer to diametrically opposite sides of the diffraction pattern. The pattern itself is shown in Blaurock and Wilkins.[1] Specimen-to-film distance was 75 mm, and $D_{avg} = 275$ Å.

to the intensities measured in a radial densitometer tracing[1] were based on these observations.

Originally Gras and Worthington[7] recorded the pattern using a line-

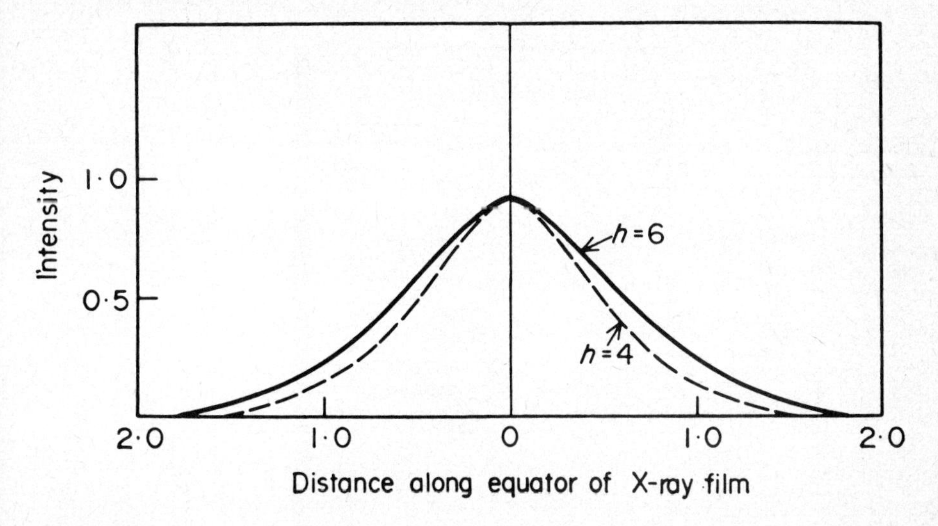

Fig. 4. Densitometer tracings of two reflections measured along the path B–B′ in Fig. 2.[3]

focussed beam, and apparently they did not know that the reflections were naturally arcs. However, Worthington[3] later published densitometer tracings made on a point-focussed pattern. He chose the path B–B′ in Fig. 2. Figure 4 shows the resultant tracings. Order 6 is longer than order 4, as expected from our own observations. However, the ratio of the lengths is not as great as the 6:4 we would find. This result is crucial to the choice of correction factor.

Too few experimental details are given for me to make critical comparison of Worthington's result with our own observations. In particular one would like to know the size of the X-ray beam in comparison to the observed arc lengths, but these data are lacking from Worthington's report. A general comment on the discrepancy is that the straight-line path he chose is not appropriate for arced reflections. That path would underestimate the length of an arc, the more so the higher the order number, which may account for the discrepancy.

As it now stands, three of us have calculated profiles (Fig. 1) which are the same to within the accuracy of the data, indicating if nothing more that we agree on the necessary correction to the observed intensities.

The diffraction from the excised retina also includes a diffuse reflection on the axis parallel to the plane of the membranes, centered at 50–55 Å.[1] We attributed the reflection to the distribution of protein molecules in the plane of the membrane. Assuming the rhodopsin predominates, this reflection relates to the arrangement of the rhodopsin molecules. The in-plane diffraction is

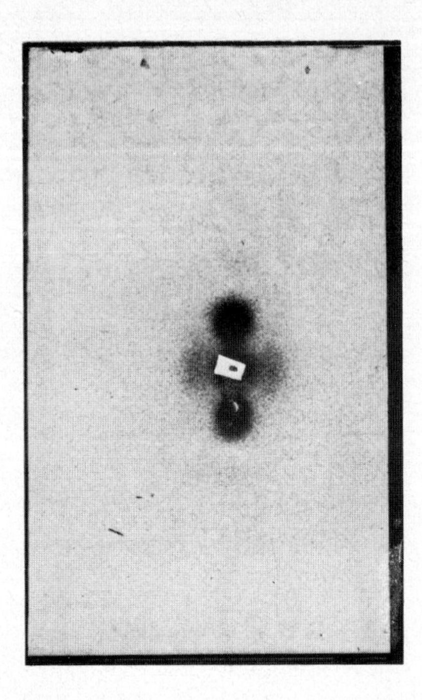

Fig. 5. In-plane reflection from the discs in the excised retina.[1] The pair of diffuse reflections on the horizontal axis are attributed to a fairly close, but irregular, packing of rhodopsin in the plane of the disc membrane.

shown in Fig. 5. We attribute the width of the reflection, in the horizontal direction, to a fairly close but not geometrically regular packing of the rhodopsin.

Blasie *et al.* (1965)[8] reported a different in-plane pattern from the isolated disc membrane, and Blasie *et al.* (1969) have analyzed this pattern in some detail.[9] The different pattern might have meant a different structure for the isolated disc membrane. However, Chabre[4] prepared specimens as reported by Blasie and Worthington[9] and concluded that the discs were not as well orientated in these specimens as had been thought. It had seemed likely to Wilkins and myself that the apparent in-plane diffraction was in fact profile diffraction from some discs which were oriented at right angles to the bulk of the membranes, and Chabre has found that this is the case. Note that a point-focus pattern is important in checking the degree of orientation of the isolated membranes. If Blasie, Worthington and myself had done this initially, some misunderstanding would have been avoided.

Interpreting the Membrane-profile and In-plane Results

X-ray workers generally agree that the structure of the disc membrane is based on a lipid bilayer. This structure was assumed initially and then proven as described above. Opinion diverges, however, as to where the protein is located in the membrane. At least five different structures have been put forward since 1969, which I will now review.

Wilkins and myself[1] estimated the mass of lipid per unit area of disc membrane by using the interference microscope and the known refractive properties of the lipid and protein, along with the stacking distance of about 300 Å. We found that there was not enough lipid to make a normal bilayer over the entire disc; I note that Moody[10] had reached a similar conclusion earlier, though with less satisfactory data. We proposed that the lipid formed a bilayer thinner than normal, with layers of protein on the two surfaces. However, studies of other membranes and of artificial lipid bilayers[11] led us to prefer a modified bilayer structure with the protein partly embedded in the core of the membrane with the lipid fatty chains. In this way the two layers of lipid headgroups would be separated by a distance, 40 Å, close to that in the purely lipid bilayers. We could also account for bands of diffraction observed at Bragg spacings of 40 Å/h, h = 2–5.[11] Similar bands have been found in the case of the pure lipids.[12]

Bearing in mind that the profile showed no significant asymmetry, we proposed that the protein molecules were each half submerged in one face or the other of the membrane bilayer; I note that Vanderkooi and Sundaralingam[13] proposed much the same structure. We found that this

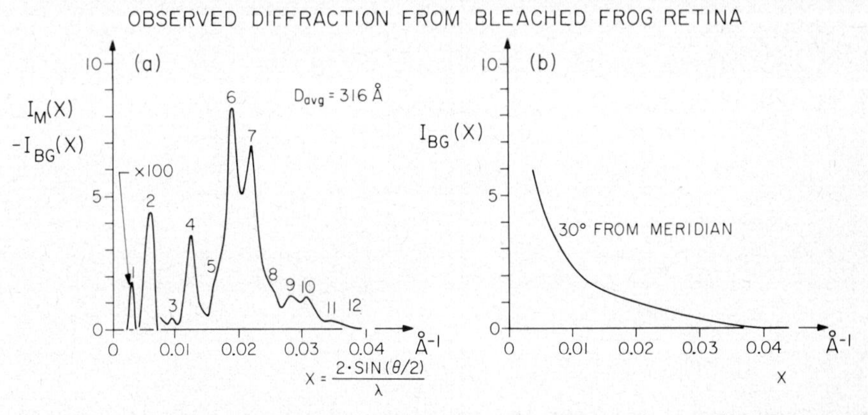

Fig. 6. Profile diffraction from the excised frog retina (Blaurock, unpublished data). The retina was bleached at 4°C and then exposed at the same temperature; D_{avg} = 316 Å. (a) The oriented diffraction obtained by subtracting the background tracing in (b) from a tracing along the axis (M) of the stacking reflections (not shown). (b) Tracing at 30° to the axis of the stacking reflections.

structure would account for the in-plane reflection (Fig. 5), assuming the then favored molecular weight of 27–28,000. The position of the diffuse reflection, the height off the in-plane axis, and the shape were all in reasonable agreement. However, two later developments have made this structure "inoperative".

An observation which influenced Wilkins and myself strongly is the freeze-fracture microscopy of outer segments. Two quite different cleavage faces are seen: most of the large particles stick to the cytoplasmic half of the membrane. This observation demonstrates an asymmetry of some kind, and two layers of protein molecules on the two sides of the membrane are very unlikely to account for the observation. Nonetheless, the X-ray profile is nearly symmetric, and therefore I am led to believe that each rhodopsin molecule spans the membrane. It appears that the rhodopsin molecules are fastened more securely on the cytoplasmic side of the membrane, possibly because of a larger area of contact with the ice on this side than on the inside of the disc. It follows that somewhat more of the bulk of the rhodopsin molecule may be on the cytoplasmic side. The results reported by Hubbell in the preceding paper also indicate that the rhodopsin spans the membrane.

Chabre[(4)] has suggested, for similar reasons, that the rhodopsin spans the membrane. In contrast Worthington[(3)] locates the rhodopsin on the inside surface of the disc membrane on the basis of his more asymmetric profile.

The second development of importance are recent reports of biochemical experiments which place the molecular weight of rhodopsin close to 40,000, and neutron-diffraction experiments of Yeager[(18)] supporting this value. The

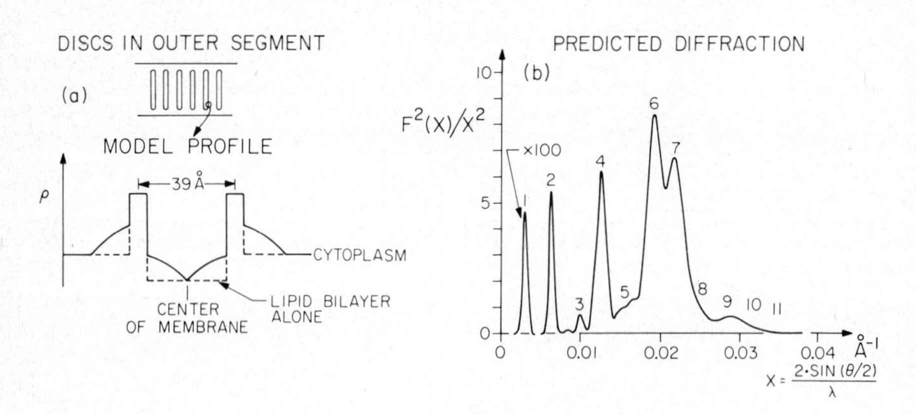

Fig. 7. A symmetric model profile and the calculated profile diffraction. Both the space between discs and the space within the disc are assumed to be variable (see text). (a) The symmetric membrane structure assumes a lipid bilayer with protein molecules centered in each headgroup layer. The more hydrophobic amino-acid residues are in the core of the membrane. (b) The calculated diffraction for comparison with Fig. 6.

consequence of the larger figure is a larger area of disc membrane per molecule than we had previously thought. We have had to recalculate the in-plane diffraction accordingly.

Some remarks about models for the membrane structure are in order since these have been useful to me in interpreting the profile. The model profiles are based on specific volumes which are calculated from the elemental composition of the lipids and protein. A good model also must take into account the disordered stacking of the membrane discs and the variable space inside the disc.[6]

A model which I developed some time ago is shown in Fig. 7. The rhodopsin molecules were assumed to be in two layers in the two halves of the membrane. The model also postulated a variable stacking distance and, independently, a variable distance between the centers of the two membranes in a disc. For simplicity both these distances were taken as constant within a given outer segment, and they were assumed to vary from outer segment to outer segment. The distributions of distances about the average values were according to two independent Gaussian functions (half heights at ±19 Å and ±7 Å, respectively). Although the model structure is now discredited, the distributions of lipid and protein are not so wide of the mark.

The calculated diffraction, on the right in Fig. 7, broadly reproduces the observed diffraction, Fig. 6, but there are small discrepancies which are

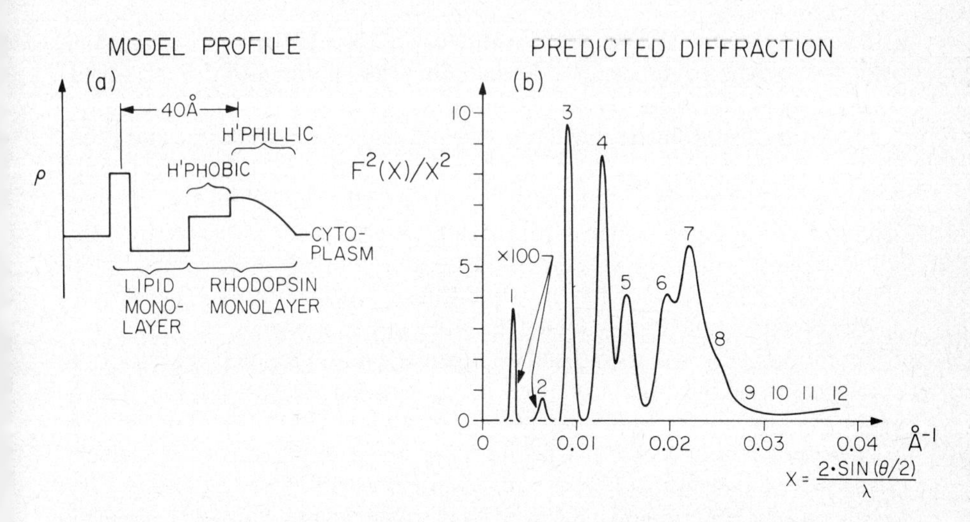

Fig. 8. A very asymmetric model profile and the calculated diffraction. (a) The membrane structure assumes that all the lipid is in a monolayer on the left side and all the protein is on the right side. (b) The calculated diffraction is in poor agreement with Fig. 6, indicating that the model is invalid.

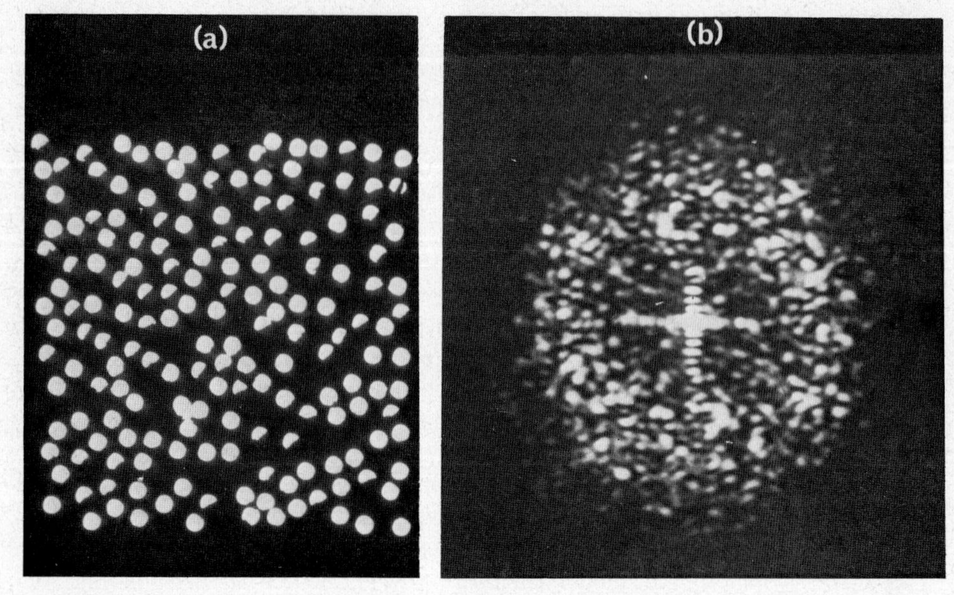

Fig. 9. An optical mask and its light diffraction pattern (Blaurock, unpublished). (a) The mask (left) was made assuming rhodopsin molecules located at random except that overlap was excluded. (b) The diffraction pattern (right) recorded using an optical diffractometer. The single broad, speckled ring resembles the intensity along the horizontal axis in Fig. 5.

sufficient to rule out this model for the disorder. Nonetheless, it is clear that both the symmetric model profile and the sizes of the variable distances between membranes are close to the truth.

I note that the problem of disorder is quite general in membrane and lipid bilayer studies, varying only in the degree to which it affects the stacking reflections. Schwartz *et al.*[5] recently have put forth a model for the variable distances in the outer segment, which finds values nearly the same as those above but for which the calculated diffraction is in better agreement with observation. Nelander and myself,[14] working independently, have developed much the same model for a similar disorder in nerve myelin.

The model in Fig. 8 places all of the lipid on one side of the membrane and all of the protein on the other side. There is very poor agreement with Fig. 6, and minor changes in the model clearly will not save it. On the basis of this calculation I cannot accept a structure based on freeze-fracture alone, which would place most of the rhodopsin molecule in one half of the membrane. The freeze-fracture observation alone leads one to overemphasize the asymmetry of the disc membrane.

The in-plane arrangement of the rhodopsin also has been studied. Fig. 9 shows an optical mask and the corresponding light diffraction pattern. The

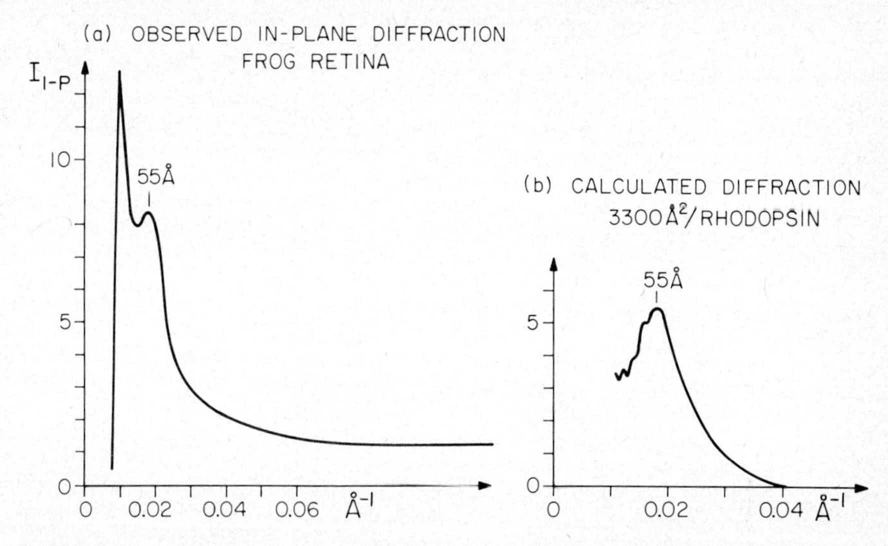

Fig. 10. Calculated in-plane diffraction for comparison with observation. (a) Densitometer tracing along the horizontal axis in Fig. 5. (b) Diffraction calculated from a model of the same kind as Fig. 9. The area per rhodopsin molecule is 3300 Å^2. The distance of nearest approach is 45 Å. The molecule is arbitrarily assumed to be a cylinder with an effective diameter of 27 Å.

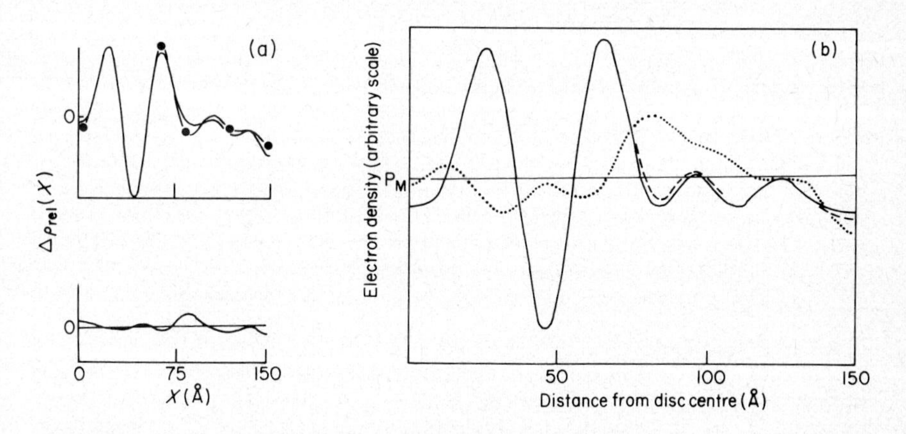

Fig. 11. The effect of bleaching on the disc membrane profile. (a) From Corless.[2] The profiles before (—●—) and after (———) bleaching are shown above, and the difference profile (bleached minus unbleached) is shown below. (b) From Chabre.[4] Both the unbleached (———) and the bleached (– – – –) profiles are shown. The difference profile (· · · · ·) is magnified five times.

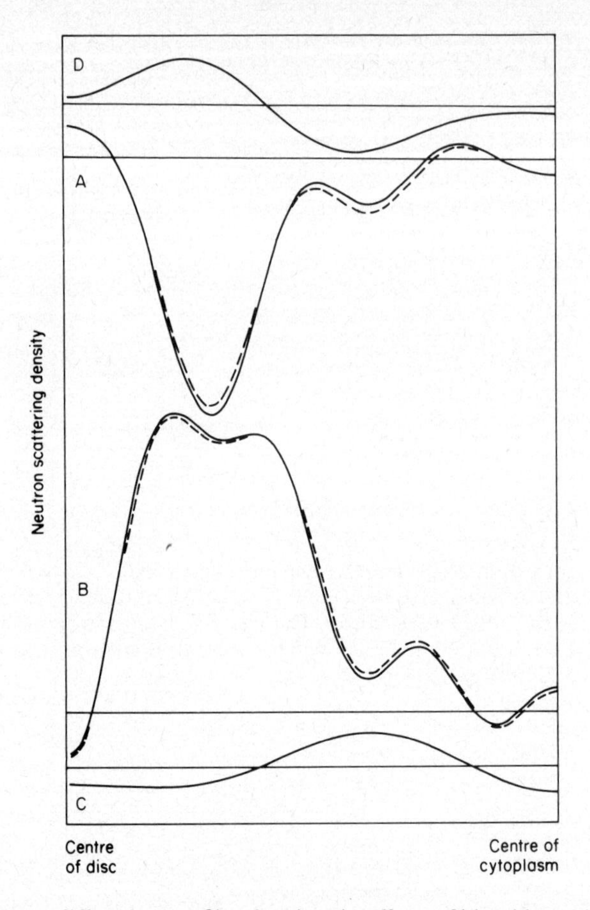

Fig. 12. Neutron-diffraction profiles showing the effects of bleaching. (a) In D_2O (A). The small difference curve (D), which is magnified five times compared to the profile, is based on significant changes in the reflection intensities. (b) In H_2O (B). The increase in density at the outer surface of the membrane after bleaching, compared to the decrease there in (A), is consistent with a small shift of protein into that region. The difference curve (C) is magnified five times compared to the profile.

model locates the protein molecules randomly in the plane of the membrane except that they cannot overlap. The single diffraction band shows that this model is a reasonable picture of the in-plane arrangement. The model also is consistent with free diffusion in the plane as found by Brown[15] and by Cone.[16]

Recently I have been doing the analogous operation by computer, using the larger area per rhodopsin molecule, as required by the larger value for the molecular weight. Fig. 10 shows the predicted diffraction in comparison to a densitometer tracing along the in-plane axis of Fig. 5. The agreement is encouraging. However, I note that the choice of a shape for the rhodopsin molecule is largely arbitrary, limiting the usefulness of this calculation.

Structural Changes due to Bleaching

X-ray and neutron diffraction experiments both have shown a small structural change when the rhodopsin bleaches. I note that the bleaching experiments involve an unphysiologically thorough bleaching of the pigment. This must be regarded, like stage make-up and stage gestures, as intended to bring out what would otherwise pass unobserved.

Corless[2] found a small change in the structure of the membrane in the first thirty minutes after bleaching (Fig. 11a). That the change is significant was shown by obvious changes in the intensities of certain reflections. Chabre[4] more recently has confirmed that similar changes occur within the first minute after bleaching the isolated outer segments (Fig. 11b).

Recent neutron-diffraction experiments have carried the study of the structural changes further. Saibil, Chabre and Worcester[17] have shown a similar, small change by this method (Fig. 12). An advantage of the method is that the scattering power of the cytoplasmic fluid can be changed dramatically by substituting D_2O for H_2O. Their observations confirm that much of the rhodopsin molecule is in the core of the membrane. They find a small but significant change in intensities in the first minute after bleaching. As in the X-ray profile, there is a change at the cytoplasmic surface of the membrane. Comparison of the H_2O and D_2O data indicate that something with the density of protein is involved. Their interpretation is that protein shifts outward from the disc membrane after bleaching.

I would suggest, as a possible alternative interpretation of the change near the cytoplasmic surface, that some constituent between the discs shifts toward the membrane. Polysaccharide is thought to be present between the discs, and this might move toward the disc after bleaching. I note that changes of comparable magnitude occur at other points in the profile, and these have yet to be explained.

References

1. A. E. Blaurock and M. H. F. Wilkins 1969, *Nature* **223**, 906–909.
   A. E. Blaurock and M. H. F. Wilkins 1972, *Nature* **236**, 313–314.
2. J. M. Corless 1972, *Nature* **237**, 229–231.
3. C. R. Worthington 1973, *Exp. Eye Res.* **17**, 487–501.
4. M. Chabre 1975, *Biochim. Biophys. Acta* **382**, 322–335.
   M. Chabre and A. Cavaggioni 1975, *Biochim. Biophys. Acta* **382**, 336–343.

5. S. Schwartz, J. E. Cain, E. A. Dratz and J. K. Blasie 1975, *Biophys. J.* **15**, 1201–1233.
6. A. E. Blaurock 1972, *Adv. Exp. Med. Biol.* **24**, 53–63.
7. W. J. Gras and C. R. Worthington 1969, *Proc. Nat. Acad. Sci. USA* **63**, 233–238.
8. J. K. Blasie, M. M. Dewey, A. E. Blaurock and C. R. Worthington 1965, *J. Mol. Biol.* **14**, 143–152.
9. J. K. Blasie, C. R. Worthington and M. M. Dewey 1969, *J. Mol. Biol.* **39**, 407–416. J. K. Blasie and C. R. Worthington 1969, *J. Mol. Biol.* **39**, 417–439.
10. M. F. Moody 1964, *Biol. Rev.* **39**, 43–86.
11. M. H. F. Wilkins, A. E. Blaurock and D. M. Engelman, 1971, *Nature N. B.* **230**, 72–76. Y. K. Levine and M. H. F. Wilkins 1971, *Nature N. B.* **230**, 69–72.
12. N. P. Franks 1976, *J. Mol. Biol.* **100**, 345–358. D. L. Worcester and N. P. Franks 1976, *J. Mol. Biol.* **100**, 359–378.
13. G. Vanderkooi and M. Sundaralingam 1970, *Proc. Nat. Acad. Sci. USA*, **67**, 233–238.
14. A. E. Blaurock and J. C. Nelander 1976, *J. Mol. Biol.* **103**, 421–431. J. C. Nelander and A. E. Blaurock 1977, *J. Mol. Biol.* in press.
15. P. K. Brown 1972, *Nature N. B.* **236**, 35–38.
16. R. A. Cone 1972, *Nature N. B.* **236**, 39–43. M. Poo and R. A. Cone 1974, *Nature*, **247**, 438–441.
17. H. Saibil, M. Chabre and D. Worcester 1976, *Nature*, **262**, 266–270.

6

Photosensitivity of Electrical Conductance of the Rod Disc Membrane

P. FATT and G. FALK

Department of Biophysics,
University College London, U.K.

A highly favoured hypothesis for visual excitation which is considered in other papers of this volume[1-4] involves the visual pigment serving as a light-activated gate for the passage of ions, especially calcium ions, across the membrane in which it is situated. Evidence in support of such a gating mechanism, though not specific for the passage of calcium ions, has been obtained from electrical measurements on rod outer segments of the frog.[5,6] The essential results obtained in this work and the analysis whereby it has been inferred that light produces a disc-membrane conductance increment of a magnitude per activated pigment molecule comparable to that obtaining for other chemically modulated membrane processes will be reviewed here. In addition, attention will be directed to certain aspects of the findings which suggest that the mechanism whereby the light-induced bleaching of rhodopsin causes a permeability increase of the disc membrane is not so straightforward as to follow directly from a configurational change of the individual rhodopsin molecule implanted in the membrane, but that it involves some essential cofactor in the cytoplasm and probably interaction between rhodopsin molecules. It is important to recognize such phenomena; for in addition to complicating the experimental work, they may be indicative of regulatory mechanisms operating at the level of the pigment-bearing membrane.

Material and Methods

In the most recent procedure[5] rod outer segments were shaken off the retina gently into a modified Ringer solution, avoiding as far as possible mechanical and osmotic disturbance. In order to prevent a gradual transfer of water, the osmotic pressure of the solution was maintained by Na_2SO_4 in place of NaCl or the NaCl of Ringer solution was partly replaced by sucrose. A colloidal substance (either bovine serum albumen (BSA) or methylcellulose) was added to improve the long-term stability of the preparation, possibly acting through the increase in viscosity of the solution. In addition, some metal-chelating substance (usually EDTA) was sometimes added with the specific aim of its removing trace amounts of copper ions, the presence of which would cause a

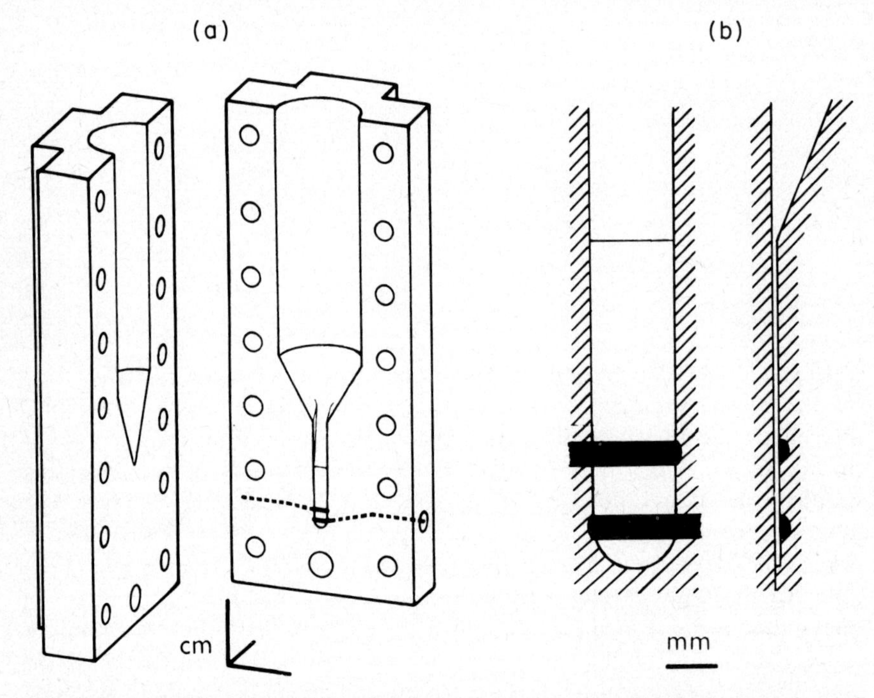

Fig. 1. Conductivity cell used for admittance measurements on packed suspensions of rod outer segments. (a) Perspective view of dismantled cell. In use, the two parts were held together by steel bolts passing through the holes along their edges. Platinum-wire leads are shown by heavy lines, dotted where leads pass through the insulating material of the cell. The major constructional material was a polyisopentene plastic (TPX, I.C.I.) which had desirable optical, electrical and mechanical properties. (b) Orthogonal sectional views of the shallow channel into which the rods were packed by centrifugation. The exposed surfaces of platinum, constituting the electrodes, were flush with the walls of the channel and were coated with platinum grey. Reproduced from Falk and Fatt.[5]

rapid disappearance of the response being investigated. The addition of millimolar concentrations of ionized calcium, magnesium, potassium, phosphate and ATP were all tried singly and in various combinations without the response being noticeably affected.

By a process of differential sedimentation, the rods were packed into a transparent conductivity cell (Fig. 1) where they formed a layer 0·1 mm deep in the direction through which the light passed. The space between and adjacent to the electrodes, through which current passed to give the measured electrical properties, had a volume of about 0·25 mm^3, which is about one-quarter the volume of rod outer segments in a single retina. The rods are assumed to take on all orientations within this space (i.e. to be randomly oriented). Electrical measurements indicate that the rods occupy a volume-fraction of about 0·8.

Measurements were made of the passive linear electrical properties of the packed rods to obtain the component of current in phase with an applied sinusoidal voltage—that is the real part of admittance, G, which is most directly related to the conductance of the various structural elements of the rod mass. These measurements eventually extended over the frequency range from 15 Hz–17 MHz. The reason for working over this wide frequency range is that it allows one to probe the electrical properties of different structures.[7,8] The basic assumption used in relating the observed electrical behaviour to structure is that the only frequency-dependent electrical elements present are the bimolecular lipid membranes of the rods having a capacitance of about 1 $\mu F/cm^2$.

Separation of Components of Response

The response to light was observed as a change in G following exposure to a 1 ms duration flash bleaching usually between 0·2 % and 10 % of the rhodopsin content of the rods. While there was no difficulty in observing a response to light within this range, the response was often of a composite nature, and problems were encountered in separating it into components and in deciding where and how each component arose. To begin with, it was necessary to distinguish those components of response that are directly dependent on light-evoked changes within the rhodopsin molecule and do not involve the structure of the rod, other than in its role of compartmentalization of aqueous spaces, and other components that involve the structural organization of the rod in a more complicated way. The magnitude of the first type of component would be in direct proportion to the amount of rhodopsin bleached. A so-called buffer component, obtained in measurements at high frequencies, with low-conductivity solutions for preparing the rods and with bright flashes, is of this kind.[8] It arises from the uptake of a hydrogen ion in the metarhodopsin I to II conversion, which in turn causes a change in ionization of any buffer present in the rod interior. Another such component, called the negative component as it appeared as a decrease in G, the same at all frequencies of measurement, was attributed to an increase in the water-excluding volume occupied by rhodopsin.[6] Further details about these components may be found in the references quoted.

Greater interest with respect to the mechanism of visual excitation is attached to other components that apparently do depend on the structural organization of the rod, and for which the amplitude has been found to vary in direct proportion to the amount of rhodopsin bleached only up to about 1 % bleached, and rapidly saturates for greater amounts bleached. With reasonably intact rods suspended in a solution with the ionic composition of Ringer solution, these are the most prominent components of response for flashes bleaching up to 3% of the initial rhodopsin content of the rods.

A component of response has been recognized under these conditions which has a time course of development that lags behind metarhodopsin I to II conversion by a few milliseconds. This response component, which is of particular interest here, has been called component II. This designation, which is perhaps unfortunate in view of the prime importance attached to it, is based on the observation that with increasing frequency of measurement this component appears only above a certain cut-off frequency, somewhere between 1–10 kHz, where it is superimposed on another, called component I, which is of invariant amplitude down to the lowest frequency of measurement. The latter has a much slower time course of development than component II. Neither of these components is reversible: each is in the form of an increase in conductance which is not followed by any noticeable return of conductance towards the initial level, at least over the course of a few minutes.

The time course and frequency dependence of the response to a 1 % bleach is shown in Fig. 2 for two experiments. The time resolution of the records extends from about a millisecond to several seconds and the frequency of measurement from 1 kHz–17 MHz. A split time-base and superimposed successive sweeps were used to obtain the extended time resolution. Separate flashes delivered at the time marked by the arrows were used to obtain measurements at different frequencies. The vertical calibration bar represents 0·5 μmho/cm for each record, this being the change in real part of admittance reduced to a sample of unit geometry. In the experiment of Fig. 2(a), represented by the upper set of 6 records, the amplitude of response varies with frequency, but the time course is invariant. There is no measurable response at 1 kHz. The response appears as a negative change in G at 1·6 kHz. It inverts to a positive change at higher frequencies, increasing to reach a plateau at 720 kHz, and then remaining constant to the highest frequency of measurement. The appearance of a reversed response, in this case a negative ΔG, in the neighbourhood of a transition frequency above which ΔG increases rapidly, is

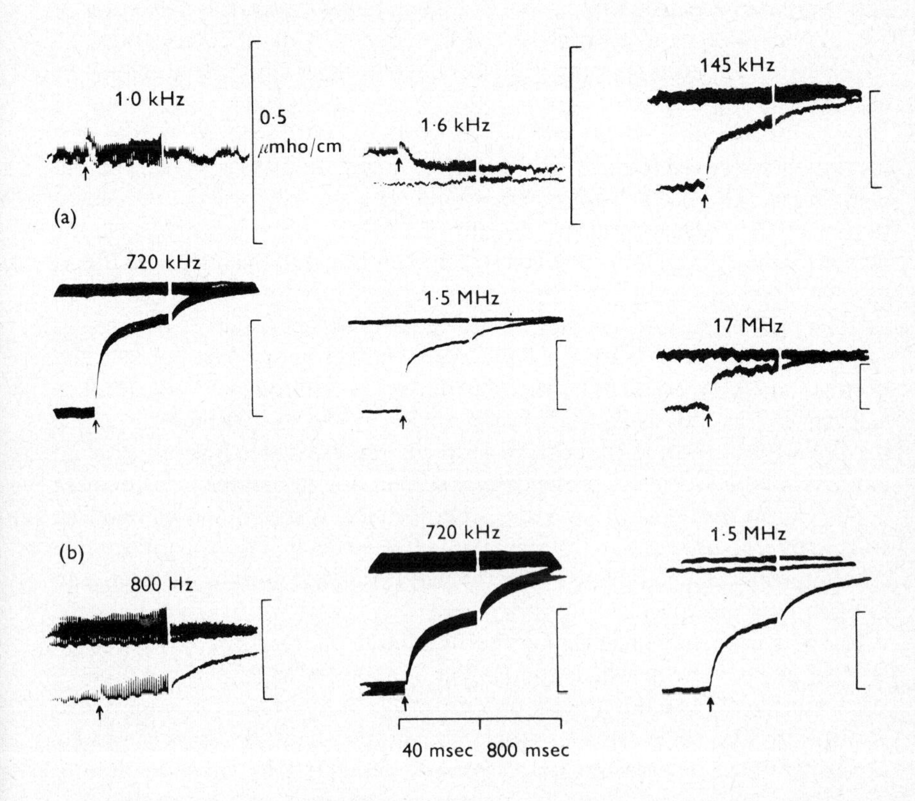

Fig. 2. Records of the change in real part of admittance (ΔG) of a packed suspension of rods in response to the sudden bleaching of 1 % of the rhodopsin content of the rods. Each record consists of 2 or 3 sweeps, commencing at intervals of 2·5 sec and utilizing a split time-base (the later part of each sweep is at a speed $\frac{1}{20}$ that of the early part). The response arises following a 1 msec-duration flash of blue-green light delivered during the first sweep of each record, at the time marked by the arrow. A separate flash was used to obtain each record, but flashes were not repeated at intervals shorter than 12 min. Two experiments are represented: (a) by the upper 6 records, (b) by the lower 3 records. The rods were prepared similarly in both experiments. An isotonic suspending medium in which, it was established, rods neither swelled nor shrank over many hours, was employed throughout the preparation. Purification of the rods was accomplished by differential sedimentation during a 15-min period of low-speed centrifugation. The suspending medium had the composition: 136 mM Na^+, 73·3 mM SO_4^{--}, 12 mM Tris, 0·01 g methylcellulose/ml (giving a viscosity of 10 cP), pH 7·2. Temperature was 18·4–18·9°C. Note the large variation in the sensitivity used for displaying ΔG at different frequencies, the vertical calibration bar having the same value for all records. In the records at low frequencies, the initial upward deflection, coincident with the flash, is an artifact associated with operation of the flash lamp. Reproduced from Falk and Fatt.[6]

not peculiar to the present type of response. It is a characteristic of the use of ΔG as a parameter to represent the change in passive electrical behaviour of any system which can be represented in terms of a model consisting of a variable resistor (either lumped or distributed) in series with a fixed capacitor, as shown in Fig. 5.

In its time course and frequency dependence, the response in Fig. 2(a) corresponds to component II, virtually uncontaminated by any other component. The buffer component would be expected to make its contribution at the highest frequencies—above 1 MHz, where it would give the response a more rapid initial rise (on the assumption that the internal buffer is anionic although the buffer present in the external solution is cationic). This may be present, but its contribution to the overall amplitude of response is probably negligible at these low light levels. If other known components had been present, they would have been seen at the lowest frequency, 1·0 kHz. The achievement of this isolation of component II clearly depends on gentle treatment of the rods. In the other experiment, represented in Fig. 2(b), a small amount of component I is present and is seen in isolation in the measurement at 800 Hz. At the higher frequencies, component II is superimposed on it. At still higher frequencies than those shown, it is expected that the contribution to the composite response made by component I would decline, as has been seen in other experiments.

The frequency-dependence of the amplitude of response, obtained in another experiment in which component II appeared nearly in isolation, is shown in Fig. 3. The initial level of G from which the response arose is also given, with the scale for this appearing at the right. In this particular experiment the rods were separated from the retina in a hypertonic solution, including 1·08 M sucrose, as is commonly employed for the purpose of purifying a rod suspension by flotation. After dilution of the sucrose to about 0·7 M, the rods were sedimented into the conductivity cell. The achievement of a response consisting almost entirely of component II with rods prepared in this way was possible only by starting with a large excess of material (rods separated from the retinas of four frogs) and, by means of prolonged low-speed centrifugation, packing a selected population of intact, undistorted rods into the conductivity cell.

It was of interest to find that, on equilibration on a bovine serum albumin gradient, rods freshly separated from the retina distributed themselves into two bands at positions of different density of the suspending medium. Measurements carried out on rods recovered from each of the bands showed component II to be relatively larger for rods equilibrating at a position of lesser density and component I to be larger at greater density. This is shown in Fig. 4(a) and (b). For the lesser-density band in (a), the response consists of a mixture of components I and II with component II predominating, as judged

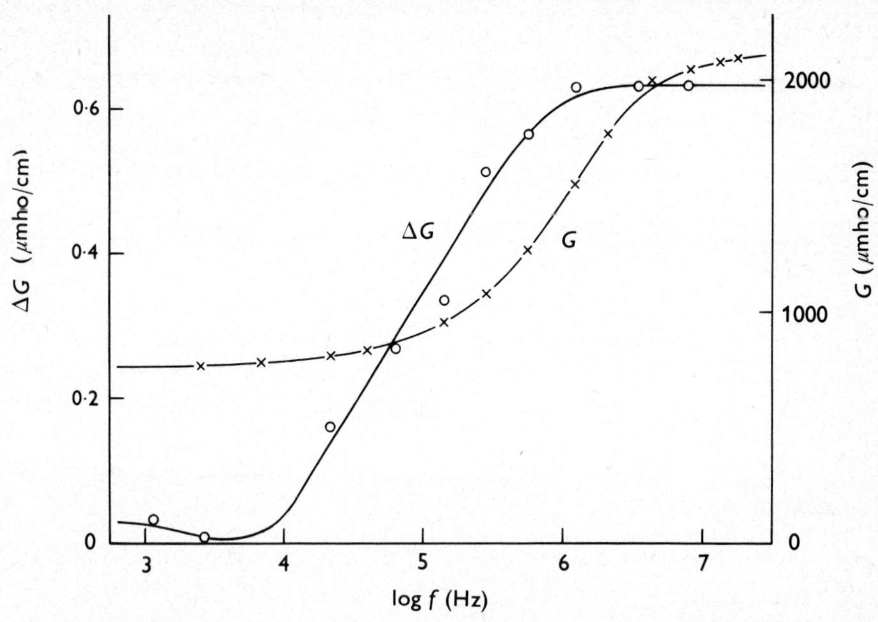

Fig. 3. Plots of G and ΔG against log frequency for a packed suspension of rods prepared in hypertonic sucrose solution. A suspension of rods was purified by flotation in a solution of 105 mM Na^+, 10 mM K^+, 108 mM Cl^-, 6 mM phosphate, 1·08 M sucrose, 0·01 g BSA/ml, pH 7·2. After dilution of the sucrose to 0·70 M, the rods were packed by centrifugation at 150 g for 20 min followed by 5000 g briefly. Each flash used to obtain ΔG (measured as the change in G from immediately before the flash to 2·5 sec later) was of an intensity which bleached 1 % of the rhodopsin content of the rods. Measurements at different frequencies were made in random order with repeats at 720 kHz every third or fourth flash. The measured values of ΔG have been divided by the fraction of the initial rhodopsin content of the rods remaining unbleached at the time of each flash. Thus the plotted values represent ΔG expected for the bleaching of 1 % of the initial rhodopsin content. It is important to note that the form of response illustrated here (consisting almost entirely of component II) was obtainable from only a small fraction of the total mass of rods prepared in the hypertonic solution. It was associated with the more rapidly sedimenting rods. Reproduced from Falk and Fatt.[6]

by the increment in amplitude of response on change of frequency from 750 Hz to 360 kHz. In contrast, for the greater-density band in (b), component II is effectively absent and the fall-off of component I (in the region of 1–3 MHz) results in there being no measurable response at 16 MHz. The interpretation placed on this separation of rods into two populations is that the more dense band consists of rods whose surface membrane has been damaged so as to allow the penetration of albumin, while the less dense band contains a high proportion of intact outer segments. It was further found that any procedure which would be expected to cause disruption of the surface membrane, such as shearing forces in the suspending medium, exposure to

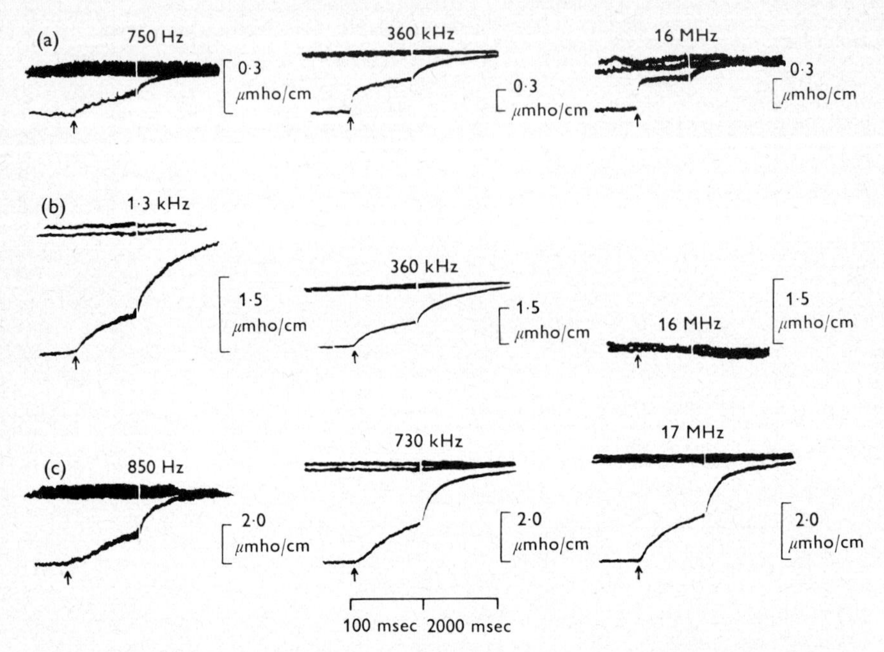

Fig. 4. Responses obtained from two fractions of a rod suspension separated by equilibration on a BSA density gradient. Rods were shaken off the retina in a solution consisting of 73 mM Na^+, 2·5 mM K^+, 62 mM Cl^-, 2 mM EDTA, 1·5 mM phosphate, 100 mM sucrose, 0·01 g BSA/ml. The rods were placed on top of a gradient of BSA extending between 0·2 and 0·3 g/ml, and centrifuged at 7800 *g* for 50 min. Records of the response of a sample of rods recovered from the band extending between 0·248 and 0·266 g BSA/ml (mean sp.gr. 1·082, the low-density band) are shown in (a); in (b) response of rods distributed in the band from 0·280 to 0·284 g BSA/ml (mean sp.gr. 1·089, the high-density band). An additional sample of rods from the low-density band was frozen and thawed and then sedimented in a third conductivity cell. The response of these freeze-thawed rods is shown in (c). Note the different calibrations for ΔG in each set of records. Each flash was of an intensity which bleached 1 % of the initial rhodopsin content. Reproduced from Falk and Fatt.[6]

hypotonic solution and freezing and thawing, led to component I increasing while component II decreased to the point of disappearance. This is shown in Fig. 4, where the same population of rods that gave the response in (a) in which component II predominated was found to give a response consisting exclusively of component I as shown in (c) after freeze-thawing. (The response of these freeze-damaged rods differs from that in (b) in its failure to fall off at high frequencies. This has been a consistent finding with grossly damaged rods and may be related to complete stripping away of the surface membrane.)

A simple explanation can be offered for these findings. Components I and II arise in different populations of rods, with component II being present only in those rods whose surface membrane is intact. It is assumed that some diffusible, but membrane-impermeant, substance is present in the cytoplasmic space of the rod, which is essential for component II and which is lost through disruption of the surface membrane. After its loss, the bleaching of rhodopsin results in a different kind of change in the properties of the rod, manifest as component I.

Mode of Origin of Component II

The frequency-dependence of component II, in particular the low-frequency cut-off in the region of 1–10 kHz, has been the source of considerable puzzlement. After much deliberation it was perceived that an explanation could be given for the frequency dependence of component II which assumed that it, like other frequency-dependent electrical characteristics of the packed rods, was determined by the surface membrane capacitance. The disparity between the characteristic frequency (or transition frequency) for the dispersion of G at 1 MHz and the low-frequency cut-off of component II, some 2–3 decades lower, is then accounted for by the shape of the rod and the expected anisotropy of the electrical properties of its interior. To begin with, one assumes that initially (i.e. preceding a response to light) no appreciable current passes across membranes, except via their capacitance. In fact, over the entire frequency range which is under consideration here, no appreciable current will flow through the disc-membrane capacitance, so that the discs will act simply as obstructions to the passage of current through the interior of the rod. There are two cases to consider. For rods oriented with their axes at right-angles to the direction of the applied electrical field, the internal conductance and surface membrane capacitance appear as lumped elements in series; while, for rods oriented with their axes in the same direction as the applied field, the internal conductance and surface membrane capacitance are distributed, with the rod behaving as a cable of finite length. These two theoretical cases show a similar form of dispersion of G and a related cut-off frequency for a response, ΔG, arising from a change in internal conductance, as is shown in Fig. 5. They differ, however, very much with regard to the frequencies at which these transitions occur and with regard to the contributions they make to G.

For rods oriented at right angles to the field, the relevant frequency will be given in Hz by $f_t = 0{\cdot}16 G_t / C_m a$, where G_t is the effective internal conductivity in the transverse direction, C_m is the surface membrane capacitance per unit area and a is the radius of the rod. For rods oriented parallel with the field, the frequency is given by $f_l = 0{\cdot}8 G_l a / C_m l^2$, where G_l is the effective conductivity of the rod interior in the longitudinal direction and l is the length of the rod. If G_t and G_l are determined by ionic conduction in the cytoplasmic space unobstructed by the discs, one would expect G_t to be about one-half the

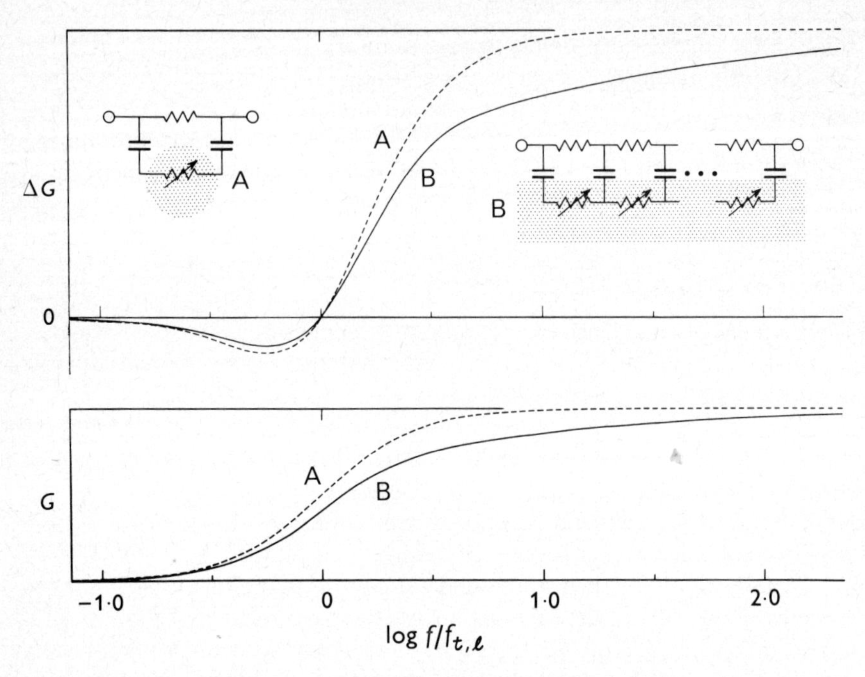

Fig. 5. Theoretical curves showing the frequency-dependence of G and ΔG for two models of the rod with electrodes in contact with the conducting medium external to the rod. (A) The case where the rod is oriented with its axis at right angles to the direction of the applied electric field so that current passes transversely through the rod interior. (B) The case where the rod is oriented parallel with the field so that current passes longitudinally through the interior. The internal conductivity of the rod is considered to change by a small fractional amount in each case, thus giving rise to ΔG. Horizontal lines extending from the left-hand edge of the graphs indicate the limiting values of G and ΔG approached at low and at high frequencies. The scale log f/f_t applies for case (A); log f/f_l for case (B). Expressions relating the transition (or cut-off) frequencies f_t and f_l to the electrical properties of the rod and the spatial dimensions of the rod are given in the text. Note that for frog rods f_t and f_l are expected to differ by a factor of about 300, with f_t in the region of 1 MHz and f_l in the region of 3 kHz. Modified from Falk and Fatt.[5]

conductivity of the fluid in the cytoplasmic space and G_l about 1/50, both of which figures are arrived at on the basis of the spatial arrangement of disc membranes as shown by electron microscopy. With the known values for the radius and length of the frog rod, one finds that the position of the low-frequency cut-off of component II can be very well accounted for on the assumption that it has its origin in a change in the effective conductivity of the rod interior in the longitudinal direction. Indeed, despite the necessity to suppose considerable scatter of the cut-off frequency f_l within the population of rods in order to account for the extended range over which component II increases, the absence of an appreciable further increase of component II above f_t can be taken as an indication that this component does not involve a change in G_t of comparable magnitude to that in G_l. One can thus eliminate the possibility that the component might arise from an increase in the conductance of the gap between the surface membrane and the disc edges or of the incisures; for, owing to the two-dimensional nature of conduction within such structures, any change within them would appear as a change in both G_l and G_t. One is thus left with the explanation for component II: that it arises from an increment in conductance of the disc membrane, which increment is appreciable compared with the initial conductance of the membrane.

Response to Activation of a Single Rhodopsin Molecule

The magnitude of this increment of disc membrane conductance may now be considered. For a 1 % bleach of a packed mass of randomly oriented rods prepared from a solution having an ionic composition similar to that of Ringer solution, component II measured on the plateau of the change in real part of admittance at about 1 MHz amounted to about 0·5 μmho/cm. Allowing for orientation, that effectively only $\frac{1}{3}$ of the rods have their axes parallel to the applied field and so are capable of contributing to the response arising from a change in G_l and, furthermore, that the disc membranes are stacked in the rod interior at a density of some 700,000 cm^{-1}, this 1 % bleach is inferred to produce within the disc membrane a conductance increment of 1·0 mho/cm^2. One is made aware by this figure that the change produced in the disc membrane by a 1 % bleach is large compared with that capable of being produced in most excitable membranes. This can, of course, be related to the high concentration of rhodopsin in the disc membrane compared with the situation obtaining for gating (or channel-controlling) molecules in other kinds of membranes. Thus if one takes the accepted figure of $2{\cdot}3 \times 10^{12}$ molecules of rhodopsin/cm^2 of disc membrane, a 1 % bleach producing a conductance increment of 1·0 mho/cm^2 represents activation of $2{\cdot}3 \times 10^{10}$ molecules/cm^2 and one arrives at the result that activation of a single rhodopsin molecule gives a disc-membrane conductance increment of 4×10^{-11} mho. This lends confidence to our interpretation of component II, as the value agrees closely with that obtained for the conductance of ionic channels controlled by the action of acetylcholine on the chemoreceptor in the post-junctional membrane of striated muscle fibres.[9,10] These channels, when open, permit the passage of a range of cation species (e.g. Na, K, Ca), but fail to permit the passage of anions. It may, on the other hand, be noted that the calculated channel conductance of the rod disc membrane is at least an order of magnitude greater than the conductance of the highly ion-selective sodium channel of excitable membranes.[11–13] This assumes significance in the light of the suggestion that the conductance of non-selective channels may in general be greater than that of more discriminatory kinds. All that can be said at present in this regard is that the rod disc-membrane conductance change is not specific for a particular ion species, such as sodium or calcium, but a specificity for cations over anions has not been ruled out. If the internal transmitter, which transmits the effect of light absorption from the disc to the surface membrane in the functionally intact rod, were calcium ions, as has been widely supposed, their release from the disc by light would not require the permeability increase of the disc membrane to be ion-specific. Ion-specificity is more likely to be associated with the metabolically driven pump whereby an electro-chemical gradient for calcium ions is created across the disc membrane so that a permeability increase then results in a net efflux of calcium ions. A subsidiary role for the increase in permeability towards ions other than calcium is that their movement would act to prevent a displacement of potential across the disc membrane, occurring as a consequence of the light-evoked efflux of calcium ions, which displacement could retard further efflux. However, for the expected value of capacitance between the inside and outside of a disc in a frog rod, this action of the other ions would probably not become important until the efflux amounted to about 20,000 calcium ions, uncompensated by inward pumping. In the case of the human or the rat rod, because of its smaller diameter, the corresponding amount of calcium efflux would be 1500 ions.

Mode of Origin of Component I

One may turn now to consider the probable origin of component I which is observed as a constant ΔG down to the lowest frequency of measurement. At high frequencies, above about 1 MHz, it falls off to a variable extent. Although this general behaviour is consistent with a surface membrane conductance change, the particular frequency region in which the fall-off occurs—above f_t rather than below it—makes it clear that the response cannot have this mode of origin. Instead, the observed behaviour can be accounted for by a reduction in the volume fraction occupied by the rods in the packed rod mass. Component I is envisaged to arise from a change of forces acting in and on the disc membrane. The initial effect would probably be a change in distribution of electrical charge, with the resultant change in electrostatic interactions producing a tendency of the disc membrane to crumple. This might occur rapidly following illumination, perhaps coincident with metarhodopsin II formation. It would lead eventually to an outflow of water from the discs and then from the rod, with the latter taking place along a time course determined by the water permeability of the disc and surface membranes. The occurrence of a conductance increase, in a situation in which there is no movement of ions into the external space, follows from the fact that for closely packed particles there is a nonlinear relation between the amount of external space and the effective conductance for the passage of current through this space.[14]

A crumpling of the disc membranes, as has been considered to be involved in the development of component I, may also be inferred from light-scattering studies on suspensions of rod outer segments.[15] These show an increase in the optical inhomogeneity of the rod interior (represented in the above reference by signal *P*) which has a similar dependence on the amount of rhodopsin bleached to that found for the conductance change. A long-range disordering of the disc repeat pattern, as would be caused by crumpling of the disc membranes, will also account for the pronounced smearing of the X-ray diffraction pattern obtained within a few minutes following bleaching of the rods.[16] This effect has been observed in the isolated retina. It is to be noted however that in this preparation, with the pigment epithelium separated from the neural retina, even though the rod outer segments remain attached to the latter, the majority of them may be damaged and leaky and thus in a condition to give rise to component I.

Response Saturation: Evidence for Clustering of Rhodopsin

While for bleaches of up to 1 % of the rhodopsin content of rods, components I and II vary in direct proportion to the amount of rhodopsin bleached, for more massive bleaches the response amplitude fails to increase proportionally. This is illustrated in Fig. 6 for a response which consisted of a mixture of 55 % component II and 45 % component I. Measurements made at different frequencies on rods prepared in isotonic solution indicate that the form of curve describing the saturation of response is similar for components I and II. Apparently saturation occurs at some stage in the generation of the response that is common to these two components. (In an early analysis[8] it was concluded that component I deviates more rapidly from a proportionality with the amount of rhodopsin bleached than does component II. This was later seen to be in error, following the discovery of the negative component of response.[6] The apparent difference in the approach to saturation of components I and II can be attributed to the fact that what was measured on osmotically damaged rods in high-conductivity solution as the amplitude of component I included a significant contribution from the negative component, which, since it varies in proportion to the amount of rhodopsin bleached and is of opposite sign to component I, accentuates the deviation from proportionality of the summed components.)

With bright flashes, bleaching 15–20 % of the initial rhodopsin content, one obtains a maximum response which is about 7·5 times as large as the response

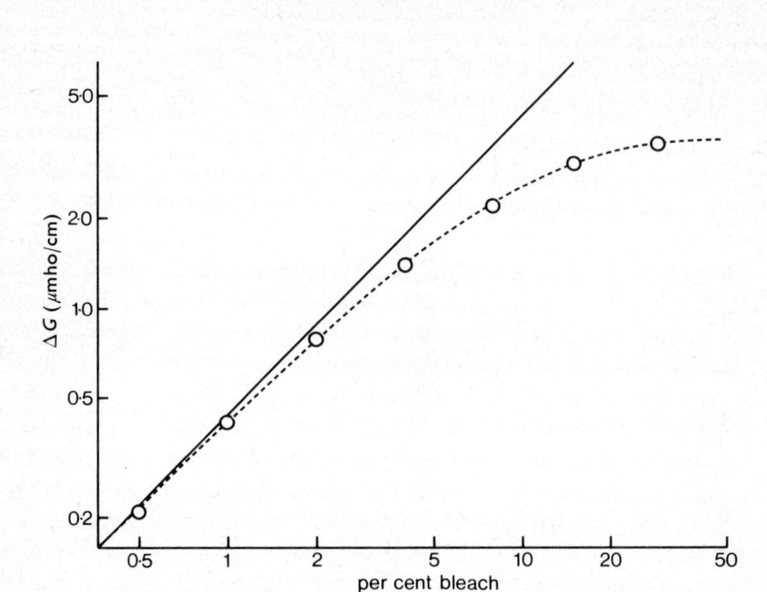

Fig. 6. Plot of amplitude of response (ΔG at 720 kHz, 2·5 sec following the flash) against percentage of rhodopsin bleached, both on logarithmic scales. Rods were prepared in a solution similar to that used for the experiments of Fig. 2, except for the addition of 2 mM EDTA and the replacement of methylcellulose by BSA. Comparison of the values of ΔG for a 1 % bleach at 720 kHz and 1·4 kHz indicate the plotted response to consist of 55 % component II and 45 % component I. In addition, there was a heat effect, marked by a development of ΔG coincident with the light flash. Its amplitude varied in proportion to the absorption of light. The plotted response amplitude has been corrected by the subtraction of this effect. The response amplitude has also been adjusted by being divided by the fraction of the initial rhodopsin content of the rods remaining unbleached prior to delivery of the flash. The position of points on the horizontal scale has been derived from flash intensity with adjustment for the effect of photoreversal (with a flash duration of 1 msec) which is expected to limit the maximum bleach to 50 %. The 45°-line represents proportionality of response to amount of rhodopsin bleached. The dotted curve represents the case in which the response is graded according to the number of clusters of rhodopsin molecules activated, with each cluster consisting of 12 molecules and the bleaching of any molecule within a cluster being sufficient for its activation.

to a flash bleaching 1 %. Still greater bleaching produces little further increase in response amplitude (up to about 8·5 times the response to a 1 % bleach). The simplest explanation that can be offered for this behaviour, on the assumption that the disc membrane consists of little more than a bimolecular lipid layer within which rhodopsin floats, is that the rhodopsin molecules are clustered and it is through the interaction of molecules within this cluster, following photo-activation of a single molecule, that a water-filled pore extending across the lipid layer is formed. The number of rhodopsin molecules required to be present in each such cluster would be in the region of 11–13 to give the observed level of response saturation. The dotted curve in Fig. 6 is drawn in accordance with this model.

Evidence against clustering of rhodopsin molecules comes from X-ray diffraction studies which show a repeat distance within the plane of the disc membranes of 55 Å, which has been interpreted as the mean distance between centres of rhodopsin molecules.(16, 17) The method, however, discriminates against the detection of greater repeat distances because of the smaller angle of scattering of the X-ray beam that would be produced by them, and the resulting interference of the undeviated beam, as can be seen in Figs 5 and 10a of chapter 5 in this volume.(18) Measurements of translational and rotational diffusion of rhodopsin within the disc membrane has also been considered to indicate that rhodopsin molecules are not clustered,(19–21) though in view of the weak dependence of diffusivity on particle volume and the uncertainty of membrane viscosity, this evidence is inconclusive.

That rhodopsin molecules are capable of interacting to some extent to form clusters is shown by molecular size determination of detergent-solubilized rhodopsin.(22) This has shown the existence of dimers and trimers. Freeze-fracturing of rods has shown the presence of particles in the disc membrane having diameters in the region of 125 Å.(23–26) It has been suggested from the size of these particles and from their density of packing, that they could each consist of 4 or 5 rhodopsin molecules. There is no reason to suppose, however, that such grouping of rhodopsin molecules exists prior to the preparation of the rod for freeze-fracturing and, apart from its showing a capability for interaction between molecules, we do not consider the phenomenon relevant to the rather larger clusters proposed for explaining the saturation of response.

An interactive photochemical behaviour of rhodopsin extending over a fairly large number of molecules, as is required to account for the present form of response saturation, can possibly be deduced from measurements of chemical reactivity. Studies of the susceptibility of rhodopsin in rods to phosphorylation by ATP following illumination indicate that, at levels of light stimulation at which only about 1 % of the rhodopsin is bleached, somewhere between 10 and 15 phosphate groups are attached for each molecule that is bleached.(22) With more extensive bleaching, the maximum number of phosphate groups attached per rhodopsin molecules bleached is limited to a value slightly less than unity. A possible interpretation is that the photochemical conversion of a single rhodopsin molecule within a cluster of interacting molecules is sufficient to cause all the molecules within the cluster to become available for phosphorylation.

Another piece of evidence of interaction of rhodopsin molecules within clusters is provided by studies of the thermal reactions of rhodopsin photoproducts following exposure of rhodopsin within the rod to light flashes of differing intensities. It has been found that the decay of metarhodopsin II is faster and the proportion of molecules going over to metarhodopsin III is less for weak flashes, bleaching only a few per cent of the rhodopsin content of the rods, than for stronger flashes.[27] The simplest interpretation that can be placed on these findings, analogous to that used for explaining the saturation of components I and II, is that within the assumed cluster of rhodopsin molecules, the first molecule to reach the stage of metarhodopsin II has a shorter mean life-time, of about 15 s, and converts directly to retinal plus opsin (the two being still bound together in some way so as to maintain the dichroism of the pigment, although the Schiff-base leakage has disappeared). It is now postulated to be this short-lived metarhodopsin II molecule that controls the opening of the ion-gate detected as component II. Additional metarhodopsin II molecules produced subsequent to the first within a cluster would follow a different reaction path, the mean life-time being 10–20 times as long, and there would be a high probability (approximately 0·5) of conversion to metarhodopsin III taking place. These additional metarhodopsin II molecules formed within a cluster are considered redundant to the process of ion-gate formation.

The depression of response sensitivity to successive flashes that is a consequence of saturation, is found to wear off over a period of about 8 min (time for $\frac{2}{3}$ recovery of response amplitude) following a bleach in the region of 2–10%. For more massive bleaches, the time of recovery increases, being about 20 min for a 40% bleach. This slow recovery of responsiveness may be dependent on the decay of some rhodopsin photoproduct. Alternatively, it might represent the time required for the separation and re-grouping of pigment molecules to form new clusters which exclude bleached molecules. The dissociation of the bleached pigment molecule (whether in the form of retinal bound to opsin or metarhodopsin III) into free retinal plus opsin may be facilitated by this process, which would thus give a correlation between photoproduct decay and the recovery of responsiveness. It seems likely, although it has not been definitely established experimentally, that in the preparation of rods used in the admittance measurements, a reversal of the disc-membrane conductance increase represented by component II takes place along the same time course as the recovery of responsiveness.

In relation to the possible relevance of component II to visual signal transmission, it may be noted that the recovery from saturation of the response follows a similar time course to a phase of visual adaptation following the bleaching of a substantial fraction of the rhodopsin content of rods in the retina. This phenomenon has been shown to have its origin in the outer segments and to occur independently of any regeneration of the pigment.[28–32] The later part of its time-course has been shown to be correlated with the decay of metarhodopsin III.

Methods for Future Work

One feature of the response, which makes it clear that the phenomena being observed in the admittance measurements on the isolated rods are at best an incomplete representation of what happens in the disc membrane during visual excitation, is the slowness or absence of reversibility of component II. Further indication of this incompleteness is the failure to observe a light-induced surface membrane conductance decrease, although the sensitivity of the measurements was sufficient for this to have been detected if it were present as indicated in other kinds of measurements. These apparent deficiencies of the response obtained with isolated rod outer segments, together with the response lability, whereby component II was replaced by component I, and the complication introduced by the random orientation of the rods, have led to the consideration of the possibility of examining the intact retina still in the eyecup for evidence of a disc membrane conductance change. A device to be used in this work has been described.[33]

Acknowledgment

Support has been provided by the Medical Research Council for work described in this paper.

References

1. F. J. M. Daemen, P. P. M. Schnetkamp, Th. Hendriks and S. L. Bonting 1977, Calcium and rod outer segments. This volume, pp. 29–40.
2. W. Hubbell, K.-K. Fung, K. Hong and Y. S. Chen 1977, Molecular anatomy and light-dependent processes in photoreceptor membranes. This volume, pp. 41–59.
3. W. A. Hagins and S. Yoshikami 1977, Intracellular transmission of visual excitation in photoreceptors: electrical effects of chelating agents introduced into rods by vesicle fusion. This volume, pp. 97–139.
4. L. H. Pinto, J. E. Brown and J. A. Coles 1977, Mechanism for the generation of the receptor potential of rods of *Bufo marinus*. This volume, pp. 159–167.
5. G. Falk and P. Fatt 1973 , An analysis of light-induced admittance changes in rod outer segments, *J. Physiol.* **229**, 185–220.
6. G. Falk and P. Fatt 1973, Isolation of components of admittance change in rod outer segments, *J. Physiol.* **229**, 221–239.
7. G. Falk and P. Fatt 1968, Passive electrical properties of rod outer segments, *J. Physiol.* **198**, 627–646.

8. G. Falk and P. Fatt 1968, Conductance changes produced by light in rod outer segments, *J. Physiol.* **198**, 647–699.
9. B. Katz and R. Miledi 1972, The statistical nature of the acetylcholine potential and its molecular components, *J. Physiol.* **224**, 665–699.
10. C. R. Anderson and C. F. Stevens 1973, Voltage clamp analysis of acetylcholine produced end-plate current fluctuations at frog neuromuscular junction, *J. Physiol.* **235**, 655–691.
11. R. D. Keynes and E. Rojas 1974, Kinetics and steady-state properties of the charged system controlling sodium conductance in the squid giant axon, *J. Physiol.* **239**, 393–434.
12. W. Nonner, E. Rojas and R. Stämpfli 1975, Gating currents in the node of Ranvier: voltage and time dependence, *Phil. Trans. R. Soc. B.* **270**, 483–492.
13. W. Almers and S. R. Levinson 1975, Tetrodotoxin binding to normal and depolarized frog muscle and the conductance of a single sodium channel, *J. Physiol.* **247**, 483–510.
14. G. Falk and P. Fatt 1972, Physical changes induced by light in the rod outer segment of vertebrates, *In* "Handbook of Sensory Physiology", vol. 7, part 1, Photochemistry of Vision (ed. H. J. A. Dartnall) pp. 200–244. Springer-Verlag, Heidelberg.
15. K. P. Hofmann, R. Uhl, W. Hoffmann, and W. Kreutz 1976, Measurements of fast light-induced light scattering and absorption changes in outer segments of vertebrate light sensitive rod cells, *Biophys. Struct. Mechanism* **3**, 61–77.
16. M. Chabre 1975, X-ray diffraction studies of retinal rods: I. Structure of the disc membrane, effect of illumination, *Biochim. Biophys. Acta* **382**, 322–335.
17. A. E. Blaurock and M. H. F. Wilkins 1972, Structure of retinal photoreceptor membranes, *Nature* **236**, 313–314.
18. A. E. Blaurock 1977, What X-ray and neutron diffraction contribute to understanding the structure of the disc membrane. This volume, pp. 61–76.
19. R. A. Cone 1972, Rotational diffusion of rhodopsin in the visual receptor membrane, *Nature New Biol.* **236**, 39–43.
20. M. M. Poo and R. A. Cone 1973, Lateral diffusion of rhodopsin in *Necturus* rods, *Exp. Eye Res.* **17**, 503–508.
21. P. A. Liebman and G. Entine 1974, Lateral diffusion of visual pigment in photoreceptor disk membrane, *Science* **185**, 457–459.
22. D. Bownds, A. Brodie, W. E. Robinson, D. Palmer, J. Miller and A. Shedlovsky 1974, Physiology and enzymology of frog photoreceptor membranes, *Exp. Eye Res.* **18**, 253–266.
23. A. W. Clark and D. Branton 1968, Fracture faces in frozen outer segments from the guinea pig retina, *Z. Zellforsch.* **91**, 586–603.
24. T. S. Leeson 1971, Freeze-etch studies of rabbit eye: II. Outer segments of retinal photoreceptors, *J. Anat.* **108**, 147–157.
25. Y. S. Chen and W. L. Hubbell 1973, Temperature- and light-dependent structural changes in rhodopsin-lipid membranes, *Exp. Eye Res.* **17**, 517–531.
26. J. M. Corless, W. H. Cobbs, M. J. Costello and J. D. Robertson 1976, On the asymmetry of frog retinal rod outer segment disk membranes, *Exp. Eye Res.* **23**, 295–324.
27. K. O. Donner and S. Hemilä 1975, Kinetics of long-lived rhodopsin photoproducts in the frog retina as a function of the amount bleached, *Vision Res.* **15**, 985–995.
28. K. O. Donner and T. Reuter 1968, Visual adaptation of the rhodopsin rods in the frog's retina, *J. Physiol.* **199**, 59–87.

29. R. N. Frank 1971, Properties of "neural" adaptation in components of the frog electroretinogram, *Vision Res.* **11**, 1113–1123.
30. W. Ernst and C. M. Kemp 1972, The effects of rhodopsin decomposition on PIII responses of isolated rat retinae, *Vision Res.* **12**, 1937–1946.
31. D. C. Hood, P. A. Hock and B. G. Grover 1973, Dark adaptation of the frog's rods, *Vision Res.* **13**, 1953–1963.
32. S. R. Grabowski and W. L. Pak 1975, Intracellular recordings of rod responses during dark-adaptation, *J. Physiol*, **247**, 363–391.
33. P. Fatt 1976, The space-clamped retina, *J. Physiol.* **258**, 52–54P.

7

Intracellular Transmission of Visual Excitation in Photoreceptors: Electrical Effects of Chelating Agents Introduced into Rods by Vesicle Fusion

W. A. HAGINS and S. YOSHIKAMI

Laboratory of Chemical Physics, NIAMDD
and *Laboratory of Vision Research, NEI*
National Institutes of Health, Bethesda, Maryland, U.S.A.

When a photon is absorbed in a vertebrate rod outer segment, the event is communicated to the rod synapse by means of a change in membrane potential of the entire receptor cell. Thus rods resemble other sensory receptor cells as well as many neurons: they conduct messages from point to point by acting as electric cables. A decade of research on the electrical events of rod excitation has provided many details about the machinery that operates during photon responses. A rough scheme of some of the principal events of rod excitation is shown in Fig. 1.

In the scheme, it is supposed that the events following absorption of a photon by rhodopsin and culminating in a sensory message at the first postsynaptic neuron can be divided into *N* steps, where *N* will turn out to be a very large number. Each step in the direct causal line is given a number (e.g. *N*-2), while parallel events that accompany a particular step but that would not prevent excitation if they did not exist are given two-number labels (e.g. *N*-5, 2). Thus excitation can be thought of as a set of events whose coordinates are the rows and columns of a matrix, the main events lying in column 1. This paper deals with steps *N*-4 through *N*-8, the release of a chemical transmitter substance from the rod disks and its subsequent binding and release at the plasma membrane covering the outer segments.

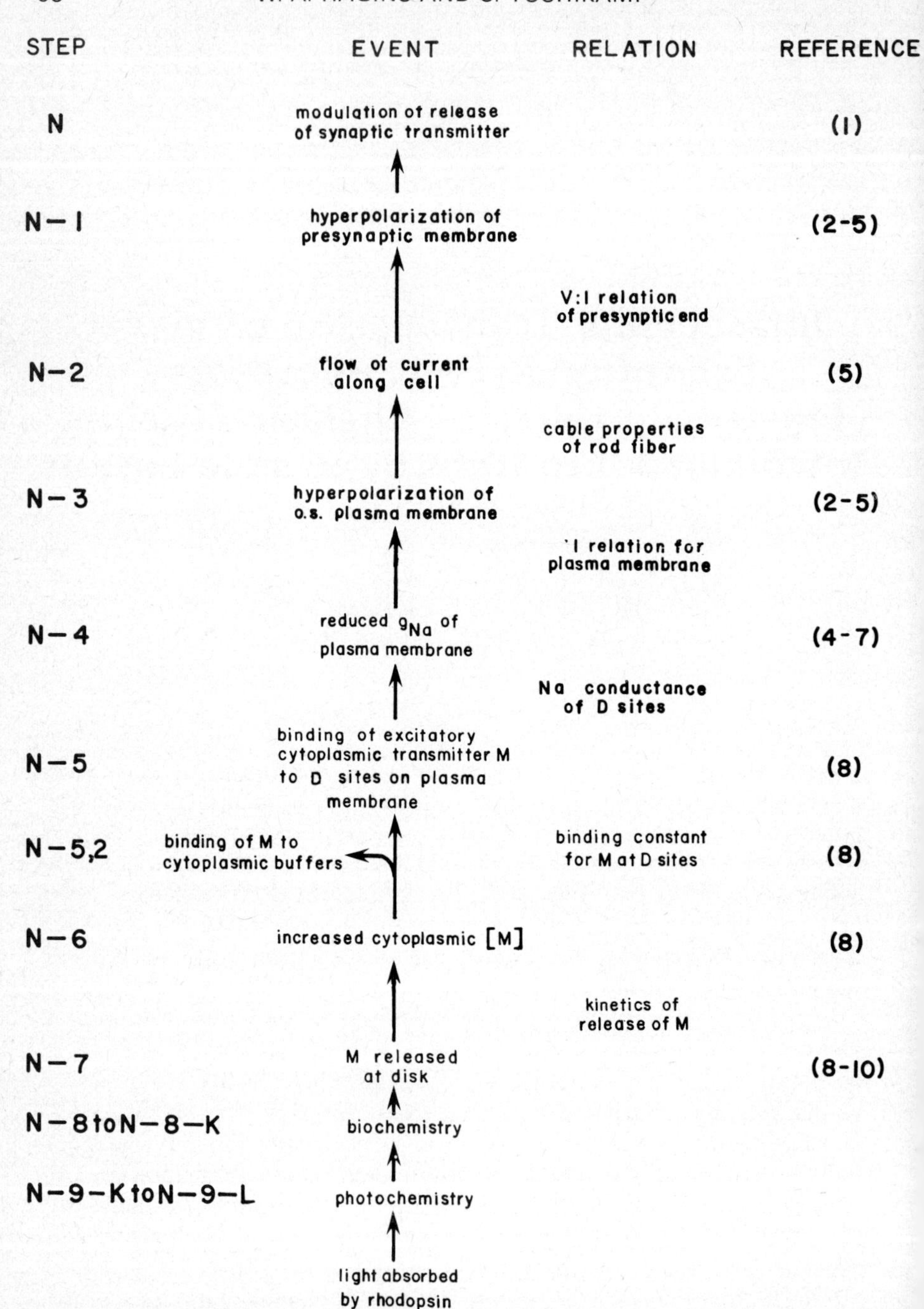

Fig. 1. Scheme of the events of a rod response to an absorbed photon. Numbers in the right hand column refer to citations in the list of references.

The existence of a chemical transmission step has been suggested to account for the kinetic complexity of electrical responses of rods and cones[3,10] and to provide a mechanism for numerical gain in the response to single photons.[9,10] The most compelling argument for the existence of a diffusible transmitter stems from the fact that the rhodopsin-bearing rod disk membranes are topologically separate from the plasma membrane[11,12,13,14] and thus cannot affect it by the usual mechanisms of cable conduction.

If there is an intracellular transmitter, what is it? In several recent papers, we have suggested that it is free calcium ions.[6,8,9,10] The evidence consists of three observations. First, increased a_{Ca} in the fluid around rods rapidly and reversibly reduces the inward-flowing dark current (step *N*-4) just as light does. Second, when the rods are treated with the Ca-bearing ionophores X-537A or A23187, externally applied Ca^{++} is at least 2000 times more effective in reducing the dark current than it is in untreated cells. Third, when the external Ca^{++} activity is reduced below 10^{-7} M, rods are greatly desensitized to light, as if an internal store of transmitter is depleted. This desensitization is reversible.

Other findings support the notion that Ca^{++} might act as an intracellular transmitter. Isolated outer segments of both frogs and rats contain at least 5 mM Ca.[10,15,16] There are several reports that isolated rod disks take up Ca^{++} in darkness and release it in light.[16-21] Cone responses of the photoreceptor layer of iguana retinas are almost completely desensitized by Ca^{++} deprivation, as if cones too use it as a transmitter but are more vulnerable to low Ca^{++} solutions because the cone disks have interiors that remain confluent with the extracellular space (Yoshikami, unpublished experiments).

While the evidence supports the excitatory role of Ca, all of it is indirect. What is needed is a set of experiments that are capable, in principle, of disproving the Ca hypothesis. In this paper we describe one such class of experiments and their results.

Design of the Experiment

In the scheme of Fig. 1, it is supposed that a transmitter substance M (or one of its precursors) is stored near rhodopsin in the rod disks. When a photon is absorbed, free M is produced near the photochemically excited rhodopsin chromophore. The free M then diffuses laterally to the plasma membrane of the outer segment where it reacts reversibly with sites D on its inside surface. While M is complexed with D, the inward Na^{+} current associated with that particular D is suppressed. After a time, the source of M ceases, M is reaccumulated in the disk or is extruded by pumps in the plasma membrane, the M bound to D is released, and the flow of the dark current through the plasma membrane resumes.

A simplified scheme of these steps is shown in Fig. 2. It contains one important approximation: longitudinal and lateral diffusion of M is assumed to be so fast as to be instantaneous on the time scale of the electrical response to light. This is the *Well-Stirred Outer Segment* (WSOS) assumption. For responses that last a few tenths of a second and involve a large enough number of absorbed photons in each outer segment, this assumption seems justified in small rods such as those of the rats used in this work.

Consider now the effect of introducing a substance B into the cytoplasm of the outer segment. Let B have the property of forming a reversible complex MB with M. In the dark, a steady state distribution of M will be formed among its various states—in the disks, free in the cytoplasm, bound to D, and so forth. Eventually, exchange of M across the plasma membrane will restore the levels of M in all of its states to the dark-adapted values except for the additional M bound to B. In darkness, therefore, the steady dark current flowing through

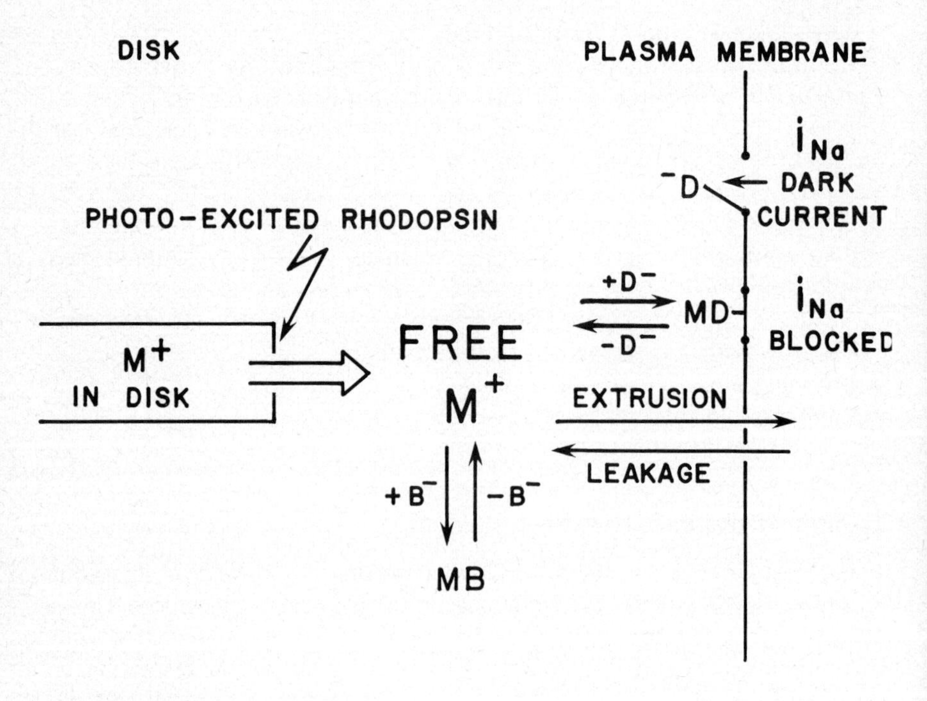

Fig. 2. Diagram of the WSOS approximation to the internal transmitter hypothesis of rod excitation. A cytoplasmic buffer B competes with the Na^+-gating D sites on the plasma membrane for the transmitter M. A steady-state level of M is present in the dark.

the plasma membrane at the D sites will be unchanged despite B's presence. When photons are absorbed from a flash of light, however, the additional Ms produced will partly complex with free Bs and relatively fewer will be available for suppression of the dark current at the D sites. Thus the light responses will be reduced. If the flash is bright enough, however, the Bs will become saturated and eventually the Ds will be also if enough Ms are released. If the binding constant of B for M and the steady-state concentration of free M are of the right orders of magnitude, the qualitative effect of the introduction of B will be to reduce the rod responses to dim light but to leave the maximal responses in bright unchanged.

Using the WSOS approximation, these qualitative statements can be reduced to equations from which effects of B on light responses can be shown.

Let

M, B, D represent the concentrations of the respective free species (moles/liter of o.s. water)
MB, MD represent the concentrations of the indicated complexes
B_0, D_0, denote the total concentrations of B and D of all forms
$K = MD/(M)(D)$, the affinity constant of D for M (moles/liter)
$R = MB/(M)(B)$, the affinity constant of B for M (moles/liter)

Now, if a flash of light causes M to be released in the o.s. cytoplasm with a time course given by, say,

$$F(t) = at^3 \exp(-bt) \tag{1}$$

and to be removed by a first-order process with a rate-constant k, a set of solutions of the differential equation for the time-course of D (representing the dark current as a time function) can be obtained for different flash energies (cf. reference 12).

This set of curves is shown in Fig. 3(a) for values of a corresponding to short flashes of energies 1–300 in 1,3,10,30,100,300 sequence. The rate constants b and k for supply and removal of M have been chosen simply to generate curves whose shapes resemble real flash responses of rat rods: they have no deeper significance. While Fig. 3(a) shows curves for the model with B absent, Fig. 3(b) shows the effect of adding a complexing agent B whose affinity constant R is 10 times that of the D sites for M and whose total concentration is equivalent to the transmitter released by 8 photons absorbed per rod. Note two effects of the buffer. First, it greatly diminishes the size of the responses to flashes of arbitrary energy 1 and 3. Second, it distorts the time courses of the low energy responses. The peak times of the curves for dim flashes lie near the time when the production of M is greatest (i.e. where dF/dt is largest). As the flashes become brighter, the peak response time becomes longer and then somewhat shorter again. The first effect is very large and depends little on the details of

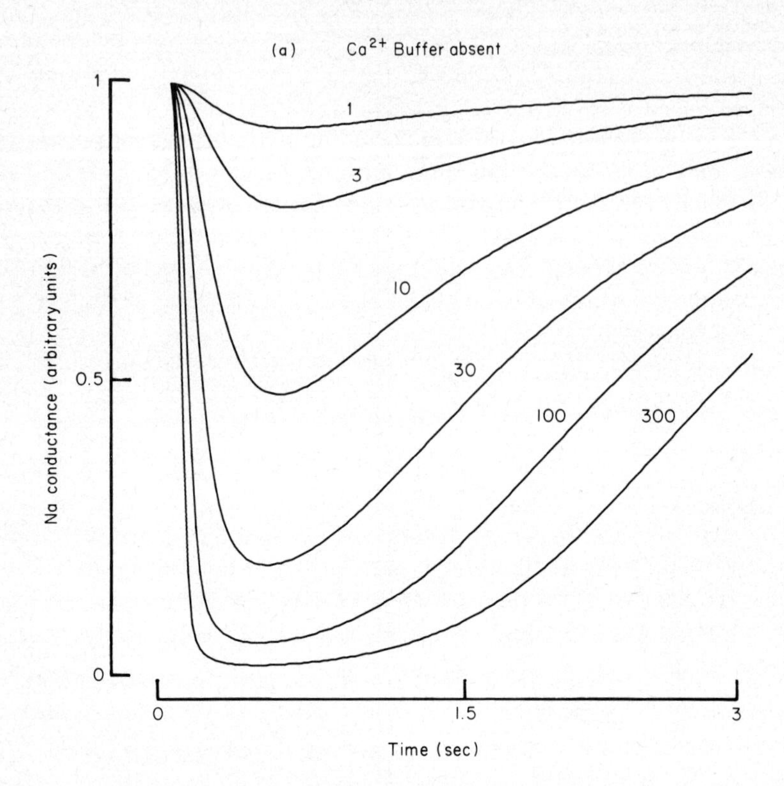

Fig. 3. Na^+ conductance *versus* time for a rod stimulated with flashes of energies of relative sizes 1,3,10,30,100, and 300 units. The transmitter is supplied by a process whose time-course is given by eqn (1) and removed by a first-order reaction. Constants have been chosen to give solutions of the WSOS model that resemble real rod responses. (a) No buffer present. (b) A buffer is present with an affinity for M ten times that of the D sites. The abscissa is in arbitrary units.

the mathematical formalism by which M is generated and removed. The second effect is smaller and much more model-dependent. In what follows, it is the first effect that will be used as qualitative evidence of the action of artificially introduced cytoplasmic buffers on the responses of rat rods.

In the experiments to be described, transient reductions in the dark current of rat rods ("light responses") are measured as functions of flash energy for a wide range of energies and in the presence and absence of a variety of hydrogen ion and metal ion buffers introduced into the rods by the phospholipid vesicle fusion technique. The wave forms and amplitude-intensity relations for the responses are compared among retinas under identical ambient conditions but differing in their history of exposure to buffer-bearing vesicles. In the

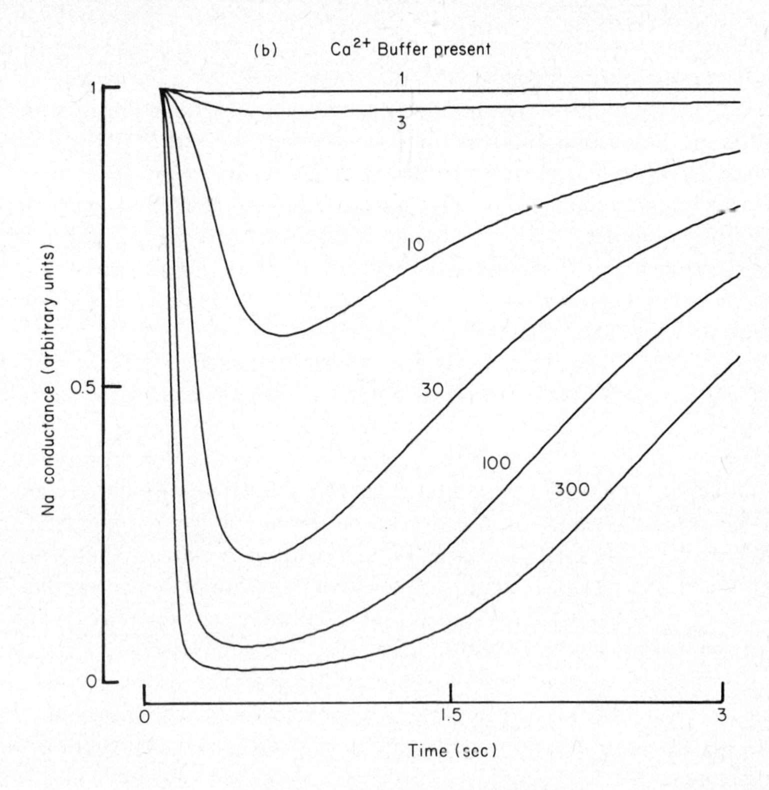

following sections the electrical recording methods and the vesicle fusion procedure are first described.

Measurement of the Rod Responses

Albino rats were dark-adapted for 1–2 h and killed by intraperitoneal injection of pentobarbital (50 mg/kgm). Their eyes were enucleated under a 25 W no. 1 ruby incandescent lamp. From this stage forward, infra-red optics and binocular image converters were used, so that the retinas were effectively in total darkness. The eyes were opened in a bath of physiological salt solution (Ringer II) containing NaCl: 136 mM, KCl: 2·7 mM, $MgCl_2$: 0·5 mM, Na_2HPO_4: 0·725 mM, $CaCl_2$: 1·36 mM, Tris: 10 mM pH 7·1, and glucose: 11 mM. The solution was saturated with O_2 by continuous bubbling and the fluid around the tissue was kept moving at all times. Scrupulous measures were taken to avoid contamination of the fluid by ferrous metals from the stainless steel dissecting instruments. The tools were soaked in EDTA solutions and lightly greased with silicone jelly before each use. The dissecting bath was lined with fresh paraffin wax.

The retinas were rapidly removed from the open eye cups and pinned near the O_2 bubbler until needed. A *circa* 2 × 4 mm section of a retina was cut and held vitreous-side-down against a Millipore type HA filter which was then lifted from the bath and blotted on its under side with filter paper until the retina was tightly flattened down and attached to its support.

The retinal fragment and its filter were quickly transferred to a small open-topped flow chamber that was continuously flushed with salt solutions of known composition, the fluid being removed by air suction at a point well past the retina. The chamber was heated to 34–37°C. Its total volume was about 0·05 ml and it could be completely flushed in 2 s.

A pair of capillary microelectrodes with tip diameters of 2–3 micrometers filled with isotonic Na isethionate solution were lowered into the retinal receptor layer under direct vision by dark-field infra-red optics that allowed the surface of the retina and the electrodes to be seen in profile at 60 × magnification. A reticle in the ocular permitted the retinal surface to be located and the depth of electrode insertion to be measured approximately. A dial micrometer on the microelectrode holder made it possible to read relative positions of the electrodes in the retina with a precision of 2 micrometers. The electrodes were connected to Ag-AgCl junctions mounted in a heavy Al block. Amplifiers with low voltage and current noise (Analog Devices type 43K) and signal-averaging allowed light responses of less than 1 microvolt to be measured with a signal to noise ratio of 5:1 when necessary. Transient retinal responses were digitized by a small computer and stored on a magnetic disk for further use.

Light stimuli were 2 μs flashes from a xenon flash tube (Edgerton Germeshausen Grier type FX 6A). These were filtered through an interference filter with a flat transmission maximum at 550 nm and a bandwidth at 1 % transmission of 60 nm. The flashes were attenuated with calibrated neutral gelatin filters. Flash exposures produced by the filter combinations were measured with a calibrated silicon photovoltaic diode over the entire range of intensities. The flashes were delivered to the retina from the side of the white Millipore filter opposite to the retina, so the rods were effectively bathed in light diffused through a 2π solid angle. The fractions of incident light absorbed by rhodopsin were calculated for two models of the rhodopsin in the rods: (a) rhodopsin chromophores randomly oriented and (b) rhodopsin chromophores oriented in the planes of the rod disks. The calculated fractions of diffuse light absorbed for the two cases differed by 20 %. This difference thus represents the minimum uncertainty in the *absolute* fraction of incident light absorbed. The *relative* energies of the stimuli were known to within ± 10 % over an intensity range of 4 decades.

All recordings were made in isotonic Ringers in which the Cl^- was replaced with the impermeant anion, isethionate$^-$. Such solutions prevented osmotic swelling of the retina in low Ca^{++} solutions so that very stable responses could be obtained for several hours. Ca^{++} and Mg^{++} activities in the bath were buffered with 1–10 mM concentrations of EGTA or other metal buffers. In all of this work the affinity constants for proton and metal buffers appearing in reference 33 were used in making Ringers and in curve-fitting. The values chosen are shown in Table 1.

Table 1

Affinity constants of buffers used in these experiments for protons, Ca, and Mg ions selected from reference 33. Figures given as $\log_{10} K(M^{-1})$ at 25°C and ionic strength 0·1.

	Protons		Mg^{++}	Ca^{++}
	K_1	K_2	K_{Mg}	K_{Ca}
EDTA	10·07	6·13	8·64	10·42
EGTA	9·46	8·85	5·21	11·00
CDTA	10·76	6·13	10·41	11·34

Vesicle Fusion Techniques

Metal ion buffers like EDTA, EGTA, and their relatives as well as many zwitterionic hydrogen ion buffers are too polar to penetrate cell membranes. Retinal rods are no exception to this rule; EDTA and EGTA have no effect on rod responses even after a three-hour incubation at 10 mM concentrations, so long as the external Ca^{++} and Mg^{++} activities in the bath are maintained at appropriate levels. Attempts to measure the penetration of ^{14}C labeled EDTA into live rat retinas (Yoshikami and Hagins, unpublished) indicate a permeability too low to measure accurately. Other work in this volume describes effects of metal buffers injected directly with microelectrodes.[34] For our purposes, however, a method of introduction is needed that permits even small photoreceptors to be loaded with known quantities of buffers and chelating agents whose effects can be assayed by the extracellular recording method. The technique we have used is that of vesicle fusion, suggested by Papahajopoulos[22-24] and others as a way to insinuate membrane-impermeable but water-soluble substances into cells. Our method permits us to measure the quantity of vesicle contents incorporated, to measure the spatial distribution of injected material and to estimate the "cytoplasmic" pH while at the same time discriminating against vesicles that have adhered to the cells without actually undergoing fusion.[25] We cannot be sure of the detailed physical mechanism by which our procedure introduces buffers into retinal

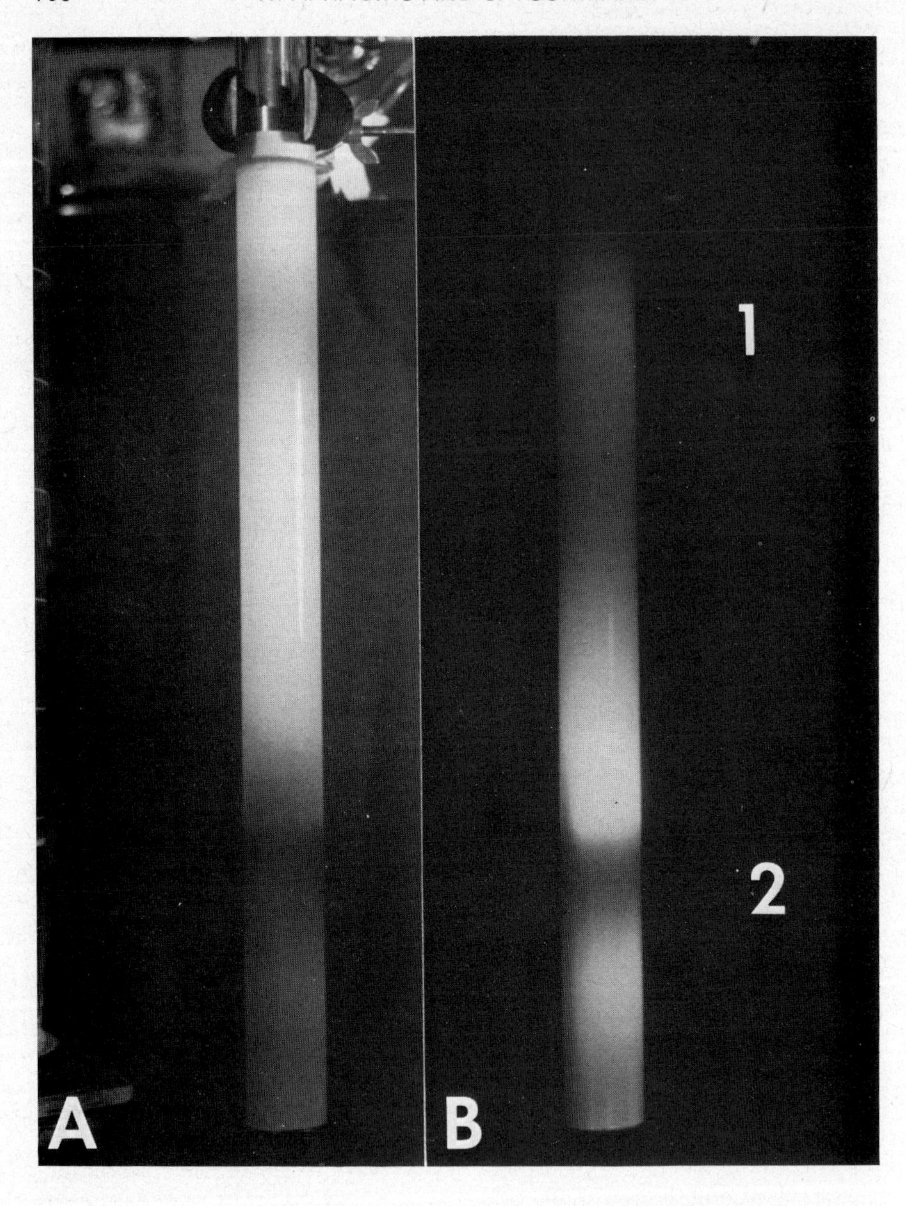

Fig. 4. DOPC vesicles containing 100 mM 6CF being separated by upward flow through a bed of Sephadex G-25. The weakly fluorescent but intensely orange band of vesicles is at 1. The free 6CF lags behind at 2, where its concentration is so high that its fluorescence is largely self-quenched, as is that of the concentrated dye in the vesicles. (A) White light illumination. (B) 365 nm excitation, fluorescence of wavelengths >520 nm recorded.

cells, but we shall refer to the process as "transfer" or "vesicle fusion" for simplicity.

Intracellular transfer of 6-carboxyfluorescein

Phospholipid vesicles were prepared by vortexing 25 mg of synthetic L α dioleoyl phosphatidyl choline (Applied Science Laboratories) vacuum-dried from benzene on the walls of a test tube with 0·5 ml of a 200 mM solution of purified 6-carboxyfluorescein (dissolved in water containing three equivalents of KOH and brought to a final pH of 7·35). The deep orange suspension was then sonicated with a 12 mm diam Ti probe for three times the clarification time at 15°C (about 20 min total sonication time). Power input to the sample was less than 15 watts and the solution temperature never rose more than 10°C during sonication. The clear suspension was loaded on a Sephadex G-25 column 1·5 × 30 cm filled with NaCl, 136 mM, HEPES, 10 mM, pH 7·35. The vesicles eluted in the void space and were collected in a total volume of 2–4 ml. Figure 4 shows a column as a separation proceeded. In Fig. 4(A) white light was used for illumination. The orange band of vesicles containing trapped dye was visible as it rose through the column ahead of the free dye band near the bottom. In Fig. 4(B), the fluorescence of the column under 365 nm excitation is shown. Note that the center of the free dye band looks dark, while its upper and lower boundaries are brilliantly fluorescent with the characteristic green emission of 6-carboxyfluorescein (excitation maximum: 490 nm, emission maximum: 520 nm in water). The vesicle band, like the center of the free dye band, is only weakly fluorescent, because the emission of 6-carboxyfluorescein at 200 mM concentration in the vesicle interiors is strongly quenched. In fact, fluorimetry shows that the dye in the vesicles has a fluorescence efficiency more than fifty times weaker than that of a dilute aqueous solution. This self-quenching serves to distinguish dye in unfused vesicles from dye that has been transferred into cells and diluted in their cytoplasm. A description of the use of this property of 6CF to study vesicle fusion in lymphocytes appears elsewhere.[25]

The vesicle suspension usually contained about 10 mg of phospholipid per ml and a total dye concentration of about 100 μM. While suspensions showed only weak orange fluorescence when fresh, they gradually developed green fluorescence on storage at 0°C for a few days. A drop of 0·1 % Triton X-100 in the suspension would immediately release the dye, as would detergent residues on any unclean glassware or utensils.

To study injection of dye into retinas or other cells, the vesicle suspension was converted into a physiological solution with the composition of Ringer II by addition of small amounts of concentrated solutions of the necessary ingredients. Three frog retinas incubated in such solutions for three hours are

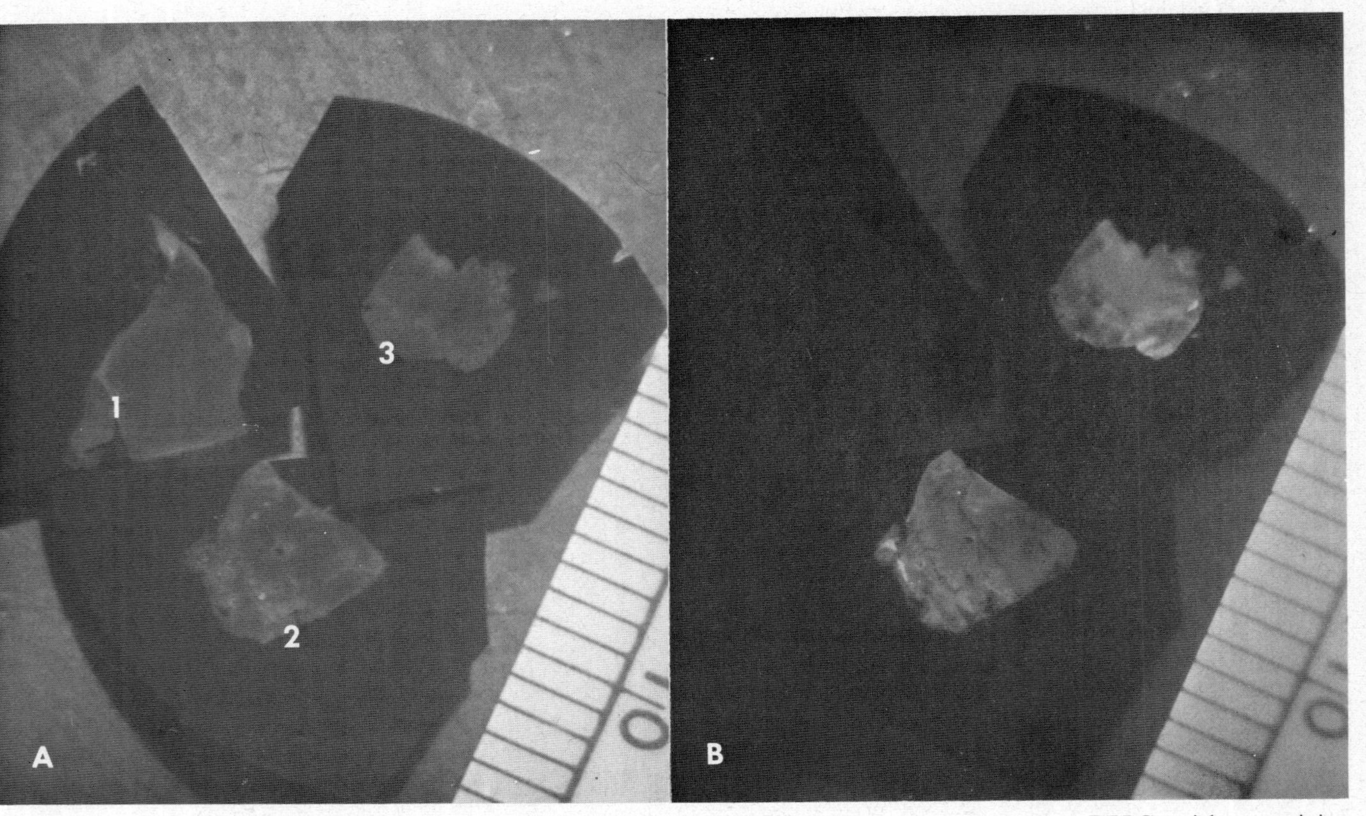

Fig. 5. Frog retinas after incubation with 10 μM free 6CF (1), DOPC vesicles containing 100 mM 6CF (2) or DPPC vesicles containing 100 mM 6CF (3). The vesicle suspensions contain 10 mg/ml of phospholipid. 3 h incubations at 25°C. (A) photographed in green light. (B) Fluorescence excitation at 436 nm, fluorescence wavelengths 515–590 nm recorded.

shown in Fig. 5. Retina 1 was exposed only to 10 μM free 6CF. It is illuminated in Fig. 5(A) by 546 nm light, but its fluorescence under 436 nm excitation while viewed through a Wratten 15 blocking filter (passing wavelengths >515 nm) is too faint to see. Retina 2 was incubated with a suspension of 6CF in di-oleoyl phosphatidyl choline, while retina 3 was exposed to similar vesicles made of di-palmitoyl phosphatidyl choline. Both show bright green 6CF fluorescence that could not be extracted by washing with Ringer for more than an hour. Clearly the dye had been transferred to the retina in the presence of phospholipid. Rod outer segments isolated from a retina incubated with DOPC vesicles containing 200 mM 6CF are shown in Fig. 6 in 546 nm dark field illumination (Fig. 6A) and under 436 nm excitation with a narrow-band interference filter transmitting at 520 nm in the path of the emitted fluorescence (Fig. 6B). Because the fluorescence is very dim in single rods, an image intensifier was used to aid the photography. Rods isolated from a retina incubated with even 0·1 mM free 6CF showed no detectable fluorescence under similar treatment. The even distribution of the dye fluorescence in the outer segments is plain. No localized accumulations were seen. Indeed, once the dye had reached the interiors of the outer segments, it could be washed out only slowly even when the plasma membranes (as indicated by the di-DANSYL cystine staining test[(14)]) were ruptured. Once inside, the dye evidently binds to structures in the outer segments.

The time course of uptake of 6CF in a live, light-adapted rat retina is shown in Fig. 7(a). The retina was mounted in a shallow, closed flow chamber close to the common end of a bifurcated fiber optic bundle. 480 nm exciting light was directed at the retina, which was mounted on a black backing consisting of a Millipore HA black filter. The resulting fluorescence was detected by a photomultiplier mounted behind a 520 nm interference filter at the other branch of the fiber optic bundle. The retina was alternately exposed to the flowing vesicle suspension or to clean Ringer, all photometric measurements being made after at least a 15 min wash to remove most adherent vesicles. On exposure to the vesicle suspension, the fluorescence of the retina increased rapidly at first but leveled off after about 2 h. The slowdown in the rate of increase of fluorescence was not due to loss of vesicles from the incubating solution or to leakage of dye from them; the measured dye content of the vesicle suspension, which was recirculated through an oxygenator and a bed of anion-exchange beads (Bio-Rad type AG1-X4 which removed free 6CF) declined by less than twofold during the experiment. Apparently a living retinal cell will permit only a few hundred vesicles of DOPC to fuse with it over these short times.

At the end of the final incubation with vesicles, the retina was briefly exposed to 0·1 Triton X-100 and the fluorescence observed. There was always a large increase, often three to five-fold. We attribute this large "Triton flash"

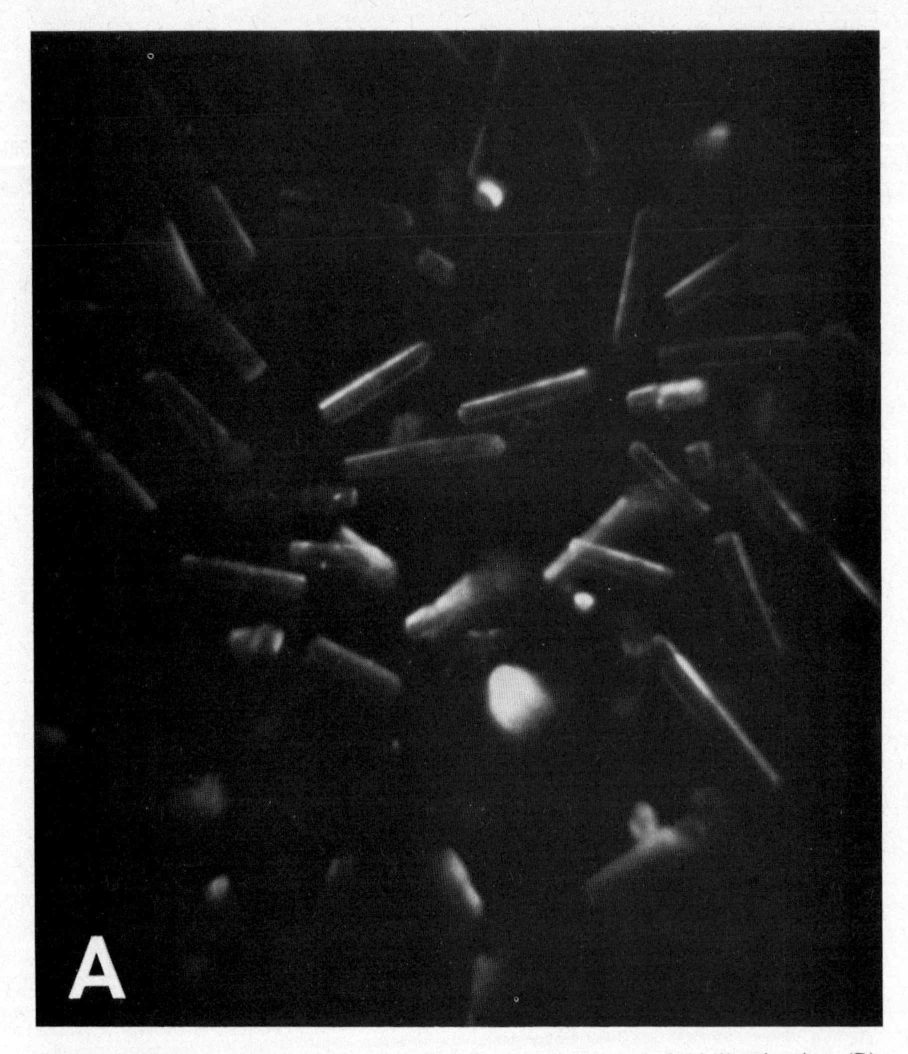

Fig. 6. Rods isolated from retina 2 of Fig. 5. (A) 546 nm dark field illumination. (B) 436 nm excitation, emissions from 515–590 nm recorded. A three-stage photoelectric image intensifier was used for photography. The outer segments are 6 μm in diameter.

to dye released from vesicles bound to the cells and trapped in the tissue interstices.

The total concentration of 6CF averaged over the volume of the retina under illumination could be estimated by flowing solutions of dye at known concentrations past the retina in the chamber. Because the chamber depth was less than 1 mm and the light pipe was more than 6 mm in diameter, a dye

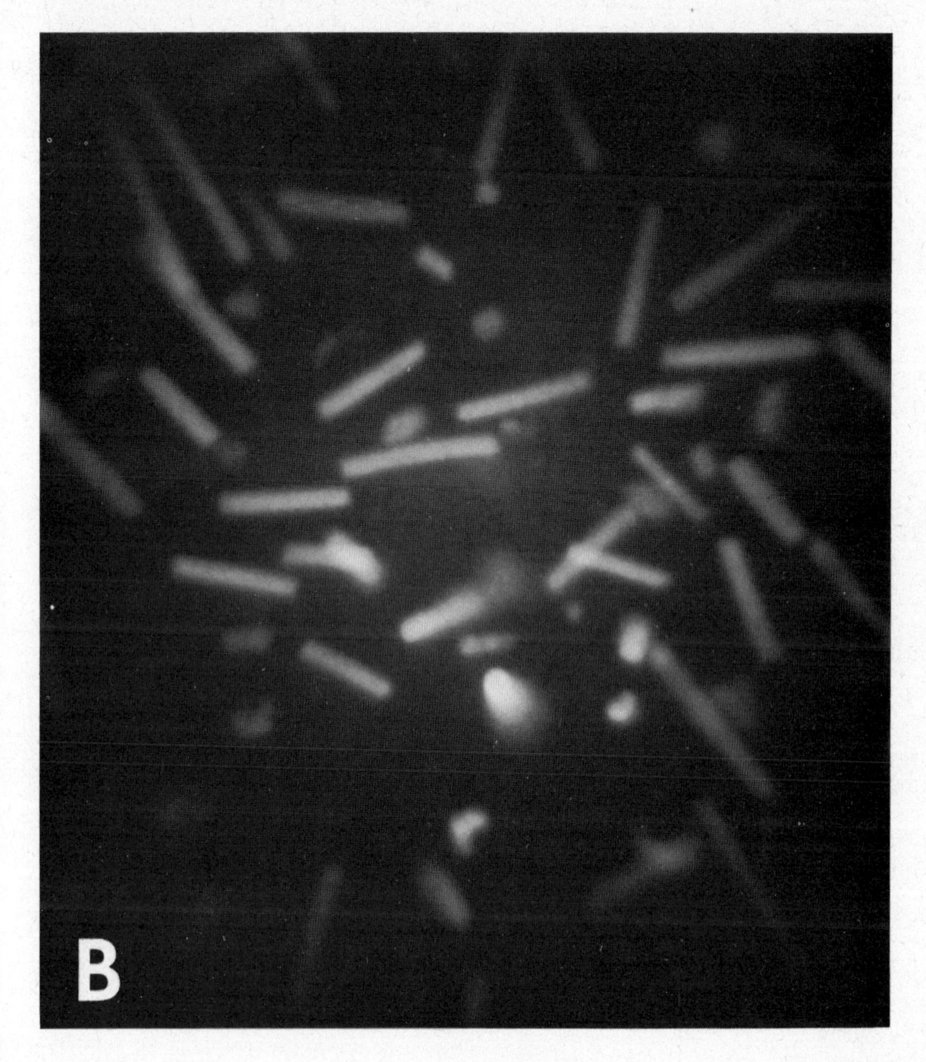

molecule at any point between retina and light was equally effective in generating a fluorescence signal. Thus liquid fluorescence standards could be compared with the retinal fluorescence on-line. In four preparations, the mean content of dye after 2 h vesicle incubation was 3 micromoles per liter of retinal volume after correction for fluorescence due to adherent vesicles.

The distribution of 6CF fluorescence with depth in the rat retina after such incubations is shown in Fig. 7(b). Here the retina was sliced into thin strips about 50 μm wide by an automatic tissue microtome. A slice was mounted

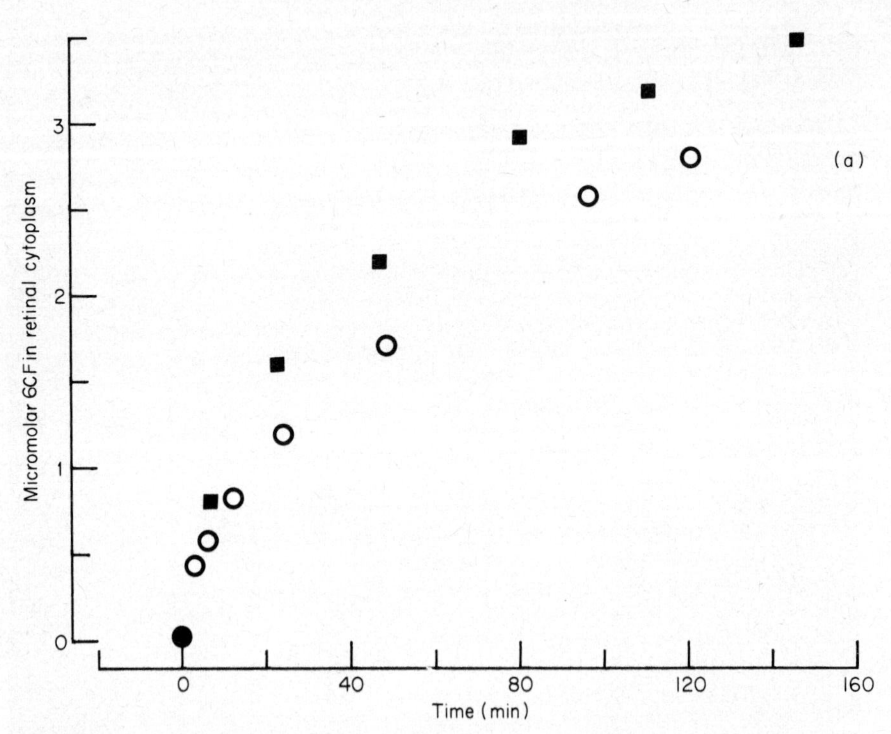

Fig. 7. (a) Time course of appearance of 6CF fluorescence in 4 × 4 mm sections of live, light-adapted rat retinas exposed to a suspension of DOPC vesicles containing 200 mM dye for times shown. Ordinate: Cytoplasmic concentration of 6CF averaged over the thickness of the retina. 480 nm excitation, 520 nm emission. Retinas were washed for 15 min after each exposure to the vesicle suspension to remove most adherent vesicles. (b) Distribution of 6CF fluorescence through the depth of a slice of rat retina like that of (a). A 50 μm wide slice of vesicle-incubated retina was cut with an automatic tissue microtome and mounted on its side in a Ringer bath. It was scanned with a slit of exciting light 50 μm long and 20 μm wide in the scanning direction. The brightest fluorescence was always in the region of rod inner segments, but all retinal layers were stained. A dark-field photograph of the slice is shown (c).

with its cells aligned as shown in Fig. 7(c). It was scanned with a slit of 480 nm exciting light from receptors to vitreal surface, the slit being 20 μm wide in the scanning direction and about 50 μm long. The resulting fluorescence signals were detected with a photomultiplier and displayed on an *X–Y* plotter.

The fluorescence was always about twice as bright in the rod layer as in the deeper parts of the retina. The inner segments were always brighter than the outer segments, probably because of the relatively low water content (*circa* 60 %) of the region containing stacks of rod disks.

Allowing for the uneven distribution of dye in the various retinal layers, the

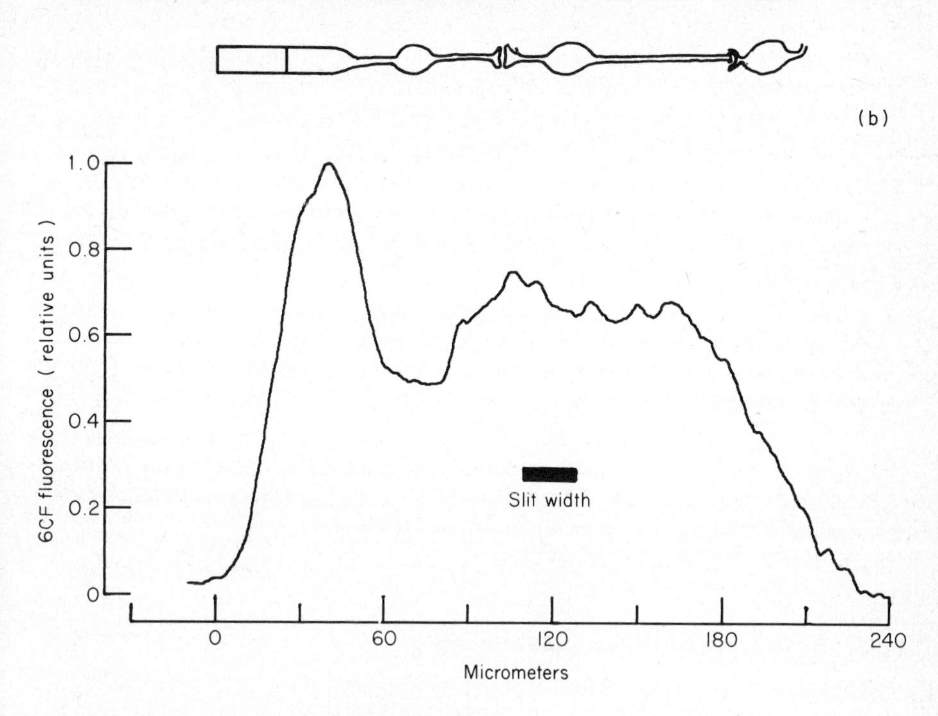
(b)
1.0
0.8
0.6
0.4
0.2
0
6CF fluorescence (relative units)
Slit width
0
60
120
180
240
Micrometers

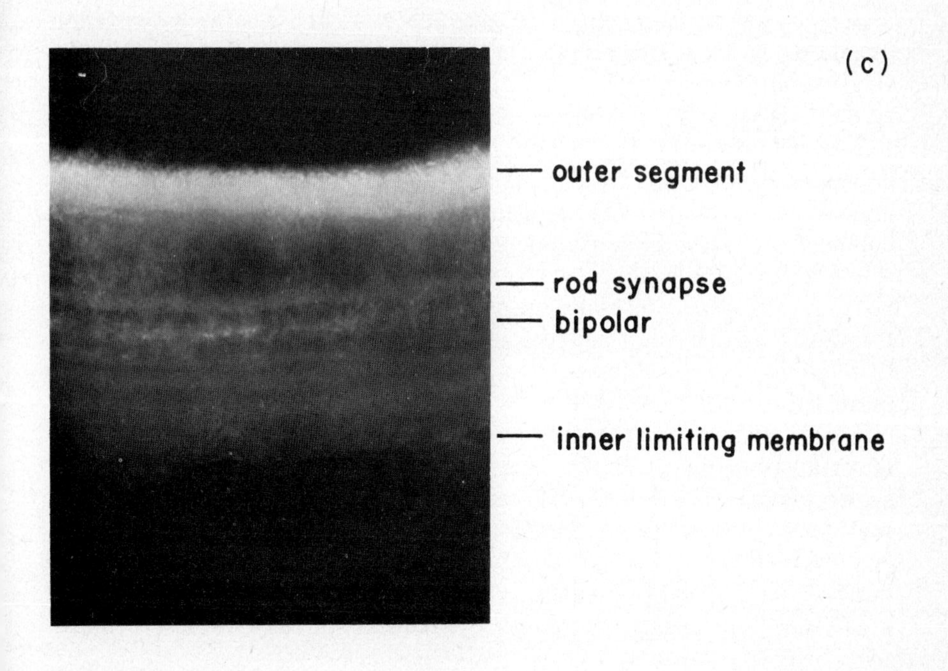
(c)
outer segment
rod synapse
bipolar
inner limiting membrane

mean concentration in the rods must exceed 4 μM. This concentration of dye could be achieved by fusion of no more than 300 vesicles per retinal rod cell, which would result in an increase of less than 1 % in plasma membrane area. This small degree of vesicle fusion is similar to that seen in lymphocytes.[25] Electron micrographs of the outer segments after vesicle treatment showed the usual disorder of disk lamellae typical of live retinas at the ends of long experiments but there were no differences in appearance between vesicle-treated cells and the controls.

Are other water soluble substances contained in the vesicles transferred to the retina in proportion to their concentration in the vesicle water? Preliminary experiments with vesicles containing both 6CF and 2-^{14}C EDTA (New England Nuclear) indicate uptakes of the two substances proportional to their concentrations in the vesicles, but the isotope technique is so insensitive compared to the fluorescence method that we can only say that the measured ratio of vesicle volumes transferred to the retinal cells lies in the range 0·8–1·2 for the two substances. In the work that follows in the rest of this paper, we shall assume that the transfer ratio for all dyes and buffers used is 1.

Estimation of the Cytoplasmic pH

Estimation of the pH in cells is a classic problem that has now been solved in many ways.[26, 27] The presence of 6CF in retinal cells permits us to add yet another way to estimate cell pH, because 6CF has a pH-sensitive fluorescence quantum efficiency. This is shown in Fig. 8(a). The titration curve indicates a single proton-binding site with a pK_a of 6·25 in aqueous media with ionic strengths of about 0·1. The fluorescence of 6CF in the retinal cytoplasm should thus be pH-dependent also. That this is so can be seen in Fig. 8(b). Here the fluorescence of a vesicle-treated and washed rat retina in the flow chamber is plotted *versus* time after various changes were made in the composition of the bathing fluid. When the membrane-impermeable zwitterion HEPES was used as a buffer in the bath, almost no change in fluorescence was seen on switching between pH 7·40 and 7·95 since the dye was not accessible to the buffer. But if the pH of the solution was maintained at 7·40 while 20 mM NH_4Cl was added to it, the fluorescence increased by a factor of almost 1·3. The same solution titrated to pH 7·95 produced a slightly larger increase. Both were rapidly reversible. We interpret these effects as being due to penetration of NH_3 into the cytoplasm from the buffer. The fact that an alkaline buffer produces no greater effect than one at pH 7·40 can be explained by supposing that 6CF is within 30 % of its maximum fluorescence efficiency at the ambient pH of the cytoplasm. If so, agents that acidify the cytoplasm should reduce the 6CF fluorescence. Curve (b) shows that this is so. A Ringer bubbled with CO_2 until its pH was 5·5 produced a reversible 50 % reduction in the fluorescence. Since

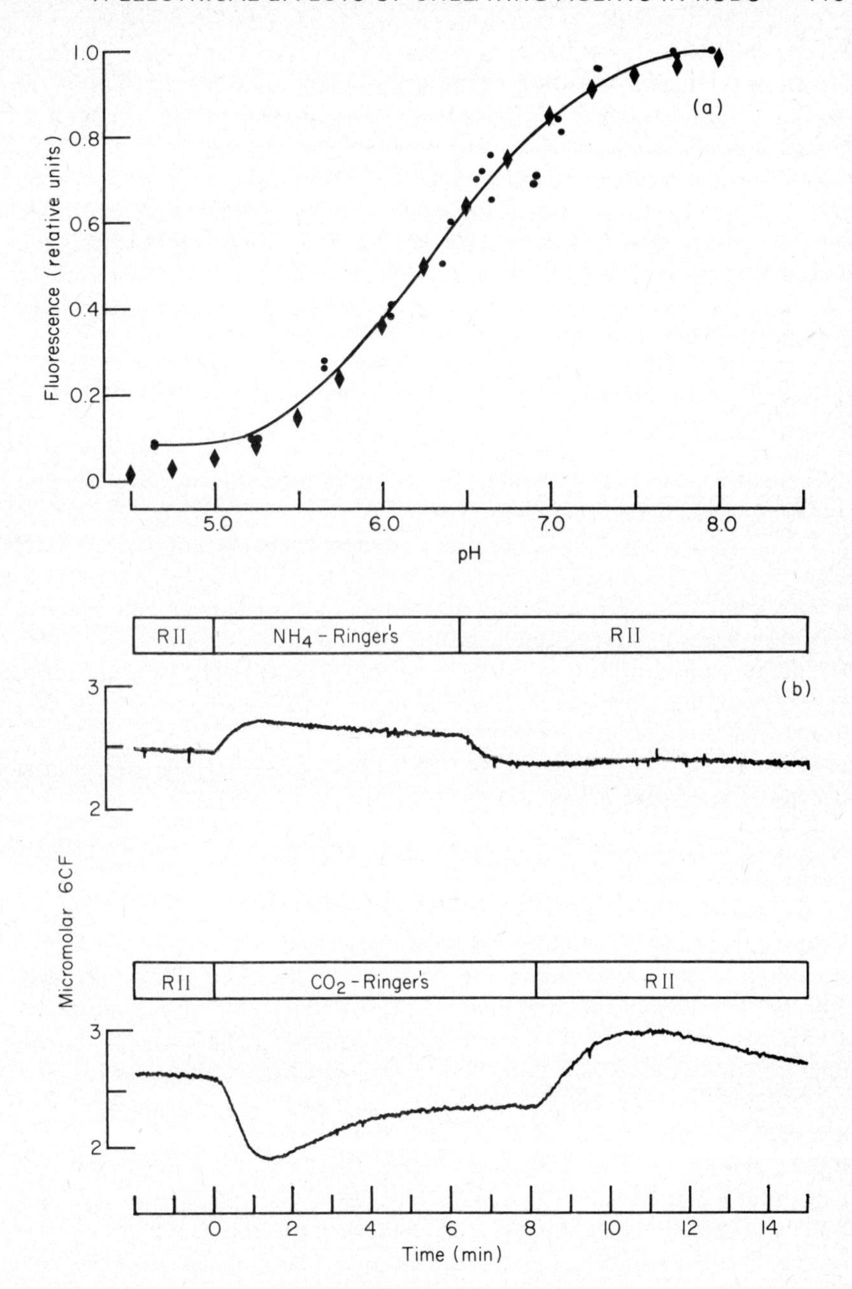

Fig. 8(a) Fluorescence of 6CF *versus* pH. 480 nm excitation, 520 nm emission. 0·15 M NaCl solution containing 10 mM buffer (HEPES, PIPES, or phosphate). (b) Changes in 6CF fluorescence of live rat retina as NH_4Cl or CO_2 is added to the perfusing fluid.

the dye should be almost non-fluorescent at this pH, evidently the cytoplasm cannot be acidified to the same degree as the buffer, or, alternatively, some of the 6CF does not respond to pH changes in the same way as does the free dye. The cytoplasmic pH can be derived on either assumption. If all the dye is titratable with the pK_a of the free species, the cell pH is about 7·1, while if some of the dye is untitratable, the calculated pH is about 6·8. In either case, rather narrow bounds are set on the pH of the space into which 6CF, and presumably other buffers and chelators, can be injected by the vesicle method. This information will be needed in assessing the effects of metal buffers on the responses of rods to light.

Summary of Studies of the Vesicle Fusion Technique

1. A membrane-impermeable dye, 6-carboxyfluorescein, can be trapped in DOPC vesicles by sonication.
2. These vesicles can introduce the dye into retinal cells after a short incubation.
3. The dye is relatively uniformly distributed both within the retinal layers and in individual rod outer segments.
4. The dye concentration in the cells exceeds 4 micromolar concentration, corresponding to fusion of about 300 vesicles of 25 nm diameter with each rod.
5. The fluorescence properties of the dye suggest that the pH of the cytoplasmic space lies between 6·8 and 7·1.

Effects of Chelating Agents on the Rod Responses

Before the effects of injected buffers are described some properties of rod responses must be considered. As we have reported previously, the dark current of rods, and the accompanying extracellular potential gradient are sensitive to the Ca^{++} activity in the bathing solution, apparently because the internal Ca^{++} tends to vary coherently with that of the external solution in the same way as in other cells.[28] The formal mechanism for this is indicated by the arrows in Fig. 2 denoting active and passive transport of M through the plasma membrane. Thus we have available a method for varying experimentally the degree to which the D sites are blocked by Ca^{++}. So, if a cytoplasmic Ca^{++} buffer with the right binding constant were present, its buffering power for Ca^{++} could also be varied by manipulating the composition of the bath. This procedure was used repeatedly in the experiments to be described, so we must consider the effects of varying external Ca^{++} activity on the rod responses of untreated isolated retinas.

When a_{Ca} in the bath is lowered from its standard value of 1·36 mM, both the dark voltage gradient and the size of the transient reduction in the gradient induced by a light flash become larger. Below $a_{Ca} = 10^{-5}$ M, however, the increase in standing dark voltage gradient cannot be maintained indefinitely. Instead, it drops back toward a value little higher than that in Ringer II after 5–10 sec. If the light responses are now tested, they are slower and usually show an overshoot, but the flash exposure necessary to produce a response of half-maximal amplitude remains at the usual value of about 30 photons absorbed rod^{-1} flash^{-1} (see reference 12). This statement applies as a_{Ca} is reduced to about 10^{-7} M. At lower Ca^{++} activities the retina becomes at least ten times less sensitive to light, and the half-saturating flash intensity increases to more than 300 photons absorbed rod^{-1} flash^{-1}.[(6)] In order to avoid this low-Ca desensitization, the present experiments were all done at $a_{Ca} > 2 \times 10^{-7}$ M.

Figure 9 shows two sets of flash responses of a dark-adapted rat retina at $a_{Ca} = 1{\cdot}36$ mM (pCa = 2·87) (Fig. 9a) and at $a_{Ca} = 2 \times 10^{-7}$ M (pCa = 6·7) (Fig. 9b). The responses were recorded as voltage differences between an electrode at the tips of the rod outer segments and a second one just past the external limiting membrane at a depth of 60 μm. The zero reference level on the ordinate scale of this and subsequent figures was obtained by rapidly withdrawing the deep electrode and noting the d.c. voltage shift just after a bright stimulus. The curves have been normalized to an arbitrary value of 1 by dividing each ordinate by the amplitude of the largest response of each group, so that the responses of different retinas could be compared. The peak response amplitude appears in each figure legend.

The stimulus energies of the 2 μs flashes are shown on each curve in units of photons absorbed rod^{-1} flash^{-1}. In both solutions the responses increased in size to a limiting amplitude as flash energy was increased. At low energies, the responses are roughly proportional to light exposure, though below 2 photons per rod the kinetics of the waveforms are somewhat slowed. This effect has been extensively studied by Hodgkin and his colleagues in intracellular recordings from cones.[(29)] Its effect is small enough in the rat retina to be separated from the buffer effects to be expected. Small oscillations due to activity of cells in the deeper layers of the retina can be seen in the 9-photon response of Fig. 9(a). These occurred in about a twentieth of the retinas and were nearly always small enough to have a negligible effect on measurements of the response amplitudes. The major differences between the two groups of responses in the two solutions are in the kinetics of the waveforms. These differences were consistent from retina to retina and were reversible. The responses were quite stable in size and shape as well, so long as the electrode position did not change and the stimuli were given at least 10 s apart. Extensive signal averaging was used for the test stimuli below the 10 photon level so that

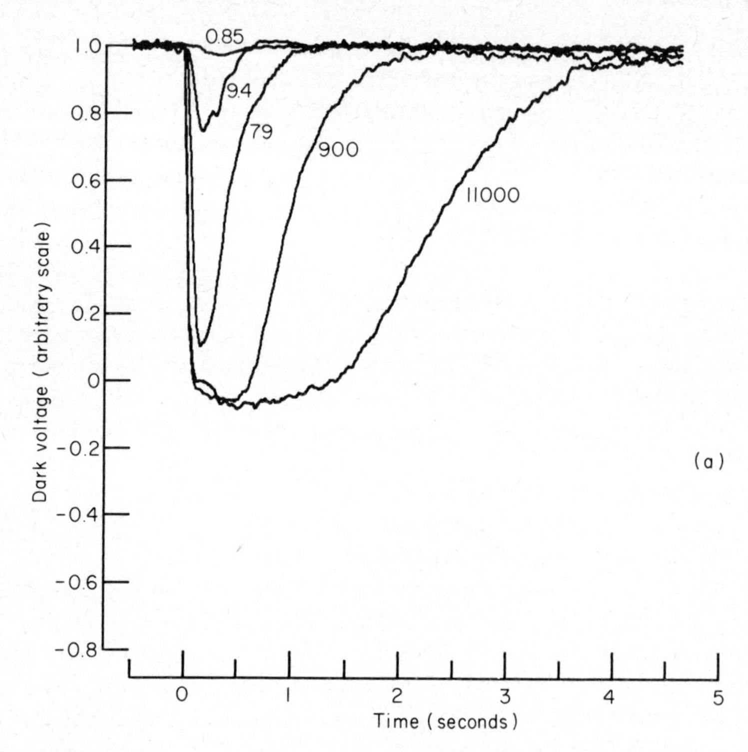

Fig. 9. Responses of dark-adapted rat retina to 2 μs flashes of 550 nm light of various flash energies. Photons absorbed per rod per flash shown as figures marked on the curves. Recording electrodes were two glass pipettes filled with 150 mM Na isethionate with tips at the distal ends of the outer segments and at a depth of 60 μm in the rod layer directly beneath. The scales of the responses have been normalized to 1 for the largest waves. Absolute dark voltage differences between the tips were measured by observing the voltage shifts on insertion and withdrawal of the intraretinal electrode. Maximal response amplitudes in this and all subsequent figures were between 130 and 250 μV. Temperature, 37°C. (a) Isethionate Ringer with pCa 2·87. (b) Isethionate Ringer with pCa 6·7. Solutions contained 1 mM EGTA and 500 μM Mg^{++}, pH 7·13. Low Ca^{++} solutions slowed the responses but had little effect on the peak amplitude: flash energy relation.

good signal: noise ratios could be had. One or two responses sufficed at energies above this level (shown on each curve in units of photons absorbed rod^{-1} $flash^{-1}$).

The effect of incubating a retina with a solution containing 10 mg/ml of DOPC vesicles formed in a solution containing 330 mM PIPES and 67 mM 6CF is seen in the responses of Fig. 10. This retina showed prominent overshoots after the responses. These are part of the current produced by the rods and are ouabain-sensitive. They have little effect on the amplitude: flash

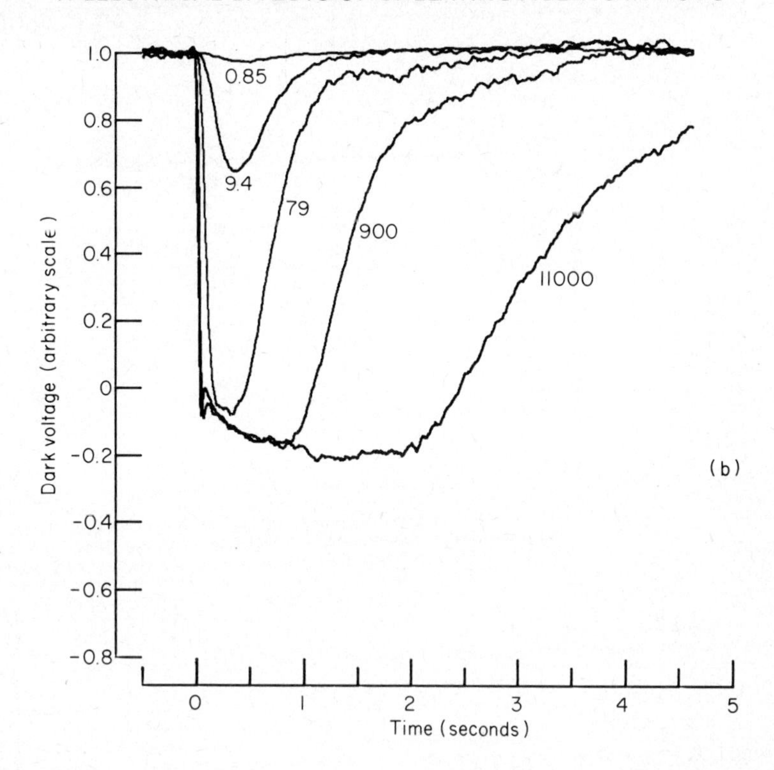

energy relation of the responses and their origin will be considered elsewhere. The retina showed essentially the same qualitative changes in response shape on going from high to low Ca^{++} activity as did the untreated preparation of Fig. 9. In both, the half-saturating flash energy was about 30 photons/rod. Since the vesicle-treated retina contained about 2·2 μM 6CF on subsequent extraction in 0·1 M NaOH after correction for 30% of the dye in bound vesicles (see above), the cytoplasmic concentration of PIPES can be estimated to be about 10 micromolar if the vesicle transfer ratio is 1·0. Clearly there was no buffering effect on the light responses. The situation was quite different when the vesicles contained 330 mM EGTA. This is shown in Fig. 11.

While at pCa = 2·87 the responses of the EGTA-vesicle treated retina were very similar to those of Figs 9 and 10, when pCa was raised to 6·7, the 9- and 79-photon responses were greatly reduced in size compared to the control retinas. In fact, intermediate stimulus intensities were found at which EGTA reduced the response amplitude at least sixfold relative to the controls. This effect of EGTA was permanent. Even if the retina was washed for two hours in EGTA-free Ringer, it remained desensitized at pCa = 6·7. Note that the responses were otherwise typical in shape, overshoot and size of the slow dark

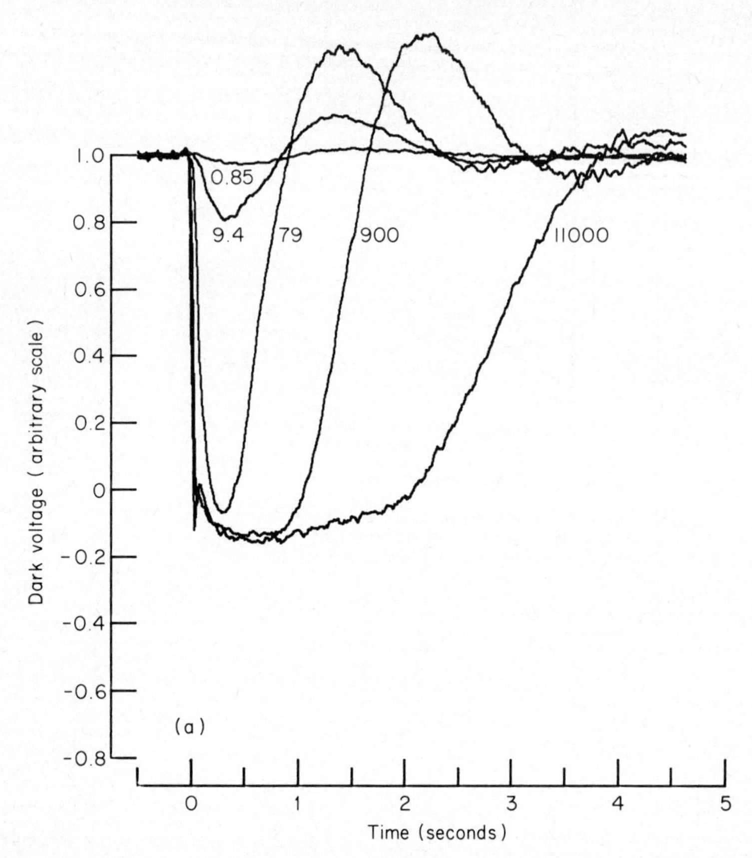

Fig. 10. Rat retina incubated for 3 h in a suspension of DOPC vesicles containing 330 mM PIPES and 67 mM 6CF. Same recording conditions as in Fig. 9. (a) pCa 4·0, (b) pCa 6·7. Incubations at 25°C, responses taken at 37°C. The responses were not desensitized at low A_{Ca}.

current reversal during responses to bright flashes. Extraction of 6CF yielded a concentration corresponding to a cytoplasmic EGTA concentration of 11 μM.

Other chelating agents were found to have similar desensitizing effects at pCa = 6·7. The responses of Fig. 12(a) show this. Here only the low Ca^{++} curves are included, but it is plain that both the 9- and 79-photon responses are reduced in comparison with the control retinas. The chelating agent in the vesicles in this case was CDTA, which complexes strongly with both Ca^{++} and Mg^{++}. This property permits a further test of the notion that the observed desensitization was due to the presence of buffer injected into the cytoplasm of the rods. Figure 12(b) shows a family of responses taken at the same pCa but with the Mg^{++} in the Ringer reduced from the normal value of 1 mM to 50 μM.

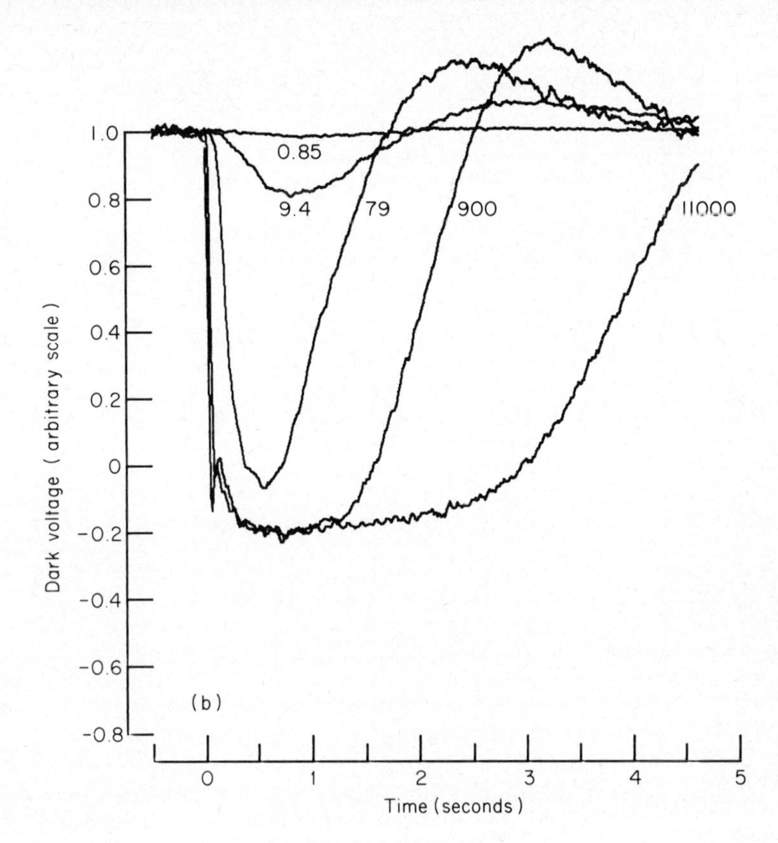

This change has no effect on the amplitude:flash energy relation for the control retinas or the EGTA-vesicle treated ones, but it produces an additional twofold desensitization when the vesicles contained CDTA. Since Mg competes strongly with Ca for this chelator, the desensitization can be put to release of additional free CDTA in the cytoplasm when the external Mg ion activity is lowered.

Similar experiments have been conducted with NTA, which has much lower Ca^{++} and Mg^{++} affinities than CDTA, and with HOEDTA, whose metal affinities are close to those of EDTA. Vesicles containing NTA did not desensitize the rod responses; those containing HOEDTA did. Thus we conclude that chelating agents that bind Ca^{++} desensitize the rod responses, whether or not they also bind Mg^{++}. Thus Mg^{++} seems ruled out as an intracellular transmitter of visual excitation, a conclusion reinforced by the lack of effect of changes in external [Mg^{++}] on the size of the dark current in normal rat retinas.[6]

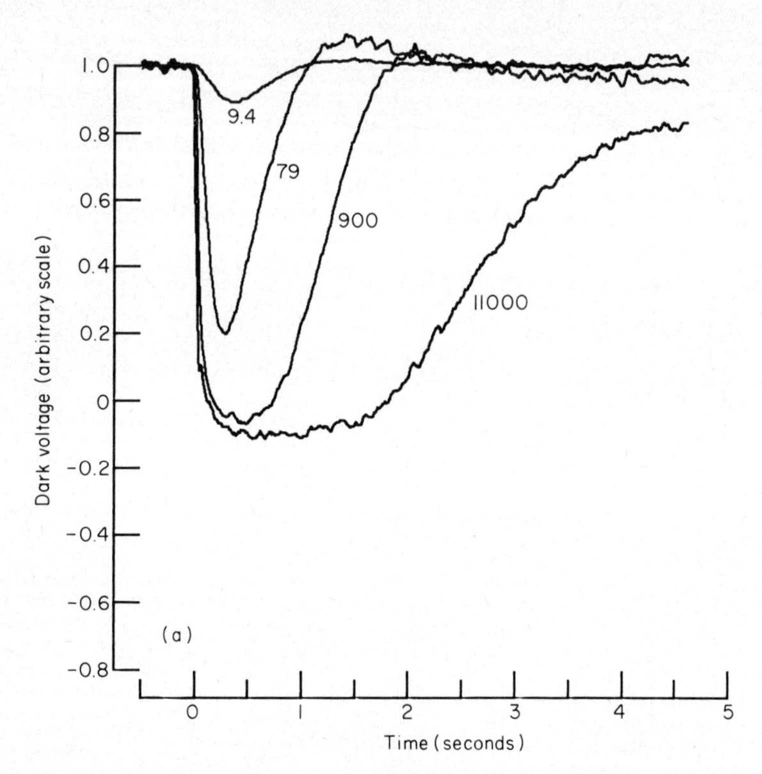

Fig. 11. Rod responses from a rat retina incubated for 3 h in DOPC vesicles containing EGTA, 330 mM, 6CF 67 mM. (a) isethionate Ringer, pCa 4·0. (b) isethionate Ringer, pCa 6·7. Recording conditions as in Fig. 9. The responses were desensitized in the low Ca^{++} solution.

Quantitative Properties of the Transmission Process

The qualitative effects of metal chelators on the rod responses just described provide a basis for estimating several interesting parameters of rod excitation. Returning to the WSOS model of Fig. 2, let

$$S = 1 - (D/D_0) \tag{2}$$

S is thus the fraction of D sites that have complexed with M and do not admit the dark current through the plasma membrane. Further, let A denote the increase in total transmitter concentration in the outer segment produced by the absorption of a single photon in a rod disk, let E denote the total concentration of M in its free and complexed forms (excluding that in the disks or bound to inexchangeable sites) from a flash in an outer segment. Then an equation can be written expressing the concentrations of the various forms of

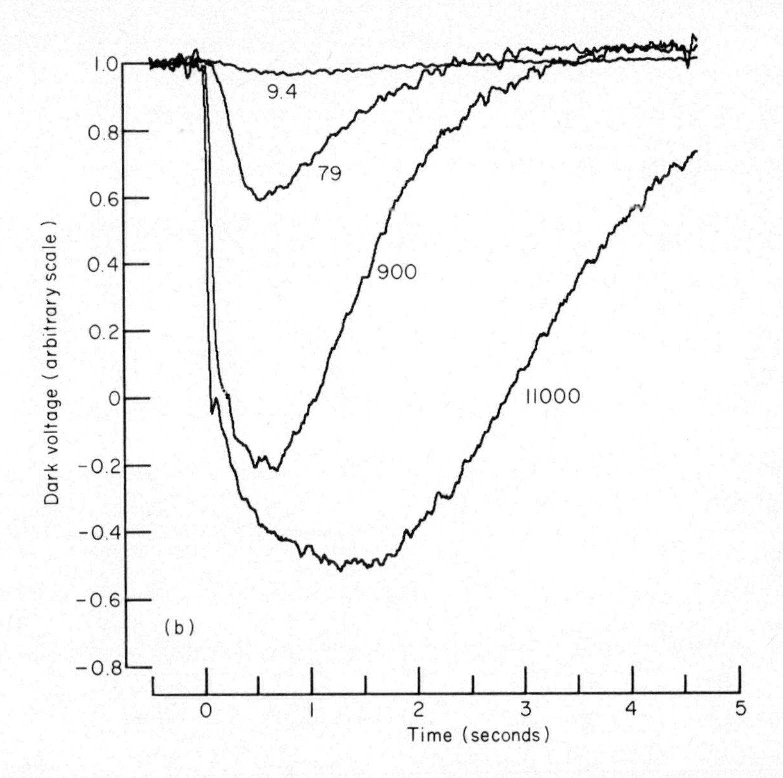

M in an outer segment in terms of the observable quantity S and the flash exposure Q (photons absorbed per rod). Thus

$$\text{Total M} = \text{Free M} + \text{M bound to D} + \text{M bound to B} \tag{3}$$

or,

$$E + AQ = \frac{1}{K}\frac{S}{1-S} + D_0S + \frac{B_0RS}{K + (R-K)S} \tag{4}$$

An important assumption in eqn (4) is embodied in A, which is treated as if it were a constant independent of light intensity over the range of values used in the experiments. However, it is actually a function $A(t)$ of time after a stimulus flash that depends upon both the kinetics of release and of removal of the transmitter. For the experiments in this paper the sampling time is always 0·4 s, so the time dependence of A is not specifically considered. But since the *shapes* of the light responses are not affected by the presence of the intracellular chelating buffers, the present analysis assumes that *$A(t)$ is the same in treated and control retinas*. A more accurate treatment of A requires a complete kinetic model of transmitter metabolism, perhaps akin to that of Hodgkin and his colleagues.[29]

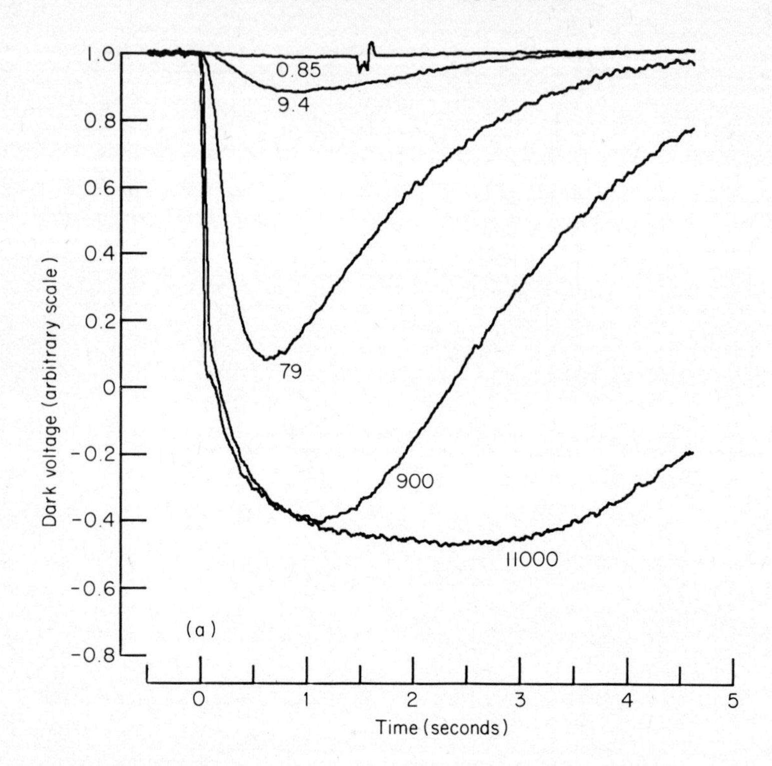

Fig. 12. Rod responses from a rat retina incubated 3 h in a suspension of DOPC vesicles containing 330 mM CDTA, 67 mM 6CF. All responses recorded at pCa 6·7. (a) pMg 3·0, (b) pMg 4·3. The responses became reversibly desensitized in the low Mg^{++} solution. Recording as in Fig. 9.

Equation (4) is an unfamiliar form of a common titration equation that can be inverted to yield S as a function of Q and the various binding constants and total amounts of D sites and buffer molecules. To spare the reader much of the messy algebra that accompanies the solution of a cubic polynomial in S, we simply write:

$$S = S(Q, D_0, K, B_0, R, E, A) \tag{5}$$

and solve for S numerically.

To make it easy to compare solutions with the responses of rods, the instantaneous amplitudes of the response waveforms 400 ms after the stimulus flashes (at or near their peaks) are measured and plotted as functions of flash energy using logarithmic scales for both axes and normalizing the response amplitudes so that the largest one has a value of 1·0. To show the general character of eqn (5) as some of the interesting experimental parameters are

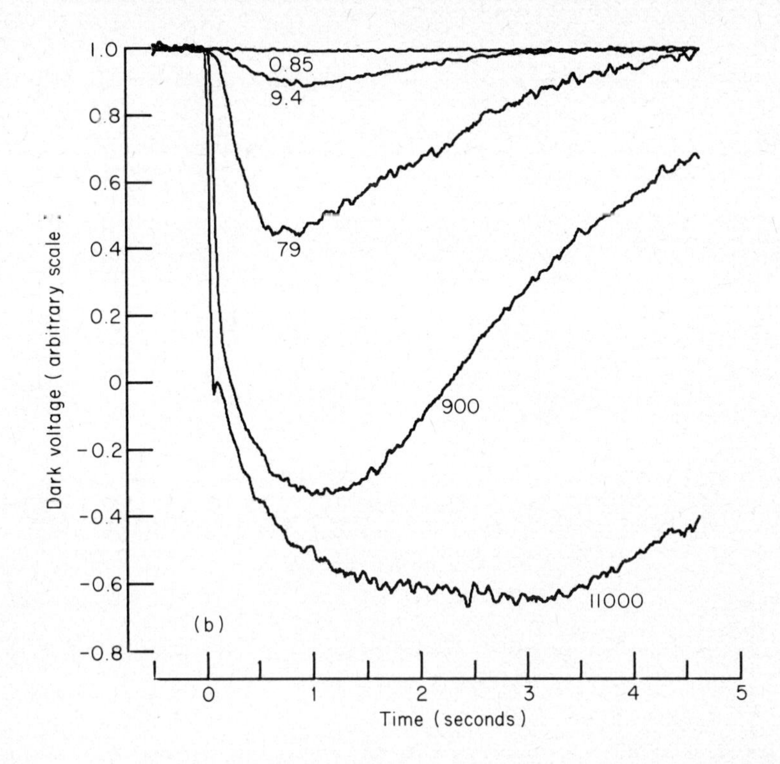

varied, representative solutions are plotted in Fig. 13. In each part of the figure, the leftmost curve is the predicted amplitude: flash energy curve for an outer segment in which the dark level of the transmitter M is negligibly small, the buffer-free half-saturating flash energy Q_1 (given by $1/KA$) is 30 photons absorbed per rod, the number of D sites is insignificant compared to the total quantity of M, and there is no chelating buffer for M present. Figure 13(a) shows the effect of varying A so that Q_1 takes values of 30, 100 and 300 photons absorbed rod^{-1} flash^{-1}. Clearly the curves maintain their shape and are simply shifted along the horizontal axis. Fig. 13(b) represents a buffer whose affinity R for M is ten times higher than that (K) of the D sites. The numbers on the curves indicate the buffer concentrations in the cytoplasm in μmoles per liter of outer segment water. As the buffer concentration is increased, the standard curve is distorted in a characteristic way. The half-saturating stimulus energy shifts to higher values, amplitude saturation sets in more abruptly, and the curve develops a region in which the slope becomes steeper than 1. At low stimulus intensities, however, the response amplitudes remain proportional to Q. Figure 13(c) shows the effect of varying the affinity R of the buffer for the transmitter M on the amplitude: flash energy curve for an outer segment containing 10 μM total buffer. The same distortion evident in

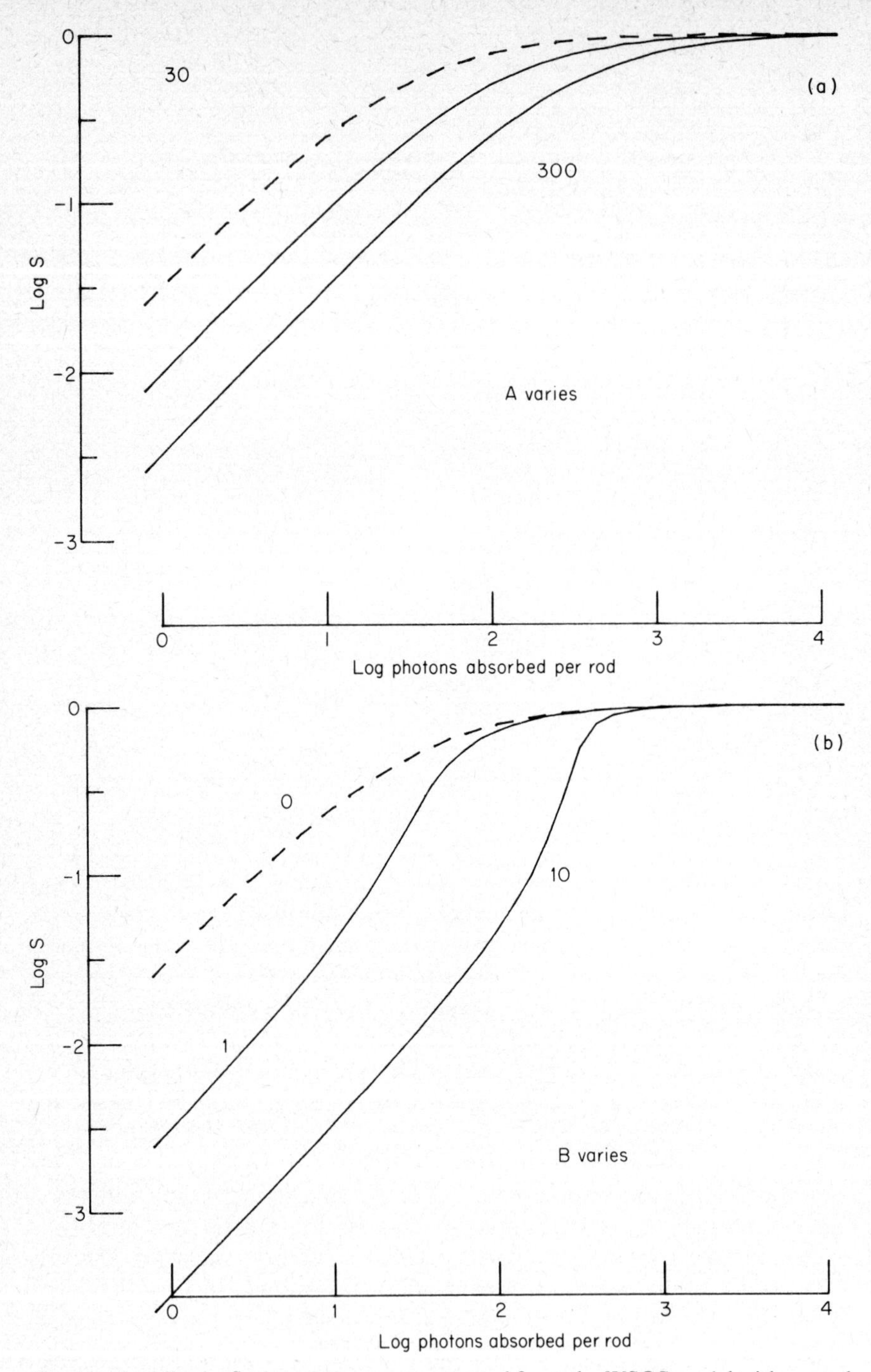

Fig. 13. Amplitude: flash energy curves computed from the WSOS model with several of the model parameters varied. Each part of the figure contains a standard curve computed with $(1/KA)$ set at 30 photons rod^{-1} flash^{-1} and with no cytoplasmic buffer present. (a) Curves computed for values of light-released transmitter A yielding half-saturating flash energies of 30, 100 and 300 photons/rod. As the quantity A of transmitter released per photon is decreased, the curves slide to the right without

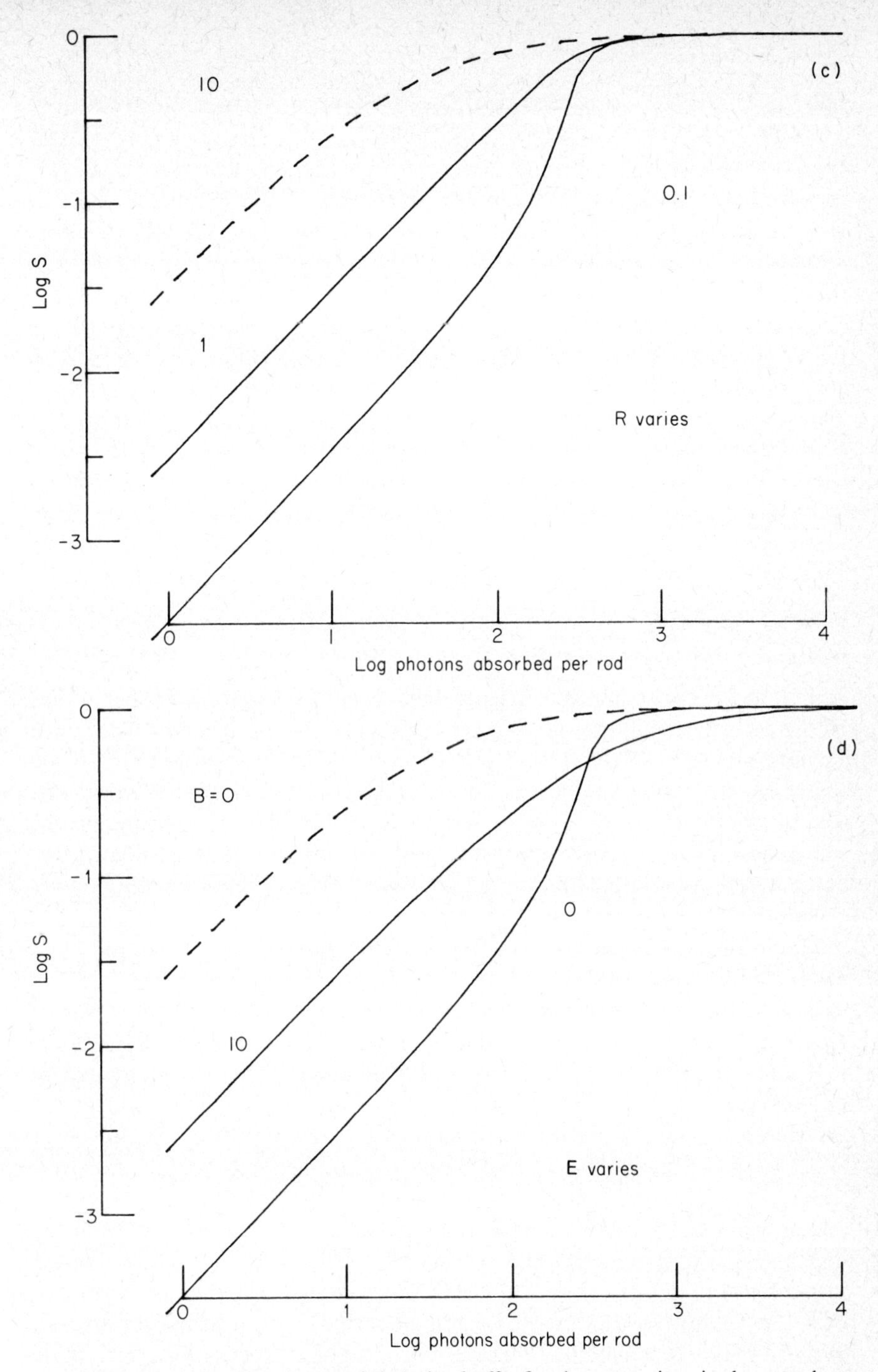

changing shape. (b) Quantity B_0 of chelating buffer for the transmitter in the cytoplasm varied. Curves for 0, 1 and 10 μM concentrations of a buffer with affinity $R = 10^7\ \text{M}^{-1}$ for M. (c) 10 μM of buffer is present in the outer segment cytoplasm. Its affinity for M is varied so that it is half dissociated at M concentrations of 10, 1 and 0·1 μM. (d) 10 μM of buffer that is half dissociated at [M] of 0·1 μM is present. Amplitude:energy curves are shown for steady-state dark concentrations E of M of 0 and 10 μM.

Fig. 13(b) is seen, but as R is increased, the upward inflexion of the curve shifts to lower and lower response amplitudes. Figure 13(d) shows the effect of varying the level E of total transmitter in darkness. This set of curves represents a model of what happens when the pCa in the bath is varied. Clearly the effects are complex, because both free D sites and free buffer molecules are depleted as E is raised. At low levels of E, the characteristic inflexion due to the buffer is visible. When E is raised, to a level that yields a free transmitter concentration of 10 μM, however, the responses to weak flashes become larger, while those produced by bright ones are reduced in size because of competition between the ambient transmitter and that released by the light. Note that all of the curves predict that at very low flash energies, the responses should always be linearly proportional to the stimulus size. This simple rule will break down when some of the approximations in the present treatment are considered. This topic will be taken up after the experimental results are analyzed.

Analysis of the retinal amplitude: energy curves

Amplitude: energy curves taken from flash responses like those shown in Figs 10–12 with flash energies at roughly equal intervals on a logarithmic scale extending from 0·85 photons absorbed rod^{-1} $flash^{-1}$ to more than 10,000 were recorded. The waveforms were sampled at 400 ms after the stimulus flashes. With averages of 8–64 responses and digital smoothing of the waveforms it was possible to measure response amplitudes of less than 1 μV. Since all retinas gave maximal responses of more than 100 μV, more than two decades of response size could be observed.

Figure 14 shows responses from a retina that had been incubated for 3 h in vesicles containing PIPES at 500 mM concentration and subsequently shown to have accepted transfer of a cytoplasmic concentration of 6CF equivalent to 10 μM of the pH buffer. The triangles and squares are from data taken at pCa = 4, while the circles are from responses at pCa = 6·7. There is no significant difference among the curves, indicating that neither the vesicle treatment nor the presence of 6CF and PIPES at micromolar levels in the cells had any large effect on excitation. When the vesicles contained EGTA at the same concentration, however, there was a large difference between the curves. This is shown in Fig. 15, in which the squares and crosses are taken from responses in Ringers whose pCa = 4 and whose pMg = 3·3. The triangles represent responses at pCa = 6·7 and pMg = 3·3, while the circles denote waveforms at pCa = 6·7, pMg = 4·3. Clearly, lowering the free Ca^{++} desensitizes the rods by at least sixfold, while a tenfold lowering of the Mg^{++} activity has little or no further effect. This result can be interpreted on the WSOS model by supposing that the cytoplasmic EGTA is mainly in the form of an inactive complex at low pCa but is unmasked when the external free Ca^{++} is lowered. The effect on the amplitude:energy curves is thus equivalent to that shown in Fig. 13(d). Evidently externally supplied Ca^{++} masks the buffering action of EGTA, while external Mg^{++} does not. Since the internal Ca^{++} activity of rods seems to increase in parallel with that of the bath, we infer that the Ca^{++} complex of EGTA in the rod cytoplasm has no buffering action on the excitatory response. *This would only be true if the excitatory transmitter forms a complex with EGTA whose stability constant is the same as or less than that of* Ca^{++}.

Competition between Mg^{++} and Ca^{++} can be demonstrated with the buffer CDTA, however. This is shown in Fig. 16, in which the triangles are taken from responses of a retina at pCa = 4, pMg = 3·3. The amplitude:energy curve is that of an untreated retina, even though it had been incubated with CDTA vesicles of 333 mM buffer concentration and containing 67 mM 6CF as a transfer marker. The dye concentration in the cells was equivalent to a cytoplasmic concentration of CDTA of 12 μM. When the pCa of the bath was raised to 6·7, little buffer effect on the responses was seen.

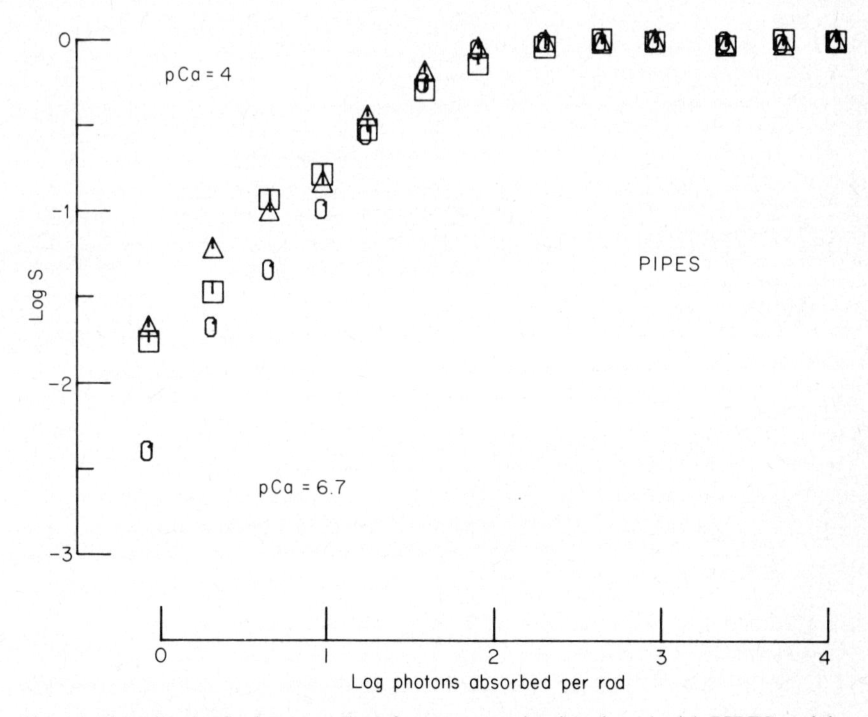

Fig. 14. Amplitude:flash energy data from a rat retina incubated with PIPES vesicles as in Fig. 10. Isethionate Ringers with pMg 3·3. Triangles and squares: pCa 4·0. Circles: pCa 6·7. Cytoplasmic PIPES concentration 10 μM, calculated from 6CF analysis of the retina.

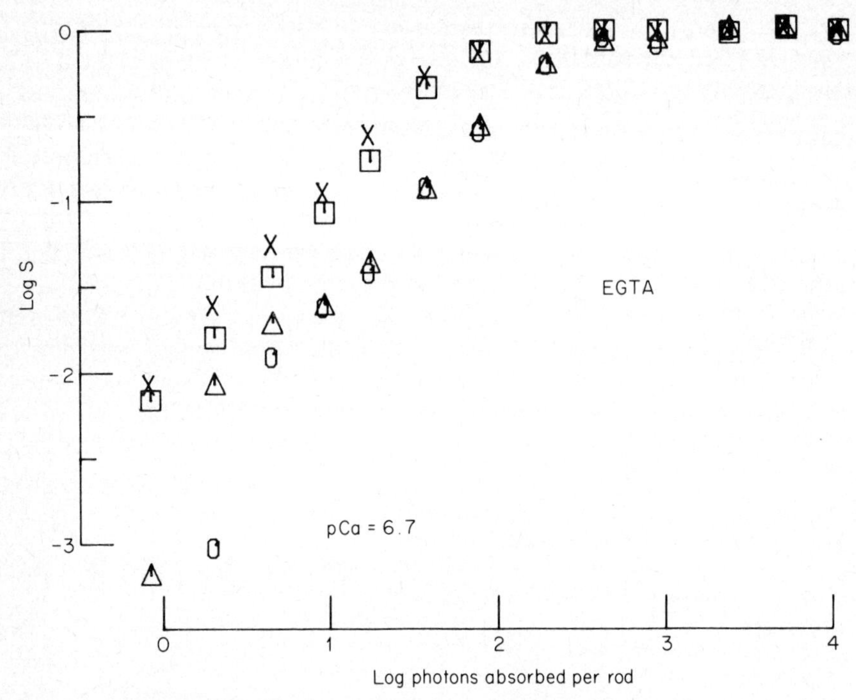

Fig. 15. Amplitude energy plots for a rat retina incubated EGTA vesicles as in Fig. 11. crosses pMg 3·3; pCa 4·0. squares pMg 4·3; pCa 4·0. triangles pMg 3·3; pCa 6·7. circles pMg 4·3; pCa 6·7. Low Ca^{++} solutions desensitized the responses. Varying pMg had no effect. [EGTA] in cytoplasm = 12 μM from 6CF analysis.

When the external Mg^{++} was then lowered to pMg = 4·3, however, characteristic buffer-like desensitization appeared. A similar experiment carried out on retinas that had not been treated with vesicles showed no buffer effects at pMg < 5. Evidently if the cytoplasmic Mg^{++} activity tracks that of the bath, the Mg complex of CDTA will not bind the excitatory transmitter at the ambient intracellular level of free Mg^{++} that obtains when pMg of the bath is <4. As the Mg^{++} in the bath is lowered, however, enough CDTA is freed to intercept the excitatory transmitter.

The Mg^{++}-Ca^{++} competition shown in Fig. 16 cannot yield an estimate of the affinity of CDTA for the excitatory transmitter, since we do not know what the actual Mg^{++} activity is in the cell. If there is a sufficient Mg^{++} activity to maintain ATP and GTP and their relatives in the biochemically effective Mg complexes, however, the intracellular pMg must not be much greater than 3 at external pMg = 3·3. At pMg 3, CDTA has an apparent affinity constant for Ca^{++} of about 5×10^5 M^{-1} at pH 7·1. The affinity increases about tenfold if pMg is raised to 4·3. At pMg = 4·3, CDTA and EGTA bind Ca^{++} about

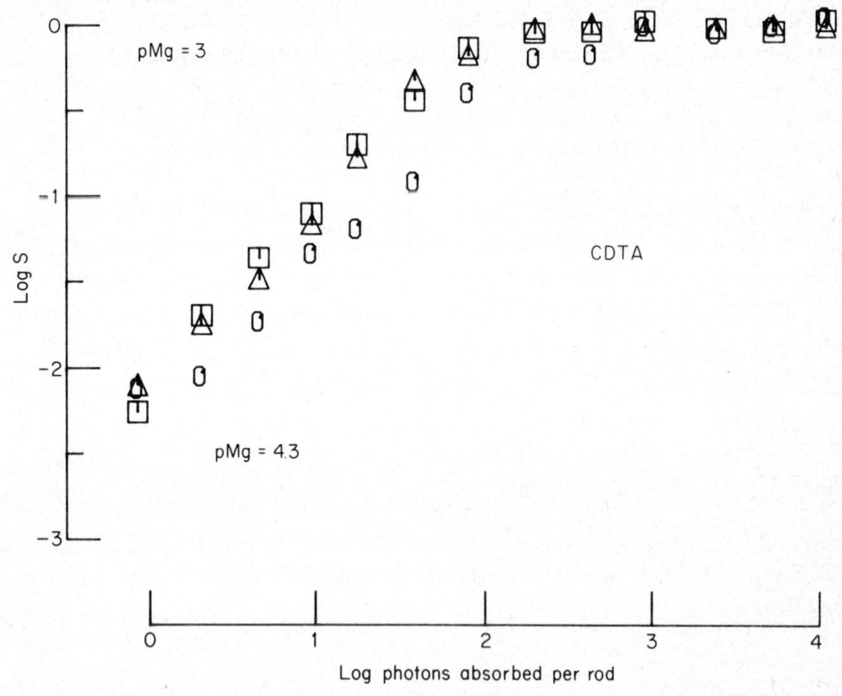

Fig. 16. Amplitude:energy relation for a retina treated with CDTA vesicles as in Fig. 12. triangles pMg 3·3; pCa 4·0. squares pMg 3·3; pCa 6·7. circles pMg 4·3; pCa 6·7. The responses desensitized only at low [Mg^{++}]. [CDTA] = 11 μM, calculated from 6CF analysis.

equally,[33] and the buffer-like desensitizations produced by the two chelators should be about equal. The results of Figs 15 and 16 confirm this. Both point to an excitatory transmitter whose affinity for the chelating buffers is of the order of that of Ca^{++}.

If the affinity constant R of EGTA is set at that for Ca^{++} at pH 7·1, and the cytoplasmic concentration of EGTA transferred from vesicles is 10 μM (previously inferred from 6CF analyses on retinas like those of Fig. 15), eqn (5) can be fitted to the observed amplitude:energy curves to yield estimates of the quantity A of transmitter released per absorbed photon and the steady-state pCa in the rod cytoplasm. The only undetermined parameter in this procedure is the affinity K of the D sites for the transmitter and their total number in each outer segment. To determine these latter two numbers, additional information is needed. A lower limit to the number D_0 of D sites can be set at 3600 by a statistical argument.[6, 10] This is equivalent to a concentration of about 0·21 μM in a rat rod outer segment with fluid volume of 28 fl. Even if this is a gross underestimate, the fraction of transmitter bound to them will be insignificant compared to that bound to EGTA or CDTA at 10 μM concentration. Thus D_0 can be safely neglected in the curve-fitting procedure. The value of K presents a more difficult problem. It sets the scale of light sensitivity for the rods, for the quantity $(1/KA)$ represents the half-saturating flash energy in the WSOS model with no buffer present. Nevertheless, K can be set within a fairly narrow range by observing that a sudden drop in Ca^{++} in the bath transiently increases the dark current of rat rods if the initial Ca^{++} activity is greater than 10^{-5} M. Below this level, the retina behaves as if all of its D sites are in the Na^{+}-conducting state and no more can be opened by further reduction in cytoplasmic Ca^{++}. Thus in the experiment of Fig. 15, at the bath pCa of 6·7, the steady state level X of free transmitter must be such that

$$KX \leq 1 \tag{6}$$

At the same time, rods treated with the Ca^{++} ionophore X-537A show an increased sensitivity to external Ca^{++} as if their plasma membranes were much more Ca^{++}-permeable than normal. Such rods show half-suppression of their dark current at external $a_{Ca} = 2 \times 10^{-6}$ M. If external and internal free Ca^{++} are in electrochemical equilibrium under this condition at rod membrane potential V(mV),

$$K < (1/[Ca^{++}]_{ext} \exp(-V/29 \text{ mV}). \tag{7}$$

Since V is not likely to be more negative than 10 mV under these conditions of low $(Ca^{++})_{ext}$,

$$K \approx 10^6 \text{ M}^{-1}. \tag{8}$$

Fitting solutions of eqn (5) to the data of Fig. 15 with relation (8) as a constraint by conventional nonlinear least-squares methods yields the curves shown in Fig. 17. The points are from amplitude:energy curves taken at pCa 4 (triangles) and pCa 6·7 (circles). The excitatory constants obtained are shown in Table 2. The values X of the cytoplasmic transmitter concentration in darkness vary with external pCa in the expected way. The quantity A of transmitter released per absorbed photon also varies in parallel with the cytoplasmic pCa, again a reasonable finding, since the store of excitatory transmitter must be accumulated through the cytoplasm. The error estimates on the numbers are due solely to scatter in the observations and are surely underestimates. Only an upper limit can be given for K at present. In a typical EGTA-vesicle experiment, if the value of K is decreased, Q must be increased by an amount that keeps their product constant. Thus the curve fit gives a lower limit for Q of about 500 transmitter molecules per absorbed photon. At the same time, the cytoplasmic Ca^{++} activity is about 1 μM or less under the conditions of the experiment. Together with previously published observations that at external pCa = 2·87, the dark current of rat rods is about 20 %

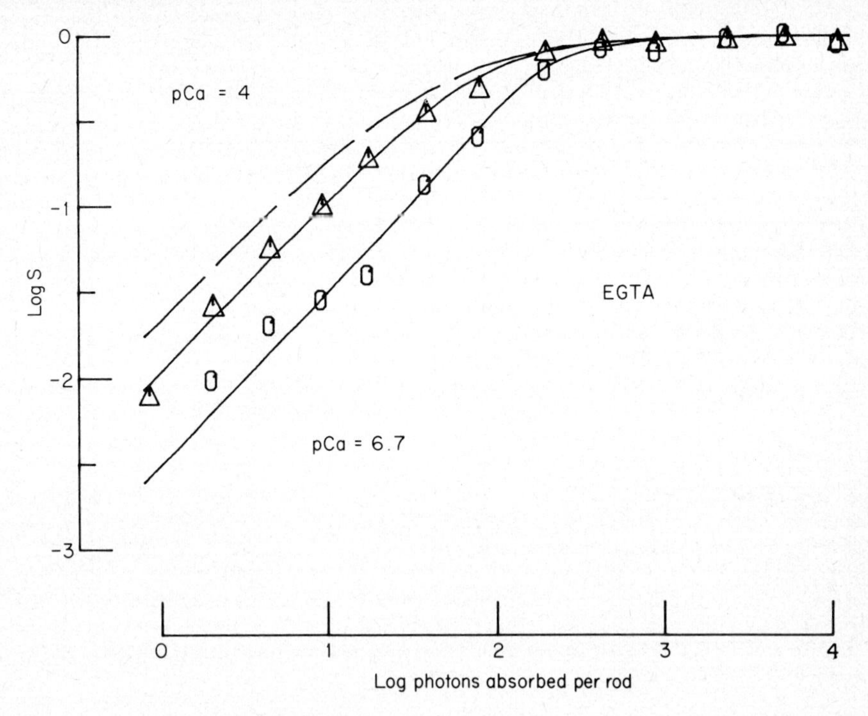

Fig. 17. Amplitude:energy data from a rat retina treated with EGTA vesicles recorded in isethionate Ringers with pMg 3·3, and pCa 4·0 (triangles) and 6·7 (circles). Curves are solutions of the WSOS model computed with the parameters of Table 1. The dashed curve is the predicted response curve that would have been obtained if the EGTA could have been removed from the rods at the end of the experiment.

of maximum, we can infer that the internal pCa is strongly stabilized against changes in external [Ca^{++}] by a control mechanism with an open-loop gain of at least 100 and ability to lower the cytoplasmic Ca^{++} level below that of the bath by at least three orders of magnitude. This finding is quite consistent with properties of many other cells (see review, reference 28).

The total transmitter released per absorbed photon depends upon the kinetics of supply and removal. The excess present at 0·4 s thus represents a lower limit to the total M released per photon. In any case *hundreds or thousands of* Ca^{++} *ions must be released in single photon responses.*

Critique of the Buffer Experiments

The argument accompanying the preceding experiments is open to two objections: (1) circular reasoning in identifying the transmitter and (2) neglect of important physical factors in using the WSOS approximation. We cannot

Table 2

Parameters of the WSOS model derived from least squares fits of eqn (5) to amplitude:energy curves of rat retina treated with vesicles containing 330 mM EGTA, 67 mM 6CF.

Experiment of 2 February 1976

Fixed Parameters	Notes
Cytoplasmic pH 7·1	(a)
Ca^{++} affinity R 8·5 × 10^6 M^{-1} of EGTA at pH 7·1	(b)
EGTA concentration B_0 12 μM in rod cytoplasm	(c)
Total D sites D_0 0·2 μM in outer segment	(d)
Ca^{++} affinity K 1 × 10^6 M^{-1} of D sites	(e)

Adjusted parameters:	pCa of bath 4·0	pCa of bath 6·7
X: Free Ca^{++} in o.s. in dark	1·2 ± 0·1	0·44 ± 0·06 μM
A: Extra Ca^{++} ions present in rod cytoplasm 0·4 s after stimulus flash per photon absorbed	880 ± 70	460 ± 85

Notes: (a) From 6CF fluorescence observations above. (b) From reference 33 and present pH estimates. (c) From 6CF recovered per unit dry wt. of tissue and assumption that 6CF:EGTA transfer volumes are the same. (d) Lower limit estimated from statistical arguments in references 6, 10. (e) From eqn (9) of text.

uniquely identify the excitatory transmitter from its affinity constant for EGTA. Yet few other metal ions bind to the chelator with affinities low enough to be displaced by low levels of cytoplasmic Ca^{++}. Only Ba^{++} and Mg^{++} bind more weakly than Ca^{++}, both so weakly that it is doubtful that EGTA could affect the response curves at a cytoplasmic concentration of only 10 μM. There are two alternatives to Ca^{++} that we cannot wholly reject. First, some organic cation with the required EGTA and CDTA affinity might be the transmitter, not Ca^{++}. We think this unlikely. Second, some other transmitter not bound by metal chelators might control the Na conductance of the plasma membrane secondarily through an increase in cytoplasmic Ca^{++} activity. We invite enthusiastic model-builders to think of other possibilities.

It is easier to deal with questions of the adequacy of the WSOS approximation. The experimental observations that are fitted to the model treat the ratio v/v_m as the equivalent of a direct measure of the fraction S of D sites that are blocked by Ms, where v is the voltage difference between the two extracellular recording electrodes and v_m is the maximum response amplitude. However, since v is related to the state of the D sites through two intermediate steps in the scheme of Fig. 1, we must take these into account. Analytical solutions of the cable equation for a cylindrical cell comprised of two regions with different resting membrane potentials and with lengths in the ratio 3:1 (a crude model of a rat rod) will show proportionality between external voltage gradient and outer segment membrane conductance so long as the space constants of the two regions are at least 0·8 times the length of the cell (unpublished results). Estimates of the cable constants of rat rods from measurements of the fast photovoltage[13] indicate that this condition is met when S is between 0·1 and 1. Since most of the statistical weight in the curve fits comes from responses in this region, nonlinearity in the voltage:conductance relation are of little matter. Such nonlinearities should be evident as a steep fall in observed values of S at low light intensities if the outer segment membrane conductance exceeds 10 m.mho in low Ca^{++} solutions. We have seen no evidence of this if pCa in the bath does not exceed 7·0. At lower Ca^{++} concentrations the amplitude:energy curves do show such effects, but these will be considered elsewhere.

A second defect of the WSOS approximation is that it takes no account of the possible spatial non-uniformity of the distribution of the transmitter along the outer segments just after its release at low stimulus intensities. Again this will affect the rod responses only at low values of S, not at the high ones upon which the curve fits mostly depend. Could the slowing of the response waveforms at low intensities (Figs 9–12) be due in part to internal diffusion of the transmitter along the outer segments? The physical basis of waveform distortion in the low intensity responses of rods and cones is not yet clear.

Implications

The present buffer experiments suggest that single photons can discharge hundreds of Ca^{++} ions from single rod disks. Since the total Ca^{++} content in living outer segments is not more than 5 mM[10] (that of isolated outer segments can apparently be higher—see the report by Daemen *et al.* in this volume, pp. 29–40), a single rod disk cannot provide full-size quanta of transmitter for more than 100 photons simultaneously absorbed. This would set an upper limit of less than a million photons absorbed per rod per flash to the physiological range of rod vision unless the size of the quanta of Ca^{++} declines at high light intensities.

How is the transmitter (which we will presume to be Ca^{++}) retained in the disks in darkness? If it is passively distributed across the disk membranes, its expulsion in light would require a parallel flux of another charge carrier driven by free energy or by a light-activated pumping of Ca^{++} itself. Although light does stimulate nucleoside triphosphate activity in rods under some conditions[30, 31] the effects are presently difficult to reconcile with the requirements of excitation. Unless some unsuspected source of free energy is available to the excitatory process, we must suppose that Ca^{++} is accumulated in the disks by means of free energy expended in the dark. Its release is then a passive downhill flux. Studies of ATPase activity of isolated outer segments show more than enough hydrolysis for charging of the disks with Ca^{++}.[32] Curiously, however, the observed ATPase was *inhibited* by Ca^{++} at normal cellular levels of Mg^{++}.

Despite the biochemical loose ends in the present story, the way now seems open to study the metabolism of transmitter release and diffusion in rod outer segments with metallochromic dyes introduced by the vesicle technique. We look forward to learning what they have to say.

Acknowledgments

We thank William Robinson, John Weinstein and Robert Blumenthal for discussions, Luther Barden, Arthur Schultz, Lawrence Showkeir, Cristina Suszynski and William Jennings for help with the computerized data collection and Gary Knott for advice about the curve fitting procedures using the MLAB modeling program.

Abbreviations

ATP Adenosine 5′ triphosphate.
DOPC L-α-dioleoyl phosphatidyl choline.
DPPC L-α-dipalmitoyl phosphatidyl choline.
6CF 6-Carboxyfluorescein.
EGTA Ethylene glycol-bis(β-aminoethyl ether) *N,N′* tetra-acetic acid.
CDTA trans-1,2 diaminocyclohexane-*N,N,N′,N′* tetra-acetic acid.
PIPES Piperazine *N,N′*-bis(2-ethane sulfonic acid).
NTA Nitrilotriacetic acid.
HOEDTA *N* hydroxyethylethylenediamine *N,N′,N′* triacetic acid.
HEPES *N*-2-hydroxyethyl piperazine-*N′*-2-ethanesulfonic acid.
EDTA Ethylenediamine *N′,N′,N,N*-tetra-acetic acid.

References

1. G. Falk and P. Fatt 1972, Physical changes induced by light in the rod outer segment of vertebrates, *In* "Handbook of Sensory Physiology" vol. 7, part 1, Photochemistry of Vision (ed. H. J. A. Dartnall) pp. 200–244. Springer-Verlag, Heidelberg.
2. T. Tomita 1970, Electrical activity of vertebrate photoreceptors, *Quart. Revs. Biophys.* **3**, 179–222.
3. D. A. Baylor and M. G. F. Fuortes 1970, Electrical responses of single cones in the retina of the turtle, *J. Physiol.* (*Lond.*) **207**, 77–92.

4. D. A. Baylor and A. L. Hodgkin 1973, Detection and resolution of visual stimuli by turtle photoreceptors, *J. Physiol.* (*Lond.*) **234**, 163–198.
5. W. A. Hagins, R. D. Penn and S. Yoshikami 1970, Dark current and photocurrent in retinal rods, *Biophys. J.* **10**, 380–412.
6. S. Yoshikami and W. A. Hagins 1973, Control of the dark current in vertebrate rods and cones, *In* "Biochemistry and Physiology of Visual Pigments" (ed. H. Langer) pp. 245–255, Springer-Verlag, New York.
7. J. E. Brown and L. H. Pinto 1973, Ionic mechanism for the photoreceptor potential of the retina of *Bufo Marinus*, *J. Physiol.* (*Lond.*) **236**, 575–591.
8. W. A. Hagins 1972, The visual process: excitatory mechanisms in the primary receptor cells, *Ann. Revs. Biophys. Bioengr.* **1**, 131–158.
9. W. A. Hagins and S. Yoshikami 1974, A role for Ca^{++} in excitation of retinal rods and cones, *Exp. Eye Res.* **18**, 299–305.
10. W. A. Hagins and S. Yoshikami 1975, Ionic mechanisms in excitation of photoreceptors, *Ann. N.Y. Acad. Sci.* **264**, 314–325.
11. A. I. Cohen 1970, Further studies on the question of the potency of saccules in outer segments of vertebrate photoreceptors, *Vision Res.* **10**, 445–453.
12. R. D. Penn and W. A. Hagins 1972, Kinetics of the photocurrent of retinal rods, *Biophys. J.* **12**, 1073–94.
13. H. Rueppel and W. A. Hagins 1973, Spatial origin of the fast photovoltage in retinal rods, *In* "Biochemistry and Physiology of Visual Pigments" (ed. H. Langer) pp. 257–61, Springer-Verlag, New York.
14. S. Yoshikami, W. E. Robinson and W. A. Hagins 1974, Topology of the outer segment membranes of retinal rods and cones revealed by a fluorescent probe, *Science* (*Wash.*) **185**, 1176–9.
15. E. Szuts 1975, Ph.D. Thesis, Johns Hopkins University.
16. P. A. Liebman 1974, Light-dependent Ca^{++} content of rod outer segment disk membranes, *Invest. Ophthal.* **13**, 700–702.
17. E. Szuts and R. A. Cone 1974, Rhodopsin: light activated release of calcium, *Fed. Proc.* **33**, Abst. 1403.
18. E. W. Abrahamson, R. Fager and W. T. Mason 1974, Comparative properties of vertebrate and invertebrate photoreceptors, *Exp. Eye Res.* **18**, 51–67.
19. A. Darszon and M. Montal 1976, Light increases the ion-permeability of rhodopsin-phospholipid bilayer vesicles, *In* "Mitochondria: Biogenesis, Structure and Function" (eds. A. Gomez-Puyon and L. Packer), Academic, New York, London.
20. W. Hubbell, K.-K. Fung, K. Hong and Y. S. Chen 1977, Molecular anatomy and light-dependent processes in photoreceptor membranes. This volume, pp. 41–59.
21. F. J. M. Daemen, P. P. M. Schetkamp, Th. Hendriks and S. L. Bonting 1977, Calcium and rod outer segments. This volume, pp. 29–40.
22. D. Papahadjopoulos, G. Poste and E. Mayhew 1974, Cellular uptake of cyclic AMP captured within phospholipid vesicles and effect on cell growth behavior, *Biochem. Biophys. Acta* **363**, 404–418.
23. G. Gregoriadis, P. D. Leathwood and B. E. Ryman 1971, Enzyme entrapment in liposomes, *FEBS Letters* **14**, 95–99.
24. Y. E. Rahman, M. W. Rosenthal and E. A. Cerny 1973, Intracellular plutonium: removal by lipsome-encapsulated chelating agent, *Science* (*Wash.*) **180**, 300–302.
25. J. E. Weinstein, S. Yoshikami, P. Henkart, R. Blumenthal, R. and W. A. Hagins 1977, Lipsome-cell interaction: transfer and intracellular release of a trapped fluorescent marker, *Science* (*Wash.*), **195**, 489–492.

26. P. C. Caldwell 1958, Studies on the internal pH of large muscle and nerve fibers, *J. Physiol.* (*Lond.*) **142**, 22–62.
27. W. F. Boron and P. de Weer 1976, Intracellular pH transients in squid giant axons caused by CO_2, NH_3 and metabolic inhibitors, *J. Gen. Physiol.* **67**, 91–112.
28. P. C. Caldwell 1972, Transport and metabolism of calcium ions in nerve, *Prog. Biophys. Mol. Biol.* **24**, 177–223.
29. D. A. Baylor, A. L. Hodgkin and T. D. Lamb 1974, The electrical response of turtle cones to flashes and steps of light, *J. Physiol.* (*Lond.*) **242**, 685–728.
30. W. E. Robinson, S. Yoshikami and W. A. Hagins 1975, ATP in retinal rods, *Biophys. J.* **15**, 168a.
31. A. Carretta and A. Cavaggioni 1976, On the metabolism of the rod outer segments, *J. Physiol.* (*Lond.*) **257**, 687–698.
32. T. J. Ostwald and J. Heller 1972, Properties of a magnesium or calcium-dependent adenosine triphosphatase from frog photoreceptor outer segment disks and its inhibition by illumination, *Biochemistry* **11**, 4679–86.
33. The Chemical Society, London. 1964, "Stability Constants of Metal Ion Complexes", Special Publication 17.
34. L. H. Pinto, J. E. Brown and J. A. Coles 1977, Mechanism for the generation of the receptor potential of rods of *Bufo marinus*, This volume, pp. 159–167.

Discussion

R. Meech. Dr Hagins reports that incorporation of proton buffers into rods of isolated rat retinae are without effect on the light responses. Barnacle photoreceptors are sensitive to changes in intracellular pH (Brown and Meech, *J. Physiol.* in press). This raises the possibility that protons may act as "intracellular transmitters". Can Dr Hagins exclude this possibility in the rat retina or is the amount of proton buffer incorporated insufficient to appreciably alter the buffering power of the rod cytoplasm?

It appears that much of the dye incorporated by the phospholipid vesicle fusion technique goes to the *inner* segment of the rod. Is it possible that the reduction of rod responses to dim light following the incorporation of EGTA and CDTA is the result of conductance changes in the inner rather than in the outer segment?

W. A. Hagins. You are quite right to conclude that one cannot rule out protons as possible excitatory transmitter particles from the lack of effect of PIPES buffer introduced into our rods by vesicles. The cytoplasmic concentration of PIPES we can achieve is no more than 10 μM, if one can judge from the amount of 6CF transferred to the cells from the same liposomes. Yet the outer segments already contain much higher concentrations of ATP, ADP, GTP and GDP. Direct analyses of both frog and rat rods by W. E. Robinson of our group indicate that the total nucleotide phosphate concentration in frog rods is more than 5 mM. Rough calculations indicate that the pH-buffering effect of these substances alone should reduce the pH increment produced by a quantum of protons released in an outer segment by a factor of more than 5000 relative to that in an unbuffered medium. The added buffering effect of 10 μm PIPES would be negligible on this scale. Nucleoside triphosphates also act as Mg^{++} and Ca^{++} buffers, of course, but as much weaker ones. At a cytoplasmic [Mg^{++}] of 5 mM and pH 6·8, the ambient level of ATP + GTP in an outer segment should only buffer an increment in Ca^{++} by a factor of 2–3. This effect is not specifically considered in our paper, but it does not increase our estimates of the amount of Ca^{++} that must be released by an absorbed photon by more than twofold. This alone suggests that protons would be a poor choice for an excitatory transmitter in rods, while Ca^{++} would be a much better one.

The complex pH and buffer effects on photoreceptor responses that you and others have described have not been fully explained. But since nearly every biochemical reaction in cells is pH dependent, a full account of the matter will likely be complicated.

We have no evidence that the distribution of 6CF in rod cell cytoplasm (Fig. 7b) is not uniform. Unfortunately we cannot see the accompanying buffers in the micrographs, so their distribution is simply assumed to be uniform.

8

Three Components of the Photocurrent Generated in the Receptor Layer of the Rat Retina

G. B. ARDEN

Department of Visual Science, Institute of Ophthalmology, London, U.K.

Introduction

It has been shown that light causes hyperpolarisation in photoreceptors, associated with a decrease of membrane conductance.[2,10] Recordings of extracellular current flow[7,8] have demonstrated an axial "dark current" which has its sink in the outer limb region of the rods and is reduced by light. This finding extends the model derived from intracellular recordings, since it localises the light sensitive conductance to the outer limb. However, Arden[1] reported that the axial extracellularly recorded photovoltage could exceed the dark voltage, implying that in illumination the dark current was not only reduced, as found by Hagins *et al.*,[7] but reversed in sign, at least in regions vitreal to the outer limb.[1] This finding cannot be explained by a model in which the rod membrane voltage is represented by a single EMF, and light only reduces a conductance located in the outer limbs.

More recently, other conductance changes have been reported in association with both rod and cone responses, in particular a voltage sensitive conductance which increases during membrane hyperpolarisation.[3,11,9] This paper reports complexities in the extracellularly recorded photoresponses, which might be attributable to such a mechanism, and distinguishes a third component of the response of the photoreceptor layer, which is produced by the associated Müller (glial) cells.

Methods

These have been described in detail in a previous paper.[1] An entire isolated retina was mounted in an incubation chamber. The normal anatomical relationships were maintained by floating the ganglion cell side of the retina

Fig. 1. The retinal surface and micropipettes, viewed through the microscope. Photograph taken with 660 nm light, 40 s exposure. Calibration bar 200 μm. The outer limb tips may just be seen. The shafts of the pipettes are out of focus since they enter the image plane from above.

onto a dome of Millipore filter paper. The retina was stimulated by 508 nm light flashes which passed through the millipore paper, ensuring even illumination. The incubation chamber was placed on the stage of a microscope, and the tips of the receptors were seen in tangential view, at a magnification of *circa* 160, using 660 nm light (see Fig. 1). The smooth, slightly curved surface formed a very sharp edge, and individual outer limbs were visible. The incubation chamber contained silver-silverchloride wire electrodes for passing current radially through the retina (0·192 or 0·064 mA cm^{-2}) and an additional reference electrode which faced the outer limb surface of the retina. Warmed, oxygenated solution flowed past the receptor surface at 8–10 ml min^{-1}. The standard Ringer, which supported a full e.r.g., was modified by lowering calcium to 0·1 mM, and increasing to 10 mM either sodium glutamate or magnesium: these procedures abolished postsynaptic neuronal activity in the retina, leaving the PIII fraction of the e.r.g.[5] Details of solution composition are given in the figure legends.

The equipment described in a previous paper has been modified in several ways. The retinal mount has been simplified, by omitting the plastic rings. The microadvancer has been modified so that a pair of saline-filled micropipette electrodes can be advanced together without relative motion. Two pipettes were fabricated and held in an adjustable Narashige micromanipulator; they were then glued together with cyanoacrylate and epoxy cements. The pair was then filled and mounted in a PTFE chuck on a single oil-filled microadvancer (see Fig. 1). The spacing between the tips varied from 8–24 μm in a direction parallel to the long axes of the receptors, and 0–30 μm at 90° to this axis. The electrodes advanced obliquely downwards, at an angle of about 18° to the horizontal. They additionally penetrated the retina obliquely, since they were not advancing along a line normal to the retinal surface. To allow for this, in estimating electrode tip positions, the distances indicated by the microadvancer have been multiplied by a factor of 0·8. The tip resistances of the micropipettes varied, in different experiments, from 10–3 MΩ. The advancer had no backlash or drift. The electrode tips were repeatedly aligned with the margin of the retina (as in Fig. 1). The S.D. of the advancer readings was less than 2 μm. The electrodes penetrated the retina without disturbing the adjacent retinal edge, viewed through an eyepiece micrometer, and the edge adjacent to the electrodes did not move more than 3 μm throughout a satisfactory experiment. Amplifier bandpasses were between 0–20 and 0–100 Hz in various experiments. The differential balance of the amplifiers was checked before each experiment, with the electrodes in position, and artefacts due to lack of channel symmetry are 40 dB less than the minimal recorded responses. Light intensities were measured with a Tektronix radiometer, and recalculated in terms of quanta absorbed per rod per flash. All experiments were recorded on an FM tape recorder for subsequent analysis.

Control experiments

Photovoltages were measured across a known lamina of the retina, the boundaries of which were determined by the electrode tip spacings. The voltages produced across the same lamina by a known current pulse were recorded in the same experiment (see Fig. 2). Thus the photovoltages recorded could be scaled to photocurrents. Changes in photocurrent between successive laminae give the photocurrent divergence, i.e. the position and magnitude of the sources and sinks.[7] However, the assumption must be made that both the applied current, and that generated by the photoreceptors, flows uniformly parallel to the long axes of the receptors. In a previous paper it was shown that

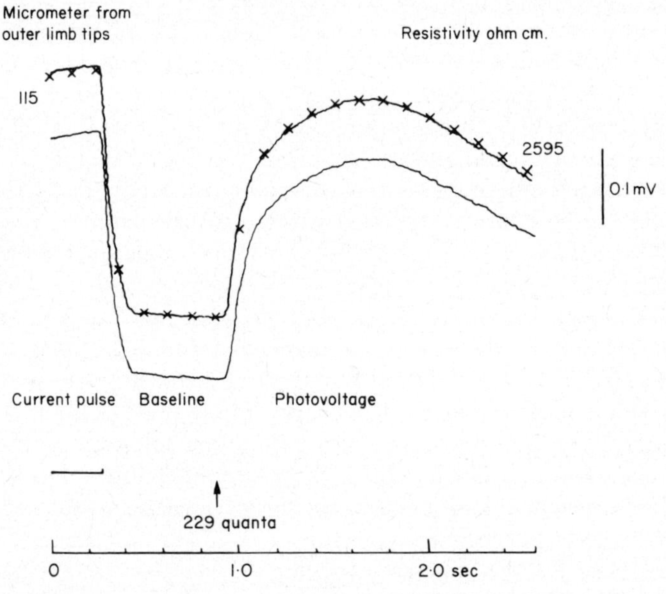

Fig. 2. Voltage records indicating stability of preparation. All these and subsequent traces are of the output of an X-Y recorder. The upper record shows the PD across a 16 μm retinal lamina, centred 115 μm from the tips of the rod outer limbs. Electrode resistances, 3 MΩ. Average of 8 responses on the Biomac. Current pulse, 0·064 mA cm^{-2}. After the upper record had been obtained, the electrodes were withdrawn 30 μm, and then withdrawn and advanced several times and further responses were averaged before the electrodes were repositioned at the same microadvancer reading as for the upper trace. The lower record was then obtained. The crosses are the lower record transferred to the upper. The region is one where photocurrent sinks are found, and a 30 μm withdrawal of the pipettes caused the photovoltage to increase by 50% and the resistivity to drop by 50%. Solution composition NaCl 109 mM; KCl 3·4 mM; Na_2HPO_4 0·8 mM; NaH_2PO_4 0·1 mM; $NaHCO_3$ 21 mM; $MgSO_4$ 2·4 mM; Glucose 15 mM; Sodium glutamate 10 mM; $CaCl_2$ 0·1 mM. Solution bubbled with 95% O_2, 5% CO_2.35°, pH 7·2.

this condition was fulfilled throughout the receptor layer, but in the present work measurements were made in all retinal layers. Since the geometry of the chamber was unaltered it seems likely that the currents continued to flow radially through the additional 80 μm, and this is supported by the following observations:

(a) Analysis of different experiments showed that the amplitude of the voltages recorded varied with axial separation of the pipette tips, but not with lateral separation (except for very small lateral separations, when the measured resistivities were low, see below).
(b) The sum of the photovoltages recorded across the successive laminae of the retina (differential recordings between the two micropipettes) was equal to the voltage recorded between the leading micropipette and the reference in the superfusing medium (the PDs in the medium were very small).
(c) When both electrodes were withdrawn from the retina, or had penetrated it, PDs could be recorded, but scalings and subtractions showed zero current divergence.
(d) The experiments showed that the sinks of receptor photocurrent were sharply localised to the receptor layer. However, PDs attributable to receptor currents could be recorded in the inner nuclear layer, but were removed by the subtraction and scaling procedures.

Other assumptions implicit in the method are that the electrodes do not affect the responses, either by damaging the retina or changing the pathways for current flow, and that the responses do not dwindle during the experiment. Figure 2 provides some evidence on this point, and also illustrates the general features and reproducibility of the responses. Each record, like the others shown in the paper, is a tracing of an XY recorder record of the output of the Biomac averager. Between 8 and 40 repetitions were employed. The cycle was initiated by a trigger pulse, with a repetition rate of 1 in 6–8 s. The trigger pulse initiated a 400 ms constant current pulse, and during this pulse the averager sweep began so that the initial portion of the record shows the voltage displaced from its baseline value, as the current passes through the retina. At the end of the current pulse, the baseline is recorded. From the PD recorded between the electrodes due to the applied known current one can calculate the resistivity of the extracellular pathway in the particular lamina of the retina. As a check, the resistivity of the superfusing solution was measured prior to any experiment, and was *circa* 60 ohm cm. Lower values implied leakage of current round the retina, or non-uniformity of current due to retinal tears, or to bubbles in the chamber, or to a poorly mounted retina. In the following figures the current pulses are not shown, but values for resistivity are indicated against the traces.

In Fig. 2 and subsequent figures, the 508 nm flash is given at the arrow, and the intensity indicated in quanta absorbed per flash per rod. The centre of the lamina from which a record is taken is indicated in μm. Thus, in Fig. 2, the electrode tips were inserted 125 and 105 μm, respectively. The upper record in Fig. 2 was first obtained, and following this the electrodes were withdrawn and some further recordings made (not illustrated). The electrodes were then advanced once more to the same reading on the micrometer and another recording made, which is shown as the lower trace. The crosses are points from the lower record, transferred to the upper. Thus both resistivity and photovoltage remain constant with time, and are unaffected by electrode motions. Such a check was usually made once during a sequence of penetration or withdrawal. Current densities were also deliberately changed during a recording sequence to demonstrate that the applied current did not affect the photoresponses.

The measured resistivities varied considerably from experiment to experiment, but always increased systematically as the micropipettes penetrated the retina. Sodium glutamate causes the retina to swell and in retinas superfused with this medium, resistivities were higher than when magnesium was used to isolate PIII. When the two electrode tips were separated laterally by 20–30 μm, the axial resistivity was higher than if the tips were separated axially by 0–10 μm. An explanation for this finding is that, in the latter case, resistivity was measured in a pocket of extracellular fluid forming around both micropipettes, while with greater separations each micropipette was more closely invested by the retinal tissue. When measured resistivity was high, the PD's recorded were much larger, so to a first approximation radial photocurrents are unaffected by lateral spacing. But large photovoltages were not associated with a corresponding increase of noise, so lateral spacings of 20–30 μm were preferred.

Results

When a microelectrode pair was placed as in Fig. 1, photovoltages were absent or were negligibly small. When the electrodes were inserted into the region of outer and inner limbs, a simple photovoltage developed. With a flash of saturating intensity, the response rose smoothly from the baseline to attain a plateau level, which was maintained for a duration which varied with flash intensity, and then the response declined. However, with further advance of the pair of micropipettes, the photovoltages not only grew in amplitude but changed in waveform. The changes were progressive, and as Fig. 3 shows, in the nuclear and synaptic region of the receptor layer the rising phase of the photovoltage has an inflection and the plateau becomes domed. Evidently, there is more than one fraction of the photovoltage. This result is in agreement with previous work, which showed that in the inner retinal layers a slowly developing prolonged voltage ("slow PIII") could be recorded and was separable from the photovoltage of the receptors.[4] It is necessary to analyse this slow response before the photoreceptor currents can be analysed in detail.

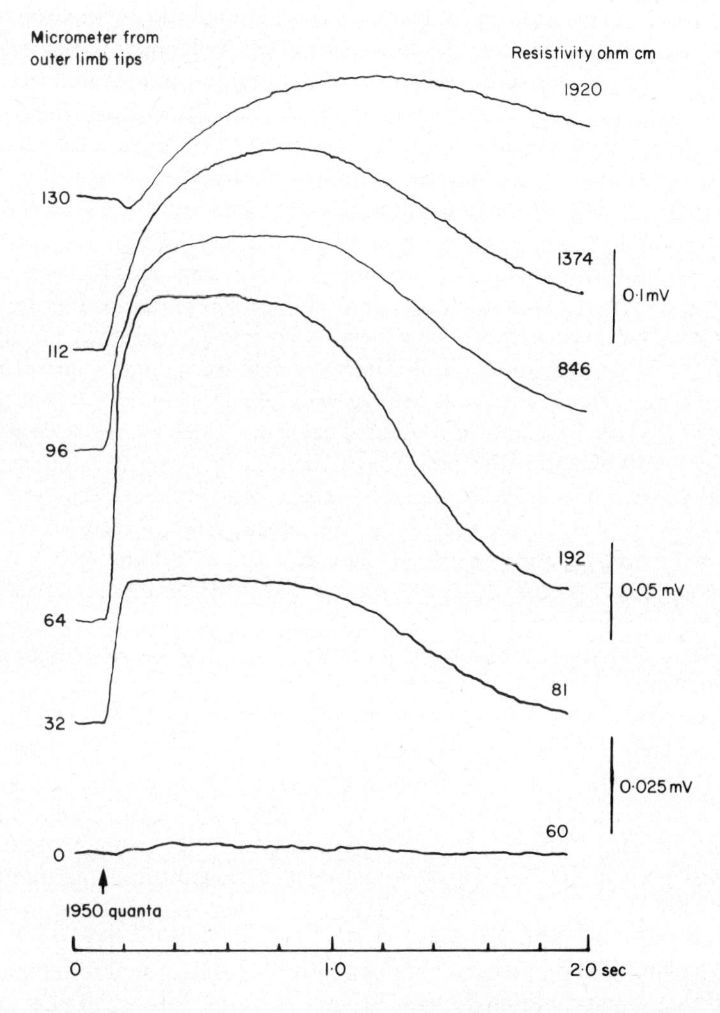

Fig. 3. Photovoltages across various retinal laminae. The lower voltage calibration bar refers to Biomac responses obtained at 0 and 32 μm. Electrode axial spacing 13 μm, resistance 3 MΩ. Note in vitreal layers, there is an inflection on the rising phase of the photovoltage and a domed response rather than a plateau. Solution composition NaCl 115 mM; KCl 3·4 mM; $NaHCO_3$ 21 mM; $MgSO_4$ 2·4 mM; $MgCl_2$ 7·6 mM; Na_2HPO_4 0·1 mM; NaH_2PO_4 0·8 mM; Glucose 20 mM; $CaCl_2$ 0·1 mM.

Sinks and sources of Müller cell current

Figure 4 shows the current divergence of successive laminae of the retina obtained from an experiment in which a pair of micropipettes was first advanced into the retina and then withdrawn in stages. The records read from top to bottom. The depths of the centre of the laminae indicate that the top 4 records were obtained from the inner plexiform and inner nuclear layer, respectively. The downwards deflection of the photoresponses indicate that they are sinks, and sinks continue to be seen for the subsequent 4 responses which were obtained from the laminae between the rod synapses and the outer limiting membrane. In the region vitread to the receptor layer, the sinks increase slowly and reach a maximum at the end of the traces. In the photoreceptor layer, the sinks rise to a maximum in less than 100 ms, and then have a maintained plateau. It is evident that two separate processes must produce these two types of sink. Since in the preparation there is no postsynaptic neuronal activity, sink currents observed in the inner plexiform layer must be generated in a cell which extends from inner plexiform to receptor layers: the Müller cells are the only possible candidate. The slow fraction of PIII has long been suspected of being generated by Müller cells: the figure shows an experimental proof of this notion. In order to complete the demonstration, it is desirable to find the sources of the Müller cell current, and these are not apparent in Fig. 4, although the records span the receptor layer to the outer limiting membrane, the Müller cells' scleral termination.

One source of this difficulty is that any source of Müller cell current would be masked by sinks of photoreceptor current. It is possible to obtain a somewhat greater temporal separation of the two responses by reducing the flash intensity(1) since under these conditions the rod response becomes briefer, there being no plateau, while the peak time of the Müller response is almost unchanged. Figure 5, left column, shows photovoltages obtained with semi-saturating flashes. In the region of inner and outer limbs, the response is simple, and rises and falls monotonically. The voltage from the nuclear region (upper record) is complex, and shows an inflection on the rising phase and a subsidiary maximum following the major peak. Disregarding, for a moment, the initial inflection, the subsidiary maximum occupies a position in time which corresponds to the glial response. From records like these, the photocurrent divergence can be obtained from the nuclear zone and from the zone of inner and outer limbs. These are shown in the right-hand part of the figure. In the more scleral laminae, a source of photocurrent is seen, which must correspond to the activity of the receptors themselves, since these are the only cells in this region of the retina. The source is inverted in the figure, so that it can be compared to the sink found in the nuclear region of the receptor layer, which contains both receptors and Müller cells. The rising phases of the two responses agree fairly well: but after the peak they diverge. It is as though the sink returned more rapidly to the baseline, because of the development of a delayed current source in the nuclear region.

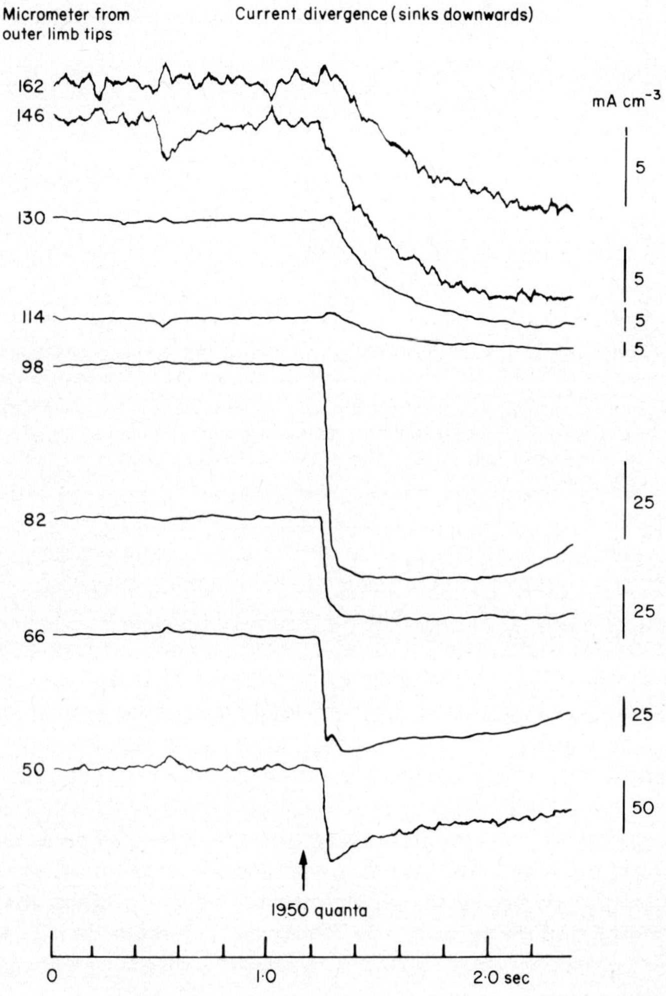

Fig. 4. Distribution of current sinks in retina. The records are of current divergence in 32 μm laminae, centred as indicated. The records were obtained by scaling and subtracting "Biomac" channels, and written out at a convenient magnification. Calibrations were calculated subsequently, and thus they all differ. These records include the initial portions (cf. Fig. 2) which contained the current pulses. The voltage deflections on the original traces have been removed by the scaling and subtracting procedures, leaving the small transients at *circa* 500 ms. Note the change in sink waveform between 114 and 98 μm. Solution composition, as in Fig. 2.

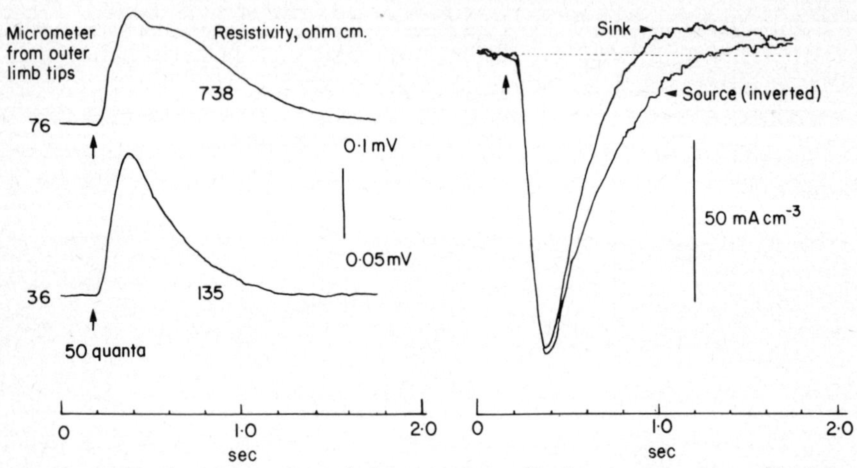

Fig. 5. Comparison of source and sink waveforms in photoreceptor layer. The left-hand column shows photovoltages recorded across 24 μm laminae centred as indicated. Note differences in waveforms. The right-hand column shows the current divergences derived from measurements such as those on the left. The scleral source (inverted) is more prolonged than the sink in the outer nuclear layer. Solution composition as in Fig. 2.

This hypothesis can be investigated more quantitatively by determining the current divergence for a series of laminae which span the entire photoreceptor layer. Such an experiment is shown in Fig. 6. In the two most scleral laminae, there are current sources, and in the more vitreal laminae, current sinks. If all the currents shown on the records were produced by receptors, the sum of source and sink currents at any moment should be equal, since the measurements span the entire receptor layer. This is indeed the case if measurements are made during the rising phase or at the peak of the photoresponse. At later times there is an excess of source current, which must sink in regions deep to the photoreceptor layer. The maximum difference between the sums of the sinks and the sources occurs *circa* one second after the flash. Attempts have been made to determine the waveform of the slow source, e.g. by subtracting the records of Fig. 6 from each other, but these "third difference" results only show a general similarity with the sink of the Müller response recorded later in the experiment, and are too noisy for valid comparisons to be made.

The source of the Müller response thus appears to be distributed from the outer limiting membrane to the synapse, and the sink from synapse to inner plexiform layer. Current divergence measurements made in a region centred on the synapse may contain both sink and source currents, and in this respect at least resemble the waveform of the photocurrent recorded scleral to the outer limiting membrane.

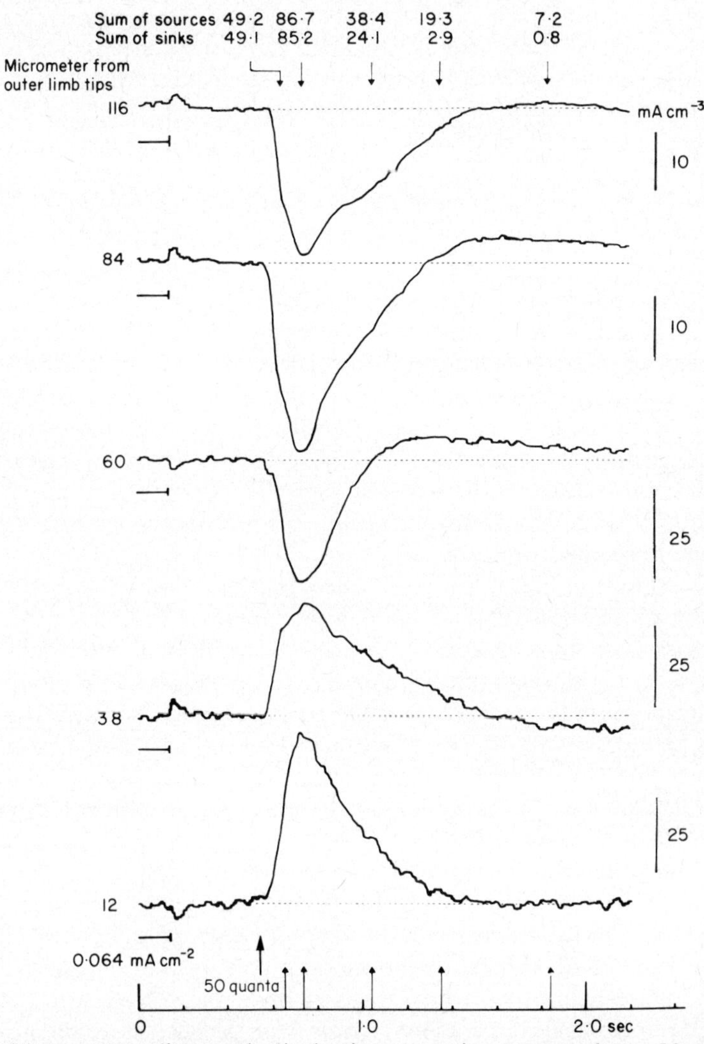

Fig. 6. Photocurrent divergencies in laminae spanning receptor layer. Note flash intensity is semi-saturating. The ending of the current pulse is indicated, and the zero photocurrent shown by dashed lines. The arrows top and bottom indicate the times after the flash when measurements of sinks and sources were made. Totals of sink and source currents are shown above. In the later portions of the response there is an excess of source current, which must therefore sink in the inner retinal layers. There are 2 scleral laminae which show sources of photocurrent. The more vitreal of these, centred at 38 μm indicates that a source of current lies between 22 and 54 μm (centre position ± electrode axial separation). The length of the outer limbs is 25 μm, so this result does not *prove* there is a source in the inner limbs of the photoreceptors, though it is difficult to reconcile with the hypothesis that the outer limb is the sole *and uniform* source of photocurrent. Solution composition as in Fig. 2.

The magnitude of the Müller current cannot be assessed with any precision but, from results such as those of Figs 4 and 6, it is small, possibly 10 % of the peak photoreceptor current. However, in transretinal recordings, the Müller cell *voltage* is larger than the receptor potential: this merely reflects the longer and more highly resistive extracellular pathway over which the Müller current flows.

A second source of photoreceptor current

Several figures have shown that the photoreceptor response appears complex, since an inflection may appear on the rising phase. This inflection has properties which strongly suggest that it is developed by a distinct process. It cannot be seen with stimuli which are so weak that they produce less than *circa* 10 % of the maximum photovoltage. The inflection becomes more prominent as flash intensity is increased much above that required for a maximum plateau photovoltage. With intense stimuli, the inflection is seen separately from the plateau and forms an initial peak or "nose" such as has been reported for some intracellular recordings.

This is shown in Fig. 7. It is also evident that the "nose" is not present on records made scleral to the outer limiting membrane (cf. Fig. 5). It appears and disappears when small motions of the electrode are made in this region, but is readily recorded throughout the nuclear layer of the photoreceptors.

The response in this region also differs in its behaviour to light adaptation. This is also illustrated in Fig. 7 where the heavy lines show responses in darkness, and the thinner lines responses in the presence of a constant background light (in this case, the red light was used intermittently to view the retina). The responses to flashes delivered in darkness were averaged both before and after the background, and the relative DC levels have been preserved. Thus it is possible to see the effect of the viewing light on the "dark voltage", and in the record made in the outer limb–inner limb region; the viewing light attenuates the photoresponse only by shifting the baseline. During the response to the intense green flash, the voltage levels are not changed by the background. In the records from the nuclear region, however, the background illumination additionally reduces the saturating photovoltage level and greatly attenuates the "nose". During recovery in darkness from an intense flash which bleaches rhodopsin, the "nose" is also selectively attenuated.

Figure 8 provides further evidence about the nature of the process which generates the "nose". Photovoltages were recorded from several retinal laminae, and are shown on slow and expanded time bases. The right-hand column shows the photocurrent divergences. In the most scleral lamina there is a source of photocurrent which rises to a plateau without any rapid overshoot:

in the subsequent lamina, a sink begins to develop, but is soon cut short and diminishes to a lower current level. In this region, the outer portion of the nuclear layer, the exact waveform is very dependent upon electrode placement. Thus, in another similar set of recordings from this retina, the sink was cut off earlier and inverted to become a source. The complexity which results in the "nose" on the extracellular photoresponse is located in the nuclear region of the photoreceptor and not in the outer limb.

The relative spectral sensitivity of the nose and plateau portions of the response are similar, so the few cone-like receptors[6] are unlikely to be the

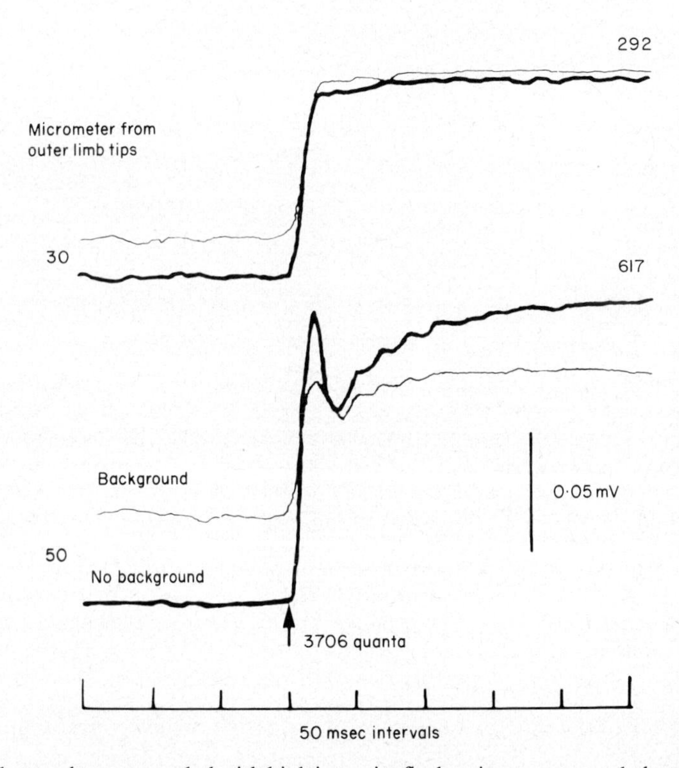

Fig. 7. Photovoltages recorded with high intensity flashes, in presence and absence of a continuous 660 nm adapting background. Note the initial peak is only seen in the outer nuclear layer. 4 records were first obtained in darkness, 8 with the uncalibrated background illumination, and then 4 more in darkness. The relative DC levels have been preserved for each pair, and during the experiment there was no DC shift not associated with illumination. For the more scleral record, the constant illumination does not change the DC level for the saturated photoresponse, but in the outer nuclear layer this is not the case and the "nose" is selectively attenuated. Note faster time base. Electrode spacings, 10 μm, 2 MΩ, bandpass 0–100 Hz. Solution composition as in Fig. 2.

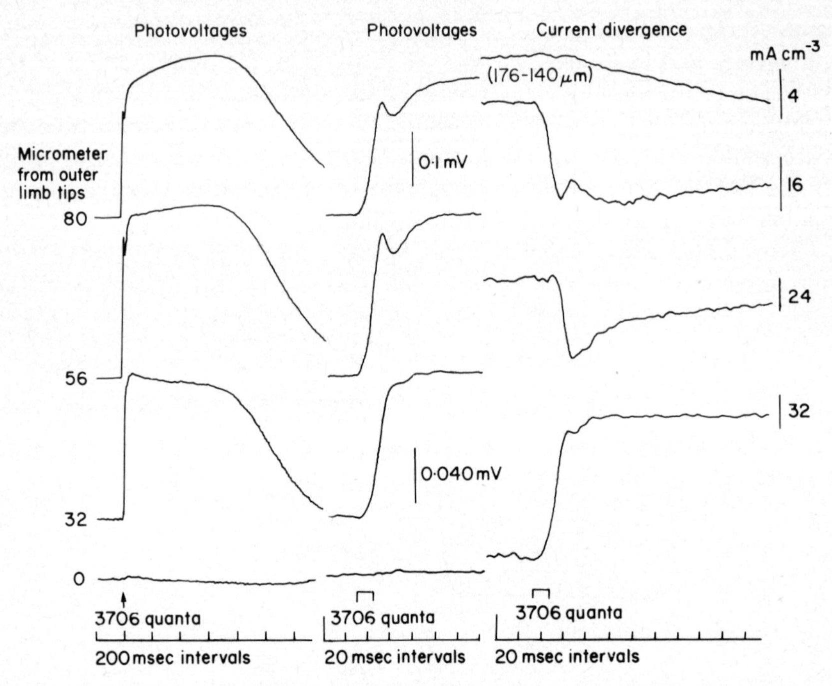

Fig. 8. Photovoltages and photocurrent divergence with intense flashes. The left-hand column shows responses across the retinal laminae indicated, with a time base similar to that in Figs 2–6. The initial portions of the photovoltages are shown with an expanded time base in the centre column. The responses on the right-hand column are the photocurrent divergencies between the laminae indicated on the left. In addition, the sink in the inner retinal layers is shown (top right-hand trace) so that the initial portion of the Müller cell response can be compared, at a fast time base, with the photoreceptor response. The source current recorded between the two most scleral laminae rises in two stages to a plateau. The sink in the adjacent more vitreal laminae is cut short after *circa* 20 ms, producing the "nose". Electrode spacing 16 μm. Solution composition as in Fig. 2.

generators of the nose. Additionally, in cat retina, which contains many more cones, it is possible to record the cone receptor potential with long wavelength light, and increase the stimulus intensity till a measurable rod response can be seen: and the nose on the response is then obviously produced by light absorbed by rhodopsin-containing rods (G. B. Arden, unpublished).

In the region where the nose develops, Müller cells are found, but it is unlikely that these cells produce the nose. Records of photovoltage or current divergence across the inner retinal layers (cf. top right-hand record in Fig. 8) show that the sink of the Müller cell responses begins after a short delay, and increases very gradually, containing none of the high frequency components which would be required to generate the nose. The experiments do not exclude the possibility that Müller cells generate rapid changes of current, but if they do, then the local circuits must be confined to the photoreceptor layer. Since additional hypotheses would be required to explain why one portion only of the glial response should be localised to the photoreceptor layer, it seems justified to assume, at this time, that the nose reflects a process which develops in the rod membrane.

Similarly, it is necessary to make an assumption about the nature of the process underlying the production of the nose. If there was only one process generating the photoresponse, the sinks and sources would be mirror images: clearly this is not the case. One hypothesis which might be advanced is that the sink in the vitreal layer consists of two components, which are both sinks. One is rather small in amplitude and mirrors the more scleral source. The other, of the same size, is very transient, and reaches a peak some 20 ms after the end of the flash, and declines to zero in less than 100 ms: the sum of these two sinks produces the observed waveform. However, no such transient process has been described for photoreceptors, and if the hypothesis is adopted, it is necessary to demonstrate a similar transient *source* somewhere in the receptor layer. This cannot be seen in the outer limb–inner limb region, or in the laminae near the synapses, where such a source would attenuate and delay the sink actually observed (Fig. 8). Thus, the alternative hypothesis is more likely, that the "nose" represents the beginning of a prolonged source of current in the scleral part of the outer nuclear layer, which subtracts from the sink current. In support of this notion is the fact that during the later portions of the "nose", there is a corresponding increase of sink current nearer the synapse, and there also is some evidence from the traces of a slightly delayed source of current in the inner limb region.

If a second source of photocurrent is developed in the portions of the receptor vitreal to the outer limb, this should influence the distribution of sources and sinks of the plateau portion of the receptor photoresponse. Figure 9 is a graph showing the distributions found in the present experiments. The dotted line is a theoretical curve, calculated by Hagins *et al.*[7] showing the expected distribution of current divergence for electrode spacings of 10 μm, and assuming that the outer limb is the sole, and uniform, source of photocurrent. Because of the different electrode spacings employed in the present experiments, exact agreement between the points and the curve is not expected so far as the sources are concerned, but it is obvious that the distribution of sinks found is not that expected from the theory outlined above. In the region of the inner limbs, the dotted curve indicates sink currents should be large, decreasing greatly towards the synapse. The results of Fig. 9 indicate that in the inner limb region sinks, if present, are small and do not appreciably diminish towards the synapse. It should be noted that Hagins *et*

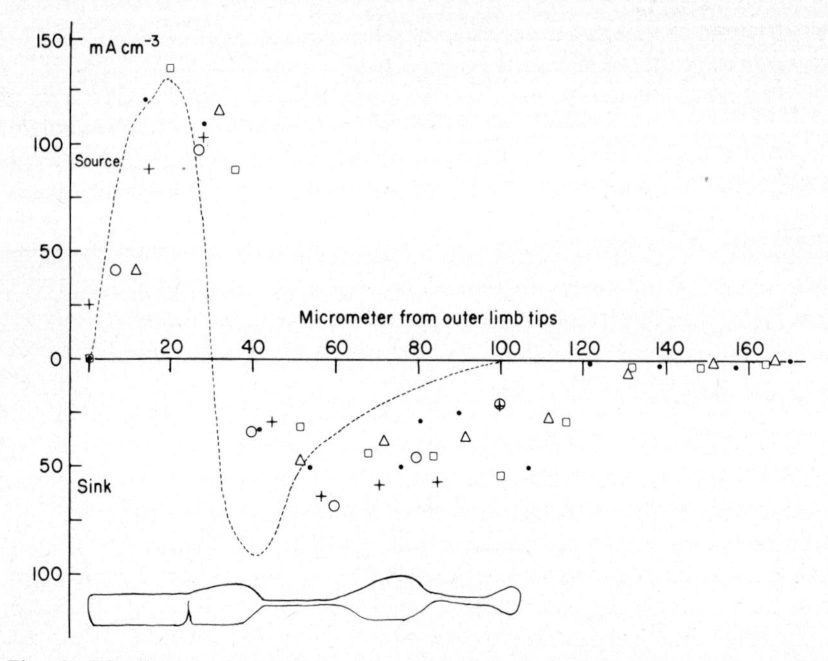

Fig. 9. Distribution of sinks and sources of photocurrent in retina. In these experiments only the "PIII" fraction of the e.r.g. was present, the light intensity produced saturating photovoltages, and measurements were made after 300–400 ms, i.e. during the plateau, but after the "nose". The experiments illustrated differ in technical details, and were chosen to demonstrate factors which do not affect the results. ○: solution, as Fig. 3 (high Mg); electrode tip separation 16 μm; resistivity at 100 μm 194 Ω cm; sequence of recording from outer limbs toward vitreous (advance). □: solution as in Fig. 2 (high glutamate—this also applies to all other experiments illustrated); electrode tip separation 11 μm; resistivity at 100 μm 592 Ω cm; sequence—withdrawal. •: tip separation 20 μm, resistivity at 1109 Ω cm; penetration. +: separation 20 μm; resistivity 1245 Ω cm; penetration. △: separation 25 μm; resistivity 305 Ω cm; withdrawal.

In the experiment with +, measurements were made with the viewing light on continuously, and the values on the ordinate scale should be divided by 6·64.

al.[7] obtained results in good agreement with their theoretical curve, but used a superfusing medium (Ringer II) which abolished the inflection on the photoresponse described above.

A model which attempts to explain the results obtained is given in Fig. 10. It is necessarily speculative, since the extracellular records may be explained in more than one way, and in the present experiments membrane voltages and resistances have not been investigated. The model is derived from the intracellular recording models of Werblin,[11] Baylor *et al.*[3] and Schwartz.[9] The light-sensitive conductance in the outer limb is in parallel with a major

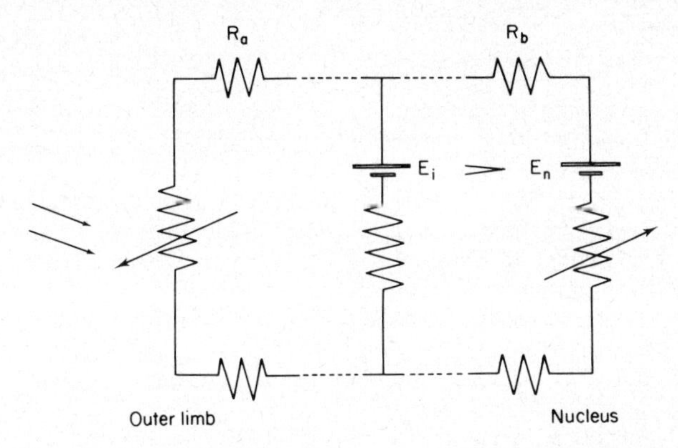

Fig. 10. A model of the photoreceptor. See text for details.

membrane EMF, E_i. In darkness, the outer limb conductance is high, and current flowing through the extracellular resistance R_a produces a "dark voltage". As a result of this current flow, the rod membrane is depolarised. On illumination, the light-sensitive conductance decreases, and the current flowing through R_a is reduced or abolished. The reduction in dark current appears, in extracellular records, as a source of photocurrent. During illumination, as the membrane hyperpolarises, a second mechanism operates with a slight delay. This is the voltage sensitive conductance, in series with E_n, which increases as the rod membrane potential rises, and thus once more shunts E_i and causes the "nose" on intracellular records. The modification suggested by the present results is that the circuit elements of the model are substantially confined to separate zones of the photoreceptor cell, as shown. Thus, when the voltage sensitive conductance increases current flows through the extracellular resistance R_b in a vitreal direction, and the membrane in that region will be a true source of photocurrent. The magnitude of any dark current flowing across R_b depends upon membrane properties which are unknown. However, there are reports[7, 12] that, under certain circumstances, a sink of dark current appears at the rod synapses. In terms of the model, this is explicable if E_i is greater than E_n. Such a conclusion has been reached from intracellular recordings.[9] The model makes the firm prediction that in the region represented by R_b the dark voltage is less than the photovoltage, and this is what is observed.[1]

Acknowledgments

Part of the equipment used in this work was purchased with a grant from the Garrick Charitable Settlement. My thanks are also due to Mr J. C. Low for technical assistance and to Dr W. J. K. Ernst for many helpful discussions.

References

1. G. B. Arden 1976, Voltage gradients across the receptor layer of the isolated rat retina, *J. Physiol.* **256**, 333–360.
2. D. A. Baylor and M. G. F. Fuortes 1970, Electrical responses of single cones in the retina of the turtle, *J. Physiol.* **207**, 77–92.
3. D. A. Baylor, A. L. Hodgkin and T. D. Lamb 1974, The electrical response of turtle cones to flashes and steps of light, *J. Physiol.* **242**, 685–727.
4. W. J. K. Ernst and G. B. Arden 1972, The separation of two PIII components in the rat electroretinogram by a flicker method, *Vision Res.* **12**, 1759–1761.
5. R. Granit 1947, "Sensory Mechanisms of the Retina," Cambridge University Press.
6. D. G. Green 1973, Scotopic and photopic components of the rat electroretinogram, *J. Physiol.* **228**, 781–796.
7. W. A. Hagins, R. D. Penn and S. Yoshikami 1970, Dark current and photocurrent in retinal rods, *Biophys. J.* **10**, 380–412.
8. R. D. Penn and W. A. Hagins 1969, Signal transmission in retinal rods and the origin of the electroretinographic a-wave, *Nature* (*Lond.*) **233**, 201–205.
9. E. A. Schwartz 1976, Electrical properties of the rod syncitium in the retina of the turtle, *J. Physiol.* **257**, 379–406.
10. J. Toyoda, H. Nosaki, H. and T. Tomita 1969, Light induced resistance changes in single photoreceptors of *Necturus* and *Gecko*, *Vision Res.* **9**, 453–463.
11. F. S. Werblin 1975, Regenerative hyperpolarisation in rods, *J. Physiol.* **244**, 53–81.
12. R. Zuckerman 1973, Ionic analysis of photoreceptor membrane currents, *J. Physiol* **235**, 333–354.

9

Mechanism for the Generation of the Receptor Potential of Rods of *Bufo marinus*

L. H. PINTO, J. E. BROWN and J. A. COLES

Department of Biological Sciences, Purdue University, West Lafayette, Indiana, U.S.A.
Department of Physiology and Biophysics, State University of New York, Stony Brook, New York, U.S.A.
and *Département de physiologie de l'Université, École de médecine, Geneva, Switzerland*

After a vertebrate photoreceptor captures quanta of light, its membrane voltage becomes more negative (the hyperpolarizing "receptor potential").[1,2,3] The hyperpolarization is accompanied by increased membrane resistance, suggesting that the receptor potential results from decreased conductance for an ionic process having equilibrium potential more positive than resting potential.[4,5] The net current that crosses the plasma membrane of rat rods has been computed using extracellular source-sink density mapping.[6,7] In the dark, current enters the plasma membrane of the outer segment and emerges from the inner segment. The effect of light is to reduce the dark current, and the removal of external Na^+ eliminates the dark current.[8] Therefore, it has been postulated that light decreases the Na^+ conductance of the membrane.[4,7,9]

The mechanism for the decrease in Na^+ conductance is of interest because the majority of the rhodopsin of the rod is located in the disks, whereas the measured decrease in conductance occurs on the plasma membrane, 100–200 nm away from the edge of the disks.[10] In addition, the capture of a single photon results in the failure of about 10^6 Na^+ ions to enter the plasma membrane.[11] An attractive hypothesis to explain the amplification and the ability of the excitation to bridge the gap between disks and plasmalemma has been proposed by Yoshikami and Hagins.[12] The hypothesis states that light induces the release of Ca^{++} from rod disks into the cytoplasm. This cytoplasmic

Ca^{++} diffuses to the plasmalemma and mediates the decrease in Na^+ conductance.

In order to test the hypothesis that light decreases the Na^+ conductance of the membrane, we have used intracellular microelectrodes to record membrane potential and membrane resistance while altering the composition of the solution bathing the rods. In order to test the hypothesis that increased intracellular ionized calcium, $(Ca^{++})_{in}$, mediates the decrease in Na^+ conductance, we have measured the effects upon the membrane potential and response to light of intracellular injections of Ca^{++} and a Ca^{++} chelating agent.

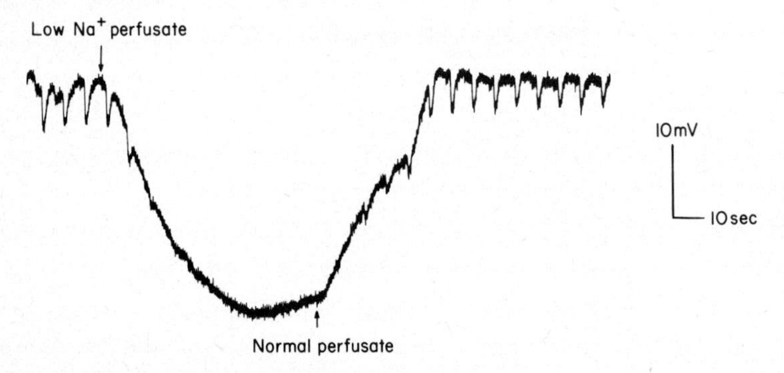

Fig. 1. The effect of substitution of choline for Na^+ in the perfusate. Stimuli of 2·0 s duration were delivered every 6·3 s and elicited responses in normal perfusate of half maximal size. Shortly after substitution (first arrow), the membrane hyperpolarized and the receptor potentials were abolished. These effects were reversed upon return to normal perfusate (1·6 mM (Ca^{++})) (second arrow). From Brown and Pinto.[13]

Impalements were made of single rods in the isolated retina of *Bufo marinus*. The retina was mounted, receptor side up, in a transparent perfusion dish. The perfusion dish was placed on the stage of a compound microscope that was equipped with an infra-red image converter. This optical arrangement allowed us to position the recording pipette in the layer of rod outer segments. Normal perfusate (unless otherwise noted) contained NaCl 111 mM, KCl 2·5 mM, $CaCl_2$ 0·8 mM, $MgCl_2$ 1·6 mM, dextrose 5·6 mM and 3 mM HEPES (N-2 hydroxyethylpiperazine N′-2 ethanesulfonic acid, titrated to pH 7·8 with NaOH). Low Na^+ perfusate was made by replacing the NaCl mole for mole with choline chloride or LiCl. The dish was perfused continuously so that its contents were exchanged every 5–6 s.

The hypothesis that the receptor potential results from a decrease of Na^+ conductance predicts that substitution of an impermeant cation for external Na^+ should result both in membrane hyperpolarization in the dark and attenuation of the responses to light. We found that when rods were bathed in low Na^+ perfusate the membrane hyperpolarized in the dark (by about 30 mV) and receptor potentials were completely eliminated (Fig. 1).[13] Both changes were reversible. These findings do not unequivocally demonstrate that the light-induced changes of conductance are perfectly selective for Na^+. If membrane potential is nearly equal to K^+ equilibrium potential when the cells are bathed in low Na^+ perfusate, then light-induced changes in K^+ conductance would not result in altered membrane potential. In order to test whether light-induced changes in conductance for ions other than Na^+ occur, we measured input impedance of cells while they were bathed in low Na^+ perfusate. We found that when rods were bathed in low Na^+ perfusate, their input impedance in the dark increased (by 15–60 MΩ) and light-induced changes in input impedance become undetectable (less than 10% of their

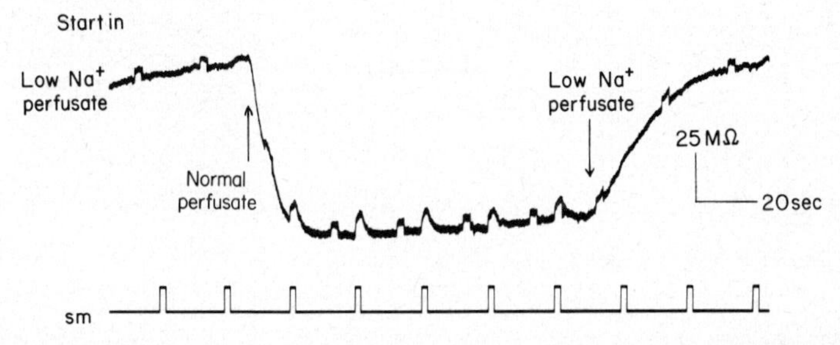

Fig. 2. Substitution of choline for Na^+ caused an increase of resting membrane resistance and abolished light-induced changes of membrane resistance. Upper trace is a continuous record of input impedance measured with an A.C. bridge. Small upward deflections preceding stimuli are (+5 MΩ) calibration pulses. SM is the stimulus monitor. From Pinto and Ostroy.[14]

normal value) (Fig. 2).[14] Thus, in the absence of external Na^+, light-induced changes in conductance for ions other than Na^+ were undetectable. This finding indicates that changes in conductance for ions other than Na^+ that normally occur[15] must be secondary to the change in Na^+ conductance.

Yoshikami and Hagins[12] examined the hypothesis that a light-induced increase in $(Ca^{++})_{in}$ mediates the decrease in Na^+ conductance. In their experiments the magnitude of the dark current was measured while the concentrations of various ions in the extracellular medium were altered. Increasing $(Ca^{++})_{out}$ suppressed the dark current, but changing $(Mg^{++})_{out}$ was without effect. By making the plasma membrane more permeable to Ca^{++}, Hagins and Yoshikami[16] showed that Ca^{++} might have an intracellular site of action. After exposing the rods to the divalent cation ionophore X-537a, changing $(Ca^{++})_{out}$ from approximately 10^{-8} M to 10^{-5} M suppressed the dark current entering the rod outer segments.

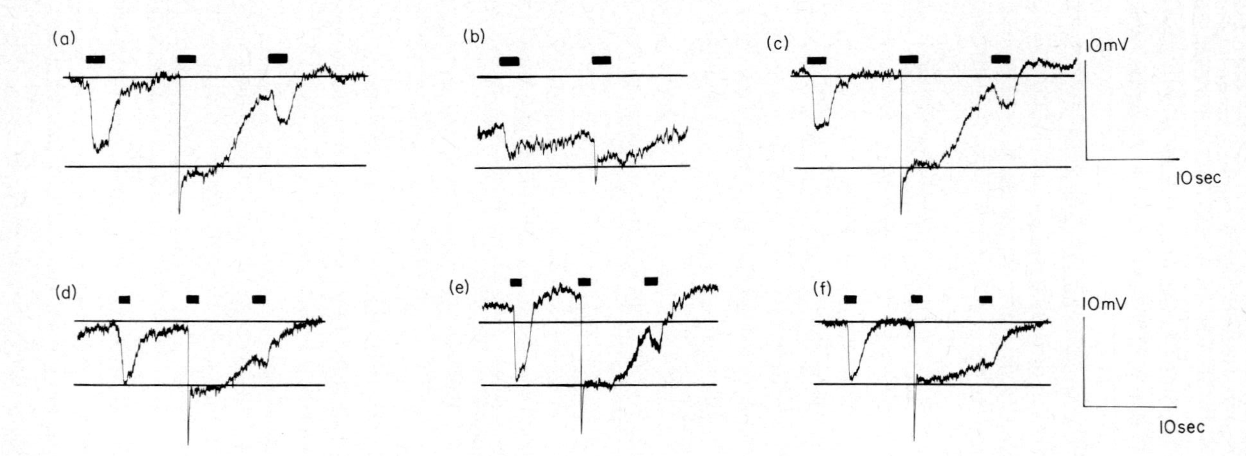

Fig. 3. Effects of increased $(Ca^{++})_{out}$ (a–c) and decreased $(Ca^{++})_{out}$ (d–f) for two different cells. Bar over each response is stimulus marker; the second response of each group was elicited by a stimulus that was 60-fold stronger than the stimuli used to elicit the smaller responses. Responses in (b) were obtained in test solution having $(Ca^{++})_{out}$ of 12·5 mM; responses in (e) were obtained in test solution having $(Ca^{++})_{out}$ of 0·6 mM. Other responses were elicited while the rods were bathed in normal perfusate (1·6 mM (Ca^{++})). Horizontal lines are drawn at two arbitrary values of membrane voltage. Note that decreased $(Ca^{++})_{out}$ causes depolarization and increased $(Ca^{++})_{out}$ causes hyperpolarization of the membrane. From Brown and Pinto.[13]

We have tested the hypothesis of Yoshikami and Hagins in three additional ways. We recorded resting potential and receptor potentials: (1) while changing $(Ca^{++})_{out}$; (2) after injecting Ca^{++} intracellularly; and (3) after buffering $(Ca^{++})_{in}$ to a low value by intracellular injections of EGTA. The hypothesis can be tested only indirectly by changing $(Ca^{++})_{out}$. Since changes in $(Ca^{++})_{out}$ are expected to yield changes in free $(Ca^{++})_{in}$ of the same sign,[17] the hypothesis predicts that increased $(Ca^{++})_{out}$ will result in hyperpolarization, and that decreased $(Ca^{++})_{out}$ will result in depolarization. We have found that increasing $(Ca^{++})_{out}$ caused membrane hyperpolarization in the dark, and lowering $(Ca^{++})_{out}$ resulted in depolarization in the dark (Fig. 3).[13] Thus, these findings are consistent with the hypothesis that increased $(Ca^{++})_{in}$ mimics the effects of light.

A more direct test of the hypothesis can be made by injecting Ca^{++} into rod outer segments. For our experiments, Ca^{++} was injected iontophoretically by passing positive current out of pipettes that contained 0·1 M calcium acetate plus 4 M potassium acetate. For several experiments this current was balanced by negative current that was passed out of the second barrel of a double barrel pipette. The second barrel was filled with 4 M potassium acetate (KAc). This balanced injection scheme was used to avoid changing membrane potential during the injection. Ca^{++} was injected in 19 cells by such balanced current injections. The injection of Ca^{++} induced an after-hyperpolarization of 2–8 mV for 10 of the 19 cells studied. For the remaining 9 cells, passage of current from the pipette that contained Ca^{++} produced no after potentials. For none of the cells did the passage of balanced current in the opposite direction (i.e. positive current from the barrel that contained only KAc) produce an after-potential.

Attempts to inject Ca^{++} by passing balanced currents through double-barrel pipettes, one barrel of which contained Ca^{++}, did not always produce an after-hyperpolarization; possibly these small barrels did not reliably transport Ca^{++}. Therefore, we also injected Ca^{++} iontophoretically using single barrel pipettes that had lower resistance than either barrel of the double barrel pipettes. Figure 4(a) shows that after the injection of Ca^{++} (by passing positive current out of a pipette filled with 0·1 M CaAc plus 4 M KAc) there was a hyperpolarization. The mean hyperpolarization after Ca^{++} injection was 15·3 mV ($\pm$1 mV S.E.) for 18 cells. The mean after-hyperpolarization that followed injection of the same positive current from control pipettes (filled with only KAc) was smaller (4·76 $\pm$ 0·47 mV S.E.). Thus, an extra hyperpolarization resulted from injection of Ca^{++}. These experiments show that a direct increase in free $(Ca^{++})_{in}$ from iontophoretic injection of Ca^{++} mimics the hyperpolarization produced by light.

The hypothesis of Yoshikami and Hagins also predicts that buffering the $(Ca^{++})_{in}$ to a low value should result in membrane depolarization in the dark

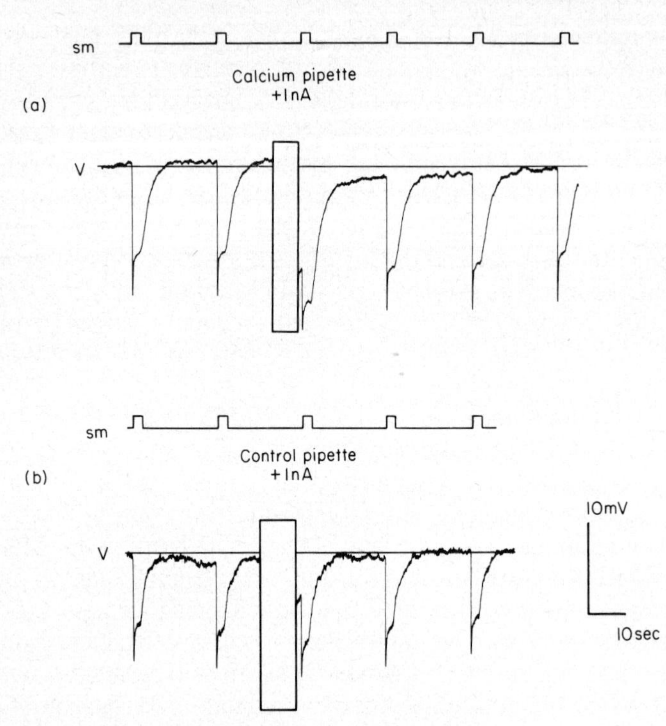

Fig. 4. (a) Injection of Ca^{++} ions from a single barrel pipette. After removal of depolarizing current (rectangle) from a pipette that contained Ca^{++}, there was an after-hyperpolarization that exceeded the plateau of the response to light. (b) After removal of depolarizing current from a pipette filled with only potassium acetate, there was an after-hyperpolarization smaller than the plateau of the response to light. Horizontal lines are drawn at two arbitrary values of membrane voltage.

and in attenuation of the receptor potentials. In order to buffer $(Ca^{++})_{in}$ to a low value, we injected iontophoretically the chelating agent EGTA (ethylene glycol *bis* (β-aminoethyl ether) N, N, N′, N′ tetracetic acid) into rod outer segments. For the injections, pipettes filled with the potassium salt of EGTA were used (pH adjusted to 7·0–9·3). After the injection of EGTA there was an after-depolarization that lasted 5–60 s (Fig. 5). When an impaled rod was stimulated by dim spots of small diameter ($\approx$30 μm), after the injection of EGTA the receptor potentials were attenuated for a period that lasted as long as the depolarization. Since depolarization of the membrane would be expected to reduce the electrochemical gradient for Na^+, the attenuation of the receptor potentials was also measured during depolarization caused by positive current passed out of the pipette. The attenuation of the receptor potentials after injection of EGTA was greater than the attenuation due to the depolarization alone. Control injections of other anions that do not bind Ca^{++} produced no depolarization or attenuation of the receptor potentials. From these experiments we conclude that decreased $(Ca^{++})_{in}$ depolarizes the membrane and that buffering light-induced elevations of $(Ca^{++})_{in}$ attenuates the receptor potentials.

In summary, the findings that the light-induced changes of both membrane potential and membrane resistance disappear when vertebrate rods are bathed

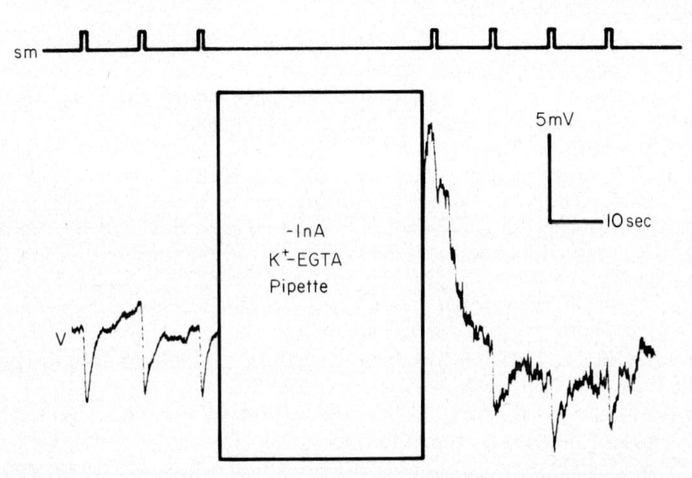

Fig. 5. Iontophoretic injection of EGTA caused an after-depolarization and attenuation of the receptor potentials elicited by small (30 μm diam.) dim stimuli.

in low Na^+ perfusate support the hypothesis that light induces a decrease in Na^+ conductance of the membrane. The hypothesis that the decrease in Na^+ conductance is mediated by a light-induced increase of $(Ca^{++})_{in}$ is supported by two direct experiments. First, increases of $(Ca^{++})_{in}$, brought about by intracellular injections of Ca^{++}, result in membrane hyperpolarization and thereby mimic the effects of light. Secondly, buffering $(Ca^{++})_{in}$ to a low value results in membrane depolarization and attenuation of the response to light. However, before accepting the hypothesis that increased $(Ca^{++})_{in}$ is a step leading to the generation of the receptor potential, we await evidence to demonstrate the following: that the hyperpolarization induced by increased $(Ca^{++})_{in}$ results from decreased Na^+ conductance, and the light does indeed induce an increase in $(Ca^{++})_{in}$ inside intact rod outer segments.

References

1. A. Bortoff 1964, Localization of slow potential responses in the *Necturus* retina, *Vision Res.* **4**, 627–635.

2. T. Tomita 1965, Electrophysiological study of the mechanisms subserving color coding in the fish retina, *Cold Spring Harbor Symp. Quant. Biol.* **30**, 559–566.
3. F. S. Werblin and J. E. Dowling 1969, Organization of the retina of the mudpuppy, necturus maculosus: II. Intracellular recording, *J. Neurophysiol.* **32**, 339–355.
4. J.-I. Toyoda, H. Nosaki and T. Tomita 1969, Light-induced resistance changes in single photoreceptors of *Necturus* and *Gekko*, *Vision Res.* **9**, 453–463.
5. D. A. Baylor and M. G. F. Fuortes 1970, Electrical responses of single cones in the retina of the turtle, *J. Physiol.* **207**, 77–92.
6. R. D. Penn and W. A. Hagins 1969, Signal transmission along retinal rods and the origin of the electroretinographic a-wave, *Nature* **223**, 201–205.
7. W. A. Hagins, R. D. Penn and D. Yoshikami 1970, Dark current and photocurrent in retinal rods, *Biophysical J.* **10**, 380–412.
8. W. A. Hagins 1972, The visual process excitatory mechanisms in the primary receptor cells, *Ann. Rev. Biophys. Bioeng.* **1**, 131–158.
9. J. T. Korenbrot and R. A. Cone 1972, Dark ionic flux and the effects of light in isolated rod outer segments, *J. Gen Physiol.* **60**, 20–45.
10. A. I. Cohen 1968, New evidence supporting the linkage to extracellular space of outer segment saccules of frog cones but not rods, *J. Cell Biol.* **37**, 424–444.
11. T. Tomita 1970, Electrical activity of vertebrate photoreceptors, *Quart. Rev. Biophysics* **3**, 179–222.
12. S. Yoshikami and W. A. Hagins 1971, Light, calcium, and the photocurrent of rods and cones, *Bioph. Soc. Ann. Meet. Abstr.* **11**, 47a.
13. J. E. Brown and L. H. Pinto 1974, Ionic mechanism for the photoreceptor potential of the retina of *Bufo marinus*, *J. Physiol.* **236**, 575–591.
14. L. H. Pinto and S. E. Ostroy 1977, Ionizable groups and conductances of the rod photoreceptor membrane, Submitted to *J. Gen. Physiol.*
15. F. S. Werblin 1975, Regenerative hyperpolarization in rods, *J. Physiol.* **244**, 53–81.
16. W. A. Hagins and S. Yoshikami 1974, A role for Ca^{++} in excitation of retinal rods and cones, *Exp. Eye Res.* **18**, 299–305.
17. P. F. Baker, A. L. Hodgkin and E. B. Ridgway 1971, Depolarization and calcium entry in squid giant axons, *J. Physiol.* **218**, 709–755.

Discussion

D. R. Copenhagen: As I understand it, the iontophoretic injections of either calcium ions or calcium chelating agents can respectively hyperpolarize or depolarize the rod potential but these effects are not observed in all cells. The rods in *Bufo marinus* are coupled together so that any potential changes induced in one rod would be dramatically reduced due to the large shunt pathways to other rods. Have you tried to estimate how large the potential changes might be if you had injected substances into an isolated rod? Also, might it not be possible that at least some of the differences you see between rods could be attributable to differences in the effective coupling between the penetrated rod and its neighbors?

L. H. Pinto: Depolarization followed injection of EGTA for all of the cells we studied.

We have insufficient data for the rods in our preparation to estimate the amplitudes of the after-potentials for an isolated cell.

The amplitudes of the responses elicited by small, dim spots of light after injection of EGTA did vary from cell to cell. It is likely that the larger light-responses were recorded from cells that were more tightly coupled to their neighboring cells. Moreover, the injection of EGTA did not attenuate the light-responses elicited by stimuli of large diameter (500 μm). In the latter case interactions probably participated in the generation of the light-responses that were recorded. However, we have no direct experimental evidence to show that shunting to neighboring rods attenuated the changes in membrane potential that were induced by injections of Ca^{++} and EGTA into the outer segments.

R. Meech: It would be useful if Dr Pinto could give an estimate of the amount of calcium injected into the rods because injection of a relatively large concentration of calcium can produce changes in intracellular pH in invertebrate neurones (Meech and Thomas, *J. Physiol.*, submitted).

Injection of calcium into central neurones causes the membrane to hyperpolarize by increasing the potassium conductance (Meech, 1976, *Symp. Soc. Exp. Biol.* **30**, 161–191). Can Dr Pinto exclude this explanation for the potential changes in rods which he has reported following calcium injection?

L. H. Pinto: We have no independent measure of the amount of Ca^{++} transported into the rods, nor have we measured membrane conductance after Ca^{++} injection into these small cells.

10

Characteristics of the Electrical Coupling between Rods in the Turtle Retina

W. GEOFFREY OWEN* and DAVID R. COPENHAGEN†

Department of Physiology, University of California at San Francisco, San Francisco, California, U.S.A.

Introduction

In their now classic paper, Hecht, Shlaer and Pirenne[1] established that as few as five photons captured by human rods within a small retinal area, and a brief interval of time, were sufficient for the perception of a stimulus. Since it was extremely unlikely that any given rod had captured more than a single photon, the clear implication of their result was that each rod is capable of signalling single photon captures to the bipolar cell and, moreover, of doing so with some measure of reliability.

On the basis of measurements of the extracellular flow of current around rods in the rat retina, Hagins, Penn and Yoshikami[2] calculated that the potential change produced at the rod's synaptic terminal by the bleaching of a single photopigment molecule was about 6 μV, more than three times larger than the mean of the fluctuations due to the thermal noise of the cell and therefore large enough to be reliably discriminated from the noise.

More recently, however, Falk and Fatt[3] considered the limitations imposed upon the reliable transmission of signals across the synapse from rod to bipolar cell by the statistical fluctuations in transmitter release (transmitter

* Present address: *Department of Anatomical Sciences, The State University of New York at Stony Brook, Stony Brook, New York, U.S.A.*

† Present address: *Department of Ophthalmology, University of California at San Francisco, San Francisco, California, U.S.A.*

noise). They concluded that potentials as small as those in the rat could not be reliably signalled to the bipolar cell and that the transmitter noise was so great that "even the complete cessation of transmitter release from a single (rod) absorbing a photon would be insufficient to produce a change in synaptic activity capable of standing out from random fluctuation in such activity".

As a way out of this dilemma, Falk and Fatt suggested that perhaps the responses of rods are not independent at low light levels, but that the light-induced hyperpolarization of one rod might induce similar hyperpolarization of its neighbours. Thus, when a single rod absorbs a quantum, the bipolar cell would receive input from many rods and the reliability of transmission would be increased.

The experiments of Baylor, Fuortes and O'Bryan,[4] and later of Baylor and Hodgkin,[5] clearly demonstrated that cones in the retina of the turtle, *Pseudemys scripta elegans*, are coupled together via both direct and indirect pathways and that the cone's light sensitivity is affected by this coupling. It has since been shown that rods of the toad, *Bufo marinus*,[6] and turtle, *Chelydra serpentina*[7,8,9,15] are also coupled together in a way that profoundly affects their sensitivity.

We were interested in how rods are coupled together, since it might be the key to clarifying how single quantal absorptions can be signalled reliably to the bipolar cell across a noisy synapse. In this paper we shall describe our measurements of the quantal sensitivity of rods in the turtle. Experiments to elucidate the characteristics and the mechanism of rod–rod interactions will then be described in some detail. Finally, we shall offer some comments upon ways in which these interactions might affect signal transmission between the rod and bipolar cell.

Quantal Sensitivity

Intracellular recordings were made from photoreceptors in eyecup preparations of the turtle, *Chelydra serpentina*. Since we were specifically interested in determining how large a potential is produced in a rod by the bleaching of a single photopigment molecule, great care was taken to ensure that all stimuli were delivered to fully dark-adapted cells. This was affected

(a) by careful light-proofing of the Faraday cage in which the eyecup was situated,
(b) by allowing adequate time for dark-adaptation to be completed following preparation of the eyecup, and
(c) by ensuring that stimuli were presented only at intervals sufficient to allow full recovery of dark adaptation after the preceding flash.

The condition of the preparation was monitored throughout the course of the experiment and any sign of abnormality or deterioration of the preparation caused the experiment to be terminated. Full details of the methods of preparation, stimulation, and recording have been given elsewhere (see ref. 7).

In theory, the quantal sensitivity of the rod can be obtained by measuring the peak hyperpolarization elicited by a stimulus whose intensity is such that a single visual pigment molecule is bleached in each of the illuminated rods. In practice we made use of the linear relation that exists between the stimulus intensity and response amplitude at low levels of illumination, to determine the flash sensitivity, S_f. This is defined by: $S_f = \Delta V / I_\lambda \, \Delta t$ where ΔV is the amplitude at the peak of the response (μV), I_λ is the stimulus intensity (quanta $\sec^{-1}$ μm^{-2}) at the stimulus wavelength, and Δt is the stimulus duration. Dividing the flash sensitivity by the effective collecting area of the rod yields the quantal sensitivity of the rod.

Measurements were made using 20 ms flashes of light of wavelength 514 nm whose intensities were adjusted to elicit a response of less than 1 mV amplitude, this being within the linear response range for all stimulus sizes tested. The effective collecting area of the rod was estimated to be 13·6 μm^2.[(7)] Before plotting, the data were adjusted to compensate for the effects of light scattering within the retina (see Appendix of ref. 7).

The circles in Fig. 1 plot the average of the values obtained from seven rods. The smooth curve was drawn by eye through the data points. It is clear that the quantal sensitivity of the rod depends strongly upon the area of the stimulus used in its measurement. Using a 25 μm diameter stimulus the rods had an average quantal sensitivity of 67 $\pm$ 12 μV. The most sensitive rod gave a value of 127 μV. Increasing the stimulus diameter to 300 μm yielded a value of 720 $\pm$ 29 μV, the most sensitive rod in this case giving 1180 μV. Increasing the stimulus diameter beyond 300 μm produced no further change in sensitivity.

Since we have adjusted our data for the effects of optical scattering upon the light intensity at the impaled rod, the observed 11-fold change in the rod's quantal sensitivity must be due to the activity of other photoreceptors within a radius of 150 μm; that is, it must result from lateral interactions between the rod and surrounding photoreceptors.

This finding raises several important questions. These include:

(1) With which classes of photoreceptors do the rods interact?
(2) What is the nature of the mechanism by which these interactions occur?
(3) What is the anatomical pathway that mediates these interactions?
(4) What effects could these interactions have upon the functional characteristics of the rods?

We will address these questions in turn.

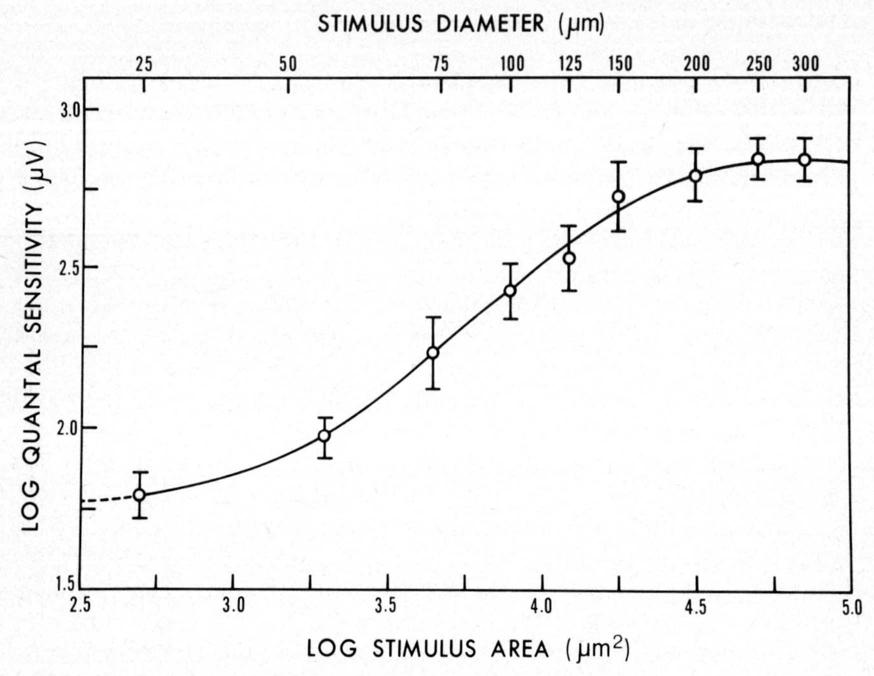

Fig. 1. The quantal sensitivity of an average rod is plotted as a function of stimulus area on logarithmic axes. Eight to ten measurements of quantal sensitivity were made at each of twelve stimulus diameters between 25 μm and 750 μm in seven different rods. The data obtained at each stimulus diameter were then averaged and the average values adjusted to compensate for the effects of light scatter within the retina. The vertical bars represent ±1 S.E. of the mean. The smooth curve was drawn through the data points by eye.

With which Classes of Photoreceptor do the Rods Interact?

To answer this question we examined the spectral sensitivity of the rod response. Criterion responses, ranging in amplitude between <1 mV and 24 mV were elicited with 750 μm diameter stimuli of twelve different spectral wavelengths. The logarithm of the relative spectral sensitivity of the rod is plotted as a function of wavelength in Fig. 2. The filled circles are averaged data from five rods in which criterion responses of either 2 mV or 3 mV amplitude were elicited. The filled squares are averaged data from three rods in which criterion responses of 0·7 mV were elicited, a signal averaging computer having been used in their measurement. The curve is the difference spectrum of a vitamin A_2 based pigment with peak absorbance at 520 nm, this being the pigment contained in snapping turtle rods.[10] The good agreement between the data and the difference spectrum is taken as a strong indication that the peak amplitudes of dark adapted rod responses are determined only by photons absorbed in the rods. One might conclude that the rod receives no input from neighbouring cones. Examination of the waveforms of responses less than 2 mV in amplitude revealed, however, that the picture is somewhat

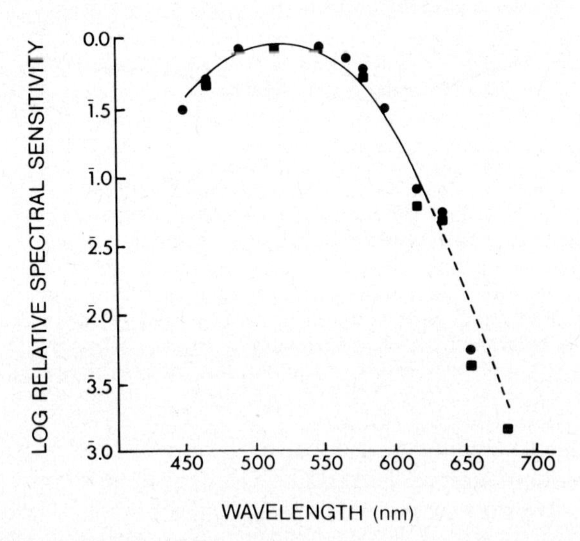

Fig. 2. The log relative sensitivity of the rod's dark adapted response is plotted as a function of stimulus wavelength. In the case of the filled circles 200 ms flashes of light, 400 μm in diameter, were used to elicit responses of peak amplitude 2 mV (3 rods) or 3 mV (2 rods) at each wavelength. The filled squares are data for 3 rods obtained by computer averaging responses, 0·7 mV in amplitude, elicited by 20 ms flashes of light, 750 μm in diameter. Data were arbitrarily normalized at 514 nm. The curve is the difference spectrum for a vitamin A_2 based photopigment absorbing maximally at 520 nm. Its form at long wavelengths (dotted line) was determined from measurements of similar pigments generously supplied by Prof. F. Crescitelli.

more complicated than this. With light of wavelengths greater than about 620 nm, an early rapid component was seen in the responses of about 70 % of the rods examined. This is illustrated in Fig. 3. As first described by Schwartz,[11] this component is due to an input to the rod from red-sensitive cones. When care was taken to ensure that the rods were fully dark adapted prior to stimulation, we found that this component never contributed to the peak amplitude of the rod response and, hence, did not affect dark-adapted rod sensitivity. We never observed this early component at wavelengths less than 620 nm and therefore conclude that the spatial variation in quantal sensitivity described in the previous section must reflect, exclusively, an interaction between rods.

Before leaving this curious cone input to the rods, it is worth mentioning a finding from experiments in which we made simultaneous recordings of the responses of rods and red-sensitive cones to the same long wavelength stimuli. We noted that in general the amplitude of the cone response was about twenty times larger than that of the cone generated component of the rod response and

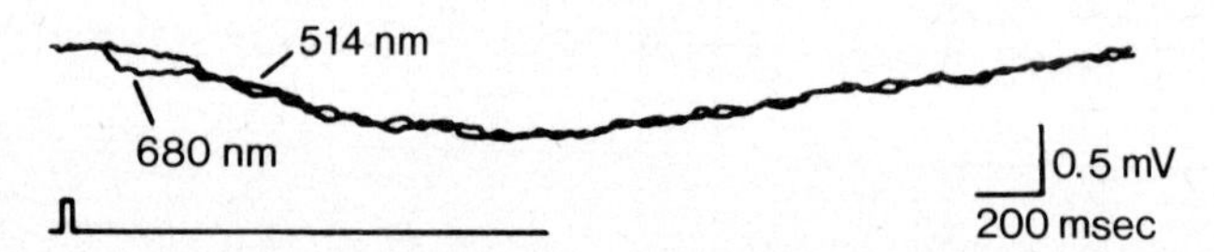

Fig. 3. Responses in the linear response range of the dark adapted rod elicited by light of wavelengths 514 nm and 680 nm. Stimuli were of 30 ms duration and 750 μm diameter. Each trace was obtained by computer averaging ten such responses. The intensities required to equalize the peak amplitudes of these responses were in the ratio 1:960, consistent with the requirement for equal quantum catches by the rod photopigment. The small, early component on the rising phase of the response to 680 nm light is believed to be due to an input from red-sensitive cones. Such a component was seen in the responses of about 70 % of the rods examined.

almost an order of magnitude larger than the response of the rod itself. At these long wavelengths, therefore, it seems likely that the turtle's visual threshold is determined not by the responses of rods but by those of red- sensitive cones. It is thus difficult to imagine that, under dark-adapted conditions, this weak cone input to the rod can have any great functional importance.

Characteristics of Rod–Rod Coupling

A number of experimental results pointed to the conclusion that rods probably interact via direct pathways between them. If this were true, it would be possible to polarize a given rod by injecting a polarizing current into a neighbouring one. This would permit a more direct approach to be adopted in studying the interaction mechanism. We examined this possibility using a double-impalement technique similar to that used by Baylor, Fuortes and O'Bryan on turtle cones.[4]

Two microelectrodes were advanced independently into the retina and simultaneous intracellular recordings were made from pairs of rods separated by a small distance. Typical results from one such experiment are shown in Fig. 4. Extrinsic currents injected into Rod 1 caused it to polarize and also produced a polarization of like sign in Rod 2. Note that hyperpolarizing currents appeared to be more effectively coupled than depolarizing ones. On reversing the situation, injecting current into Rod 2, essentially identical results were obtained. Again, hyperpolarizing currents appeared to be coupled more effectively than depolarizing ones.

Upon withdrawing the microelectrode from one of the rods (Rod 1 in this case), and injecting current into the extracellular space, it was no longer

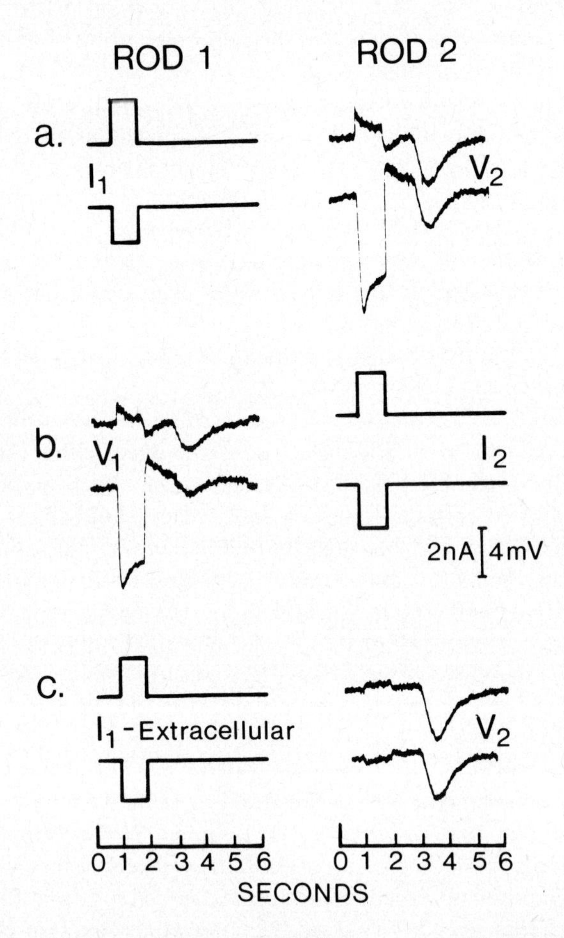

Fig. 4. Simultaneous intracellular recordings from two rods, 75 μm apart. The current records, I_1 and I_2, show the respective time courses of current injected into Rod 1 and Rod 2. An upward deflection indicates a depolarizing (outward) current, a downward deflection a hyperpolarizing (inward) current. The earlier deflections in the voltage records, V_1 and V_2, are the potentials induced in Rod 1 and Rod 2 by, in each case, currents injected into the neighbouring rod. Upward deflections indicate depolarizations, downward deflections hyperpolarizations. The slow hyperpolarizations following the current induced potentials are the rods' responses to light. In (a) an outward current of 1·8 nA and an inward current of 1·5 nA were injected into Rod 1. In (b) outward and inward currents of 1·8 nA were injected into Rod 2. In (c), as a control, the electrode was withdrawn from Rod 1 and currents of $\pm 1{\cdot}6$ nA were injected into the extracellular space while recording the potential in Rod 2.

possible to induce a potential change in the other rod. The small displacements of membrane potential seen in the records from Rod 2 were artifacts of the recording system.

From these results we conclude that the coupling pathway is both reciprocal and non-inverting. Furthermore, in view of the failure of extracellularly injected current to evoke a potential change in the rod, it seems apparent that these interactions are mediated by a specific intercellular pathway.

Since hyperpolarizing currents are more efficient in polarizing neighbouring rods than depolarizing ones, we might be tempted to conclude that the coupling pathway is electrically rectifying. However, it would be difficult to reconcile such a conclusion with the observation that the pathway is also reciprocal since this implies that current of a given sign is coupled with the same efficiency from Rod 1 to Rod 2 as it is from Rod 2 to Rod 1. It seems more likely, therefore, that the observed rectification reflects properties of the rod's plasma membrane.

Consistent with this notion were the results of several experiments in which single rods were impaled with two electrodes, advanced either independently or together in a double-barrelled configuration. The relation between extrinsically injected current (I) and the induced potential change (V) was in all cases nonlinear, rectifying strongly in the depolarizing direction. Thus both the coupled and input voltage-current relationships rectify in the depolarizing direction. Moreover, the coupling and input voltage-current relationships appear to be in constant proportion over the normal range of light induced potentials. We therefore believe that the rectification is a result of non-linearities in the rod's plasma membrane, and is not in the coupling pathway itself.

The rods described in Fig. 4 were separated by about 75 μm and were the most strongly coupled pair we examined. Of 27 pairs from which recordings were obtained, 14 pairs were found to be coupled. While coupling efficiency varied greatly, other characteristics of coupling were consistently as described above. The maximum observed separation of coupled rods was about 110 μm. Of the 13 pairs that were not coupled, 12 pairs were separated by 120 μm or more. One pair, however, were less than 50 μm apart. Curiously, there seemed to be little correlation between coupling efficiency and separation of the coupled rods. One possible explanation is that coupling efficiency was impaired by varying amounts of damage inflicted upon the rods during impalement. An alternative explanation, which will be discussed later, is that direct pathways exist between rods separated by various distances.

To summarize the results of these experiments, we find that rods are coupled by specific intercellular pathways. These pathways are reciprocal, communicating with equal efficiency in either direction. Coupling efficiency, at least over the physiological response range, appears to be independent of potential. Probably as a result of the rectifying properties of the rod's plasma membrane, however, hyperpolarizations are more efficiently communicated to neighbouring rods than depolarizations.

We performed similar experiments on pairs of cones. For comparison the results of one such experiment are shown in Fig. 5; the cones in this case being red-sensitive. The basic features are consistent with those observed in rods and agree well with the findings of Baylor *et al.*[4] from their work on cones of

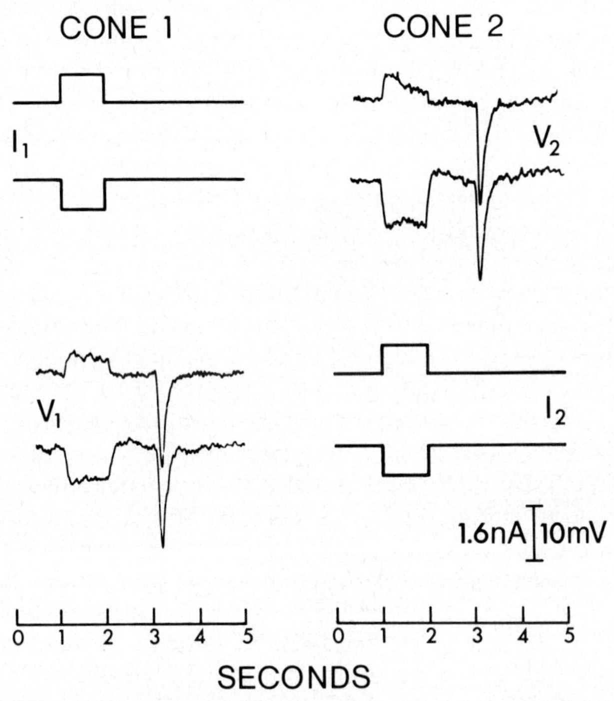

Fig. 5. Simultaneous intracellular recordings from a pair of red-sensitive cones. The records are presented according to the convention adopted in the previous figure. Thus, the upper traces show the potential recorded in Cone 2 when first, current pulses of $\pm 0{\cdot}8$ nA were injected into Cone 1 and then the cones were stimulated with a brief light flash. The lower records were obtained from Cone 1 when currents of $\pm 0{\cdot}8$ nA were injected into Cone 2 and the cones were then stimulated with light.

Pseudemys scripta elegans. Note, however, that the efficiency with which depolarizing potentials are coupled between cones is only slightly lower than the coupling efficiency for hyperpolarizing potentials. Examination of the input voltage-current relation of red-sensitive cones revealed a correspondingly smaller degree of rectification. Again we suggest that the observed rectification is primarily a property of the cone's plasma membrane and does not reflect the behaviour of the coupling pathway.

The Coupling Mechanism

Of the several possible ways in which rods might interact the most likely are:

(1) Via chemically mediated synapses;
(2) via the electrotonic spread of current through gap junctions;
(3) via an interneuron such as a horizontal cell;
(4) by any combination of these mechanisms simultaneously.

In the frog neuromuscular junction extracellular application of Co^{++} was found to inhibit the relase of transmitter.[12] Since an influx of Ca^{++} to the presynaptic terminal appears to be a normal prerequisite of transmitter release,[13] it seems likely that Co^{++} may act by competitive inhibition of Ca^{++}. If so, it should act in a similar manner upon chemical synapses that may exist between retinal rods. We therefore carried out several experiments in which Co^{++} was applied to the retina in order to distinguish between the possible mechanisms of interaction.

The results of a typical experiment are shown in Fig. 6. Prior to application of Co^{++} the retina appeared to be completely normal. Recordings from cells in the proximal layers of the retina showed normal light responses. Rods had normal receptive fields 300 μm in diameter. Figure 6(a) shows records taken from a horizontal cell. Upon penetration, the resting membrane potential was about -25 mV. A weak stimulus was presented every 5 s and it elicited a response of 5 mV. By using a weak stimulus, significant light adaptation of the rods was avoided. During the interval indicated in Fig. 6(a), drops of 6 mM $CoCl_2$ in Ringer solution were applied to the retina. The drops were touched to the shank of the electrode and allowed to run down it into the eyecup. About 10 such drops were applied, during which time the potential record was greatly disturbed. The microelectrode remained in the cell, however, and when application of $CoCl_2$ was complete, the potential returned to its previous steady value. Thereafter, the cell hyperpolarized quite rapidly by about 35 mV. When fully hyperpolarized the light response was virtually abolished. These observations are consistent with the notion that Co^{++} blocks the release of transmitter to the horizontal cells and other secondary neurons.

It might seem surprising that the light-induced response should remain fairly constant until the horizontal cell was substantially hyperpolarized. It can be shown, however, that this observation is probably a consequence of the horizontal cell's ionic mechanism and the action of Co^{++} at the presynaptic release site.

The electrode was then withdrawn from the horizontal cell and advanced until a rod was impaled (Fig. 6(b)). Comparison of the light response of this rod with that of rods in the same retina before application of $CoCl_2$ showed both the response waveform and receptive field size to be unchanged. The response waveform elicited by a large, intense stimulus (dia. 300 μm) exhibited a marked "overshoot" when compared with that elicited by a small stimulus of the same intensity. This is clearly seen in the records shown in Fig. 6(b). In an earlier paper we showed that such a difference in waveform was characteristic of the rod's response to stimuli of different diameters.[7] Attempts to elicit identical

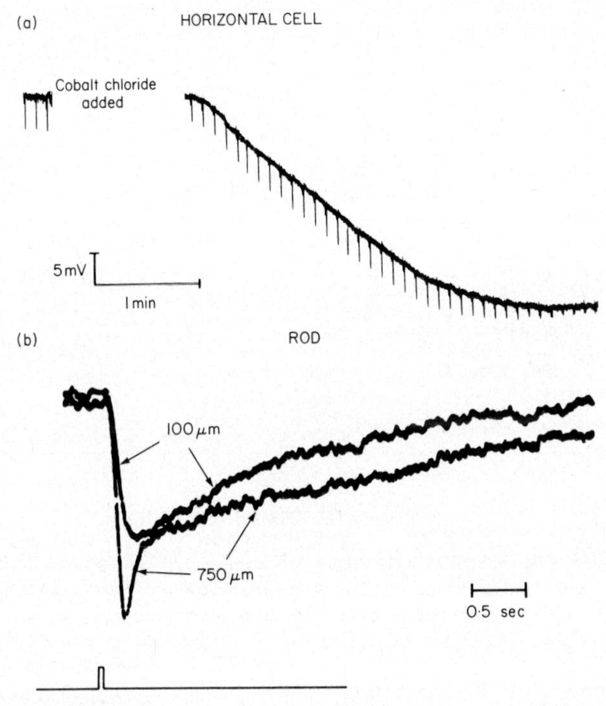

Fig. 6. (a) The effect of Co^{++} upon the membrane potential and light response of a horizontal cell. Drops of 6 mM $CoCl_2$ were applied during the indicated interval. (b) Light responses of a rod in the same retina, recorded soon after the above application of $CoCl_2$. These differences in the response waveforms elicited by stimuli of diameter 100 μm and 750 μm are typically seen when recording from rods in untreated retinae, and have been shown to be due to rod-rod interactions.

responses, either by adjusting stimulus intensities or, in later experiments, by polarizing the rod with steady extrinsic currents, were unsuccessful. We believe, therefore, that the difference in waveforms reflects a difference in the degree of interaction between the rods under the two stimulating conditions. The interactions between the rods appear to be unaffected by the application of Co^{++}. Examination of the response amplitude elicited by a stimulus of fixed intensity as a function of stimulus diameter showed that it, too, was unaffected by application of Co^{++}. This is illustrated in Fig. 7.

As a control, upon withdrawing from the rod, the electrode tip was deliberately broken and the electrode then placed in the vitreous. A strong stimulus was presented which evoked an ERG. The ERG was abnormal in that it consisted only of the pIII components and pI, the c-wave, confirming that neurons proximal to the receptors remained inactivated by the application of Co^{++}.

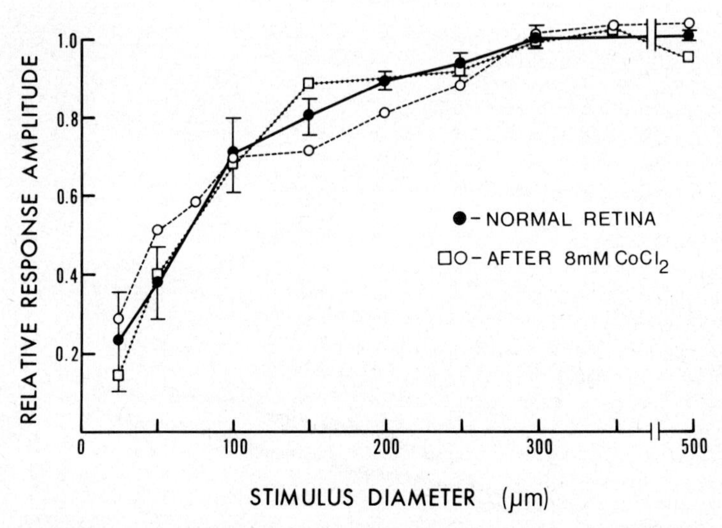

Fig. 7. Peak response amplitudes as a function of stimulus diameter. Filled circles plot the average response obtained from 4 rods in a normal retina. Unfilled symbols describe responses from two rods after application of cobalt chloride. Data points were normalized with respect to the largest response amplitude.

There are two major conclusions to be drawn from these experiments. First, the mechanism by which rods interact does not involve the activity of horizontal cells as interneurons since these were rendered unresponsive by the application of Co^{++}. Second, chemical synapses appear to play no significant role in mediating rod-rod interaction. Since electrotonic junctions are known to be relatively unaffected by *low* concentrations of divalent cations,[13] these interactions are most easily explained in terms of an electrotonic spread of current through gap junctions between the rods. This is in agreement with the conclusion drawn by Schwartz on the basis of similar experiments.[15]

The Anatomical Identity of the Coupling Pathway

One naturally asks where these gap junctions might be found. Gap junctions between photoreceptors have been identified anatomically in several vertebrate species.[16, 22, 23] Specific rod-rod interconnections between the inner segments of human rods were described by Uga *et al.*[17] In the toad, where coupling has also been demonstrated physiologically, gap junctions were identified between interlocking "fins" that radiated from the inner segments of neighbouring rods.[18] Similar junctions were described in the axolotl.[19] Such "fins" were also found on the rods of snapping turtle[20] and it is therefore possible that immediately adjacent rods may communicate in this way. However, the relative proportion of rods to cones in the snapping turtle is small. Underwood,[21] estimates that no more than 40 % of the single receptors are rods and there is, in addition, a sizeable population of double cones. The camera lucida drawing reproduced in Fig. 8 was generously supplied by Harold Leeper of the Jules Stein Eye Institute in Los Angeles. It was drawn

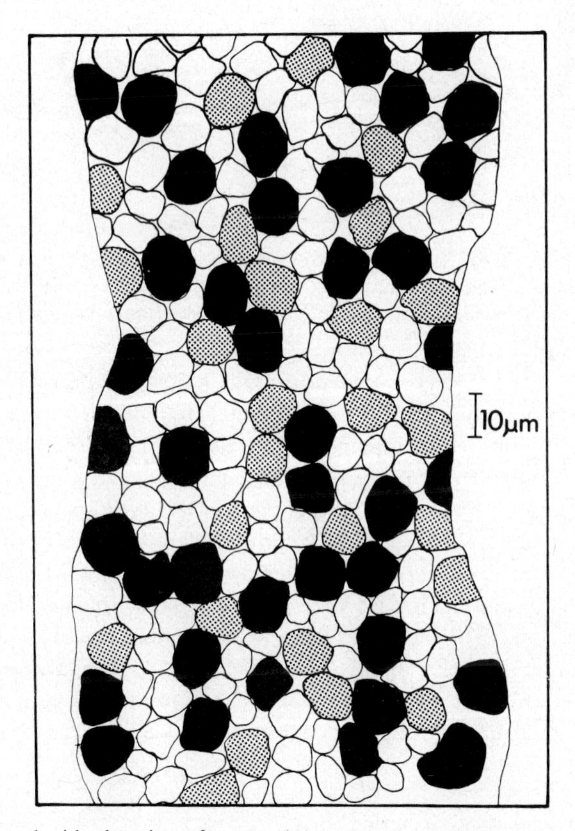

Fig. 8. Camera lucida drawing of tangential section through snapping turtle retina at the level of the inner segments. Single cones are unshaded, double cones are lightly shaded and rods are densely shaded. Photoreceptors were identified on the basis of serial sections.

from a 1 μm tangential section through a retina of the snapping turtle. The section was cut at the level of the inner segments. The cells were identified from serial sections above and below the plane of this one. The unshaded cells were single cones, the lightly shaded cells double cones, while the densely shaded cells were rods. It is clear that in this region of the retina very few of the rods are adjacent to one another. One might therefore expect them to be electrically isolated from one another. However, in recording from perhaps five hundred single rods, we have never encountered any that were not coupled with other rods over a radius of 150 μm. If this is true of the rods in Fig. 8, they must be coupled by processes that extend between them and bypass the intervening cones.

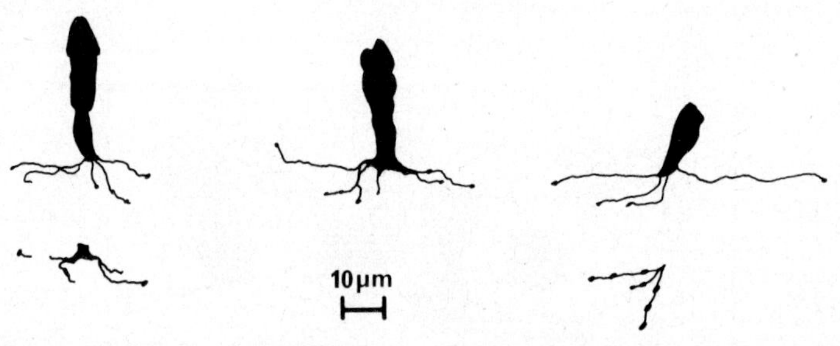

Fig. 9. Camera lucida drawings of 3 rods stained by intracellular injections of Procion Yellow dye. The drawings situated below two of the rods were stained basal processes obtained from adjacent histological sections. The scale mark represents 10 μm.

Turtle rods possess long teleodendria which spread laterally up to 40 μm from the synaptic pedicle.[7, 22] They are illustrated in Fig. 9 by camera lucida drawings of rods stained by injection of the dye Procion Yellow M4R. These teleodendria exhibit both *en passage* and terminal swellings, strongly reminiscent of synaptic structures in other parts of the nervous system. Lasansky has found that similar teleodendria emanating from the cones of *Pseudemys scripta elegans* make contact with other cones.[22] In the carp, the teleodendria of rods make contact both with the spherules and with the teleodendria of neighbouring rods.[23] While we cannot exclude the possibility of junctions between the inner segments of adjacent rods in the snapping turtle, we suggest that an important route by which these rods communicate with one another may be through their teleodendria, and that the swellings may be the sites of gap junctions.

An Electrical Model of the Rod

Several authors have proposed similar electrical models of the rod photoreceptor that are consistent both with measured light-induced resistance changes and with the extracellular current that flows during darkness from the proximal to distal portions of the rod.[2,24,26,27,28] An implicit assumption of these models has been that each rod is functionally independent of its neighbours. However, the models may be readily adapted to be consistent with the electrotonic coupling described in this paper.

Our model which includes electrotonic coupling is shown in Fig. 10. It contains only linear circuit elements and therefore will be applied only to the range of very weak stimuli to which rods respond in a linear manner. Its application will be further limited to two specific cases; the case of a single quantum absorbed in the impaled rod alone and the case when, on average, all rods simultaneously absorb a single quantum. This model is useful because it enables us to quantify some of the electrical parameters of the rod that are important in determining its response to weak stimuli.

In constructing the model, the following assumptions are made:

(1) The cytoplasmic resistance is small compared with the plasma membrane resistance of the rod.[25] Hence the outer segment, and also the proximal region comprising the remainder of the rod, will each be considered isopotential.
(2) Because the extracellular medium is of low resistance,[2] its potential will be taken as uniform and zero.
(3) Since the outer segments of snapping turtle rods are so narrow, we feel confident in assuming that our microelectrodes penetrated only the proximal region of the rod.

In the circuit illustrated in Fig. 10(a), the membrane of the rod outer segment is represented by a resistance R_o, in series with a battery, E_o. R_o represents the cumulative ionic resistances of the membrane and E_o the associated e.m.f. across the membrane. The high resistance ciliary pathway[24] between the inner and outer segment is represented by R_c. The combined surface membrane resistances of the inner segment, cell body, synaptic terminal and teleodendria are represented by R_p, and the e.m.f.'s in this region by E_p. Potentials are recorded at point A.

The effect of light is to increase the outer segment membrane resistance, R_o, thereby reducing the inward flow of current i_D, that exists in the dark. This reduction in the dark-current will be termed the photocurrent, i_p.

Since rods are interconnected by low resistance electrical pathways, current can flow laterally from rod to rod. If a single rod absorbs light, a photocurrent will be generated which will tend to flow across the rod's plasma membrane, causing it to hyperpolarize. This creates a potential difference between the stimulated rod and its neighbours. As a result, a portion of the photocurrent will instead flow across the plasma membranes of neighbouring rods via the network of connections. Under these conditions, therefore, the coupling pathways will serve to "shunt" the plasma membrane of the stimulated rod. This situation is represented by Fig. 10(b). The shunt resistance, R_s, represents the equivalent resistance of all the resistive pathways to ground via the lateral connections while E_s denotes the effective, associated e.m.f.

If all rods are illuminated equally, they may be assumed to respond in concert, their responses having equal amplitudes and time courses. Since at no time will there exist a potential difference between them, no net current can flow between them. They will respond, therefore, as though the shunt pathways do not exist. Fig. 10(a) therefore represents the rod under large-field (diffuse) illumination.

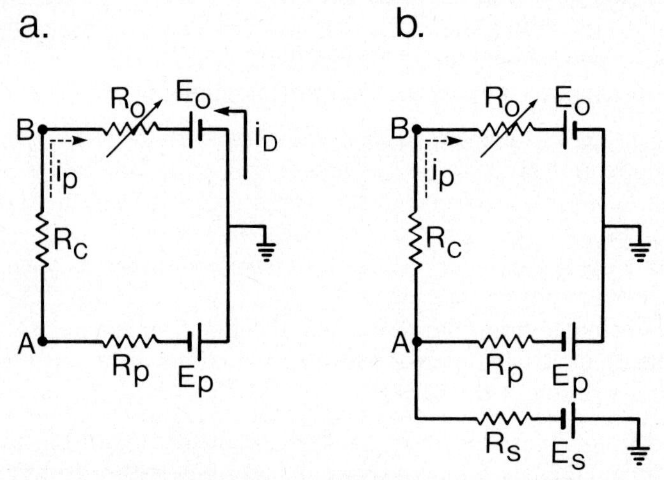

Fig. 10. (a) Circuit model of the rod for the case when all rods within the receptive field absorb a single quantum. (b) Circuit model of the rod for the case when that rod alone absorbs a single quantum. The symbols have the meanings assigned in the text.

These circuits can be used to estimate the magnitude of the photocurrent induced by the absorption of a single quantum, provided the values of the various resistances can be determined. From Fig. 10(a) we see that the under large-field illumination the photocurrent, i_p, resulting from a small change, R_o, in the outer segment membrane resistance will cause a potential change at A, of

$$\Delta V_p = R_p i_p. \tag{1}$$

If only the impaled rod is illuminated, however, the potential change recorded at A will be, from Fig. 10(b),

$$\Delta V_p^* = i_p\left(\frac{R_p R_s}{R_p + R_s}\right). \tag{2}$$

Hence

$$\frac{\Delta V_p^*}{\Delta V_p} = \frac{R_s}{R_p + R_s} \tag{3}$$

Equation (3) indicates that illuminating a single rod will always elicit a response smaller than that produced by diffuse illumination of the same intensity. The reduction will depend upon the relative values of R_p and R_s. The quantal sensitivity of the snapping turtle rod increased approximately 11-fold, from 65 μV/quantum to 720 μV/quantum, as the stimulus was enlarged from 12 μm in diameter. Thus, from eqn (3), the value of R_p must be about 10 times that of R_s to be consistent with our observations.

The input resistance of the rod was determined by injecting current at point A. Since this necessarily polarized the impaled rod relative to its neighbours, a portion of the injected current flowed between them through the lateral pathways. Hence, the circuit over which the injected current flowed to ground is that shown in Fig. 10(b). The input resistance, R_{in}, is therefore given by

$$R_{in} = \frac{R_p R_s (R_o + R_c)}{R_p R_s + (R_o + R_c)(R_s + R_p)}. \tag{4}$$

Measurements of the input resistance, defined as the slope of the steady-state input voltage-current relation at the resting potential, yielded values between 50 MΩ and 100 MΩ. The median value of 80 MΩ will be used for the purposes of calculation. The value of R_p, therefore, can be calculated, provided we have a reasonable estimate of the relative magnitudes of $(R_o + R_c)$ and R_p. Such an estimate can be obtained by considering the value of the resting potential in the dark. The circuit of Fig. 10(a) gives us expressions for the potential at A:

$$V_p = E_o - i_D(R_o + R_c) \tag{5}$$

$$= E_p + i_D R_p \tag{6}$$

where i_D is the dark current. The median value of V_p, from measurements made in some one hundred rods, was −34 mV. We will assume values of E_o and E_p, based on estimates made in the rods of *Necturus*.[24] Hence, letting $E_o = +10$ mV, $E_p = -55$ mV, in eqns (5) and (6) gives

$$R_p = \tfrac{1}{2}(R_o + R_c).$$

Substituting in eqn (a) we obtain:

$$R_p = 920 \text{ M}\Omega \qquad R_s = 92 \text{ M}\Omega.$$

(This value of R_p is seen to be reasonable when we note that the lowest possible value consistent with our data is 880 MΩ, the value obtained when $(R_o + R_c)$ is assumed to be infinite.)

From measurements made on histological sections of the snapping turtle retina we estimate the outer segment of the rod to have a surface area of 200 μm². The remainder of the rod has a surface area of about 1350 μm³. Thus, we estimate the specific resistance of the plasma membrane of the inner segment to be about 1·24 × 10⁴ Ωcm². The specific resistance of the outer segment plasma membrane is probably somewhat lower, the calculated value being about 3·5 × 10³ Ωcm² though we have less confidence in this value since it is more strongly dependent upon the assumed values of E_o and E_p. It is worth noting that if we assume a specific capacitance of 1 μF cm⁻², the membrane time constant of the rod will be close to 10 ms, in good agreement with values measured in rods of snapping turtle[15] and other lower vertebrates.[29,33] Using the above values of R_p and R_s and the appropriate value of quantal sensitivity, the magnitude of the photocurrent generated upon absorption of a single quantum (quantal photocurrent), can be calculated from either eqn (1) or eqn (2). In either case we obtain a value of 0·78 picoamps.

The fact that both these equations yield the same value of i_p illustrates an important point. The quantal photocurrent does not depend upon the spatial distribution of the light but depends only on the capture of a quantum in the rod's own outer segment. The magnitude of the potential change produced by this photocurrent, however, depends upon how much of the current is shunted through the coupling pathways. This, in turn, depends upon the extent to which the surrounding rods are stimulated.

The dark current, i_D, is believed to be carried through the outer segment plasma membrane as an influx of Na^+ ions.[2,24,26,27,28] The photocurrent, i_p, is believed to represent a light-induced reduction in this Na^+ influx. The number of monovalent ions, N, halted by a single photoisomerization is given by:

$$N = i_p \times t_i \times F$$

where F is Faraday's constant and t_i is the integration time of the rod. Dark adapted responses of 1 mV or less reach their peak amplitude about 1·6 seconds after the light flash. This implies an integration time, t_i, of 2·25 s.[5] Thus the calculated quantal photocurrent represents the halting of about 1×10^7 monovalent cations. Korenbrot and Cone[24] estimated that, in the frog rod, 10^7 Na^+ ions were stopped. Hagins *et al.*[2] estimated that $2{\cdot}3 \times 10^6$ ions were stopped as a result of a single photoisomerization in the rat rod. Despite the use of different species and experimental procedures, there is surprisingly close agreement between these three estimates.

On the basis of the relative distribution of rods and cones we earlier suggested that rod–rod coupling is probably mediated by the long teleodendria that emanate from the rod's synaptic pedicle. One naturally wonders whether the space constant of a teleodendrion is sufficiently large to ensure the condition of current along its length. From light micrographs the cross-sectional diameter of the teleodendrion is about 0·5 μm. The cytoplasmic resistivity is assumed to be about 200 Ωcm, about twice the value for frog axoplasm[30] and close to the value for frog muscle.[31] On the further assumption that the surface membrane of the teleodendrion has the same specific resistance as the rest of the proximal region of the rod, we calculate a space constant of 250 μm. Thus, the teleodendria appear to be eminently capable of conducting currents between rods separated by 150 μm.

Consequences of the Electrotonic Coupling of Rods

Being coupled together electrotonically, the rods respond not only to the light they each absorb, but also to the absorption of light by their neighbours. The response of any given rod thus provides a measure of the luminous flux falling on the area of retina that lies within a radius of 150 μm.

At very low intensities, a diffuse stimulus might provide an insufficient number of photons to allow an absorption to occur in every rod. Were the rods not coupled, a proportion of them would fail to respond to such a stimulus. As a result of the coupling, however, rods which are not directly stimulated would still respond provided photons are absorbed in neighbouring rods no more than 150 μm away. As a consequence, under diffuse illumination, the intensity range over which any given rod responds is extended downward by more than an order of magnitude. Of course, whether or not this contributes to the animal's ability to detect large, dim objects will depend on how the rod responses are integrated by the bipolar cells that contact them.

The reliability with which responses, a few tens of microvolts in amplitude, can be transmitted to the bipolar cells is likely to depend upon the level of noise inherent in the transmission process. This can be separated, according to its origin, into presynaptic noise and transmitter noise.

It was recently discovered by Simon, Lamb and Hodgkin that there exists in each receptor a presynaptic source of noise (ref. 32 and see also the papers by Simon and Lamb, p. 291, and by Schwartz, p. 323, in this symposium). Noise is random by definition, and a random fluctuation in presynaptic potential of a given rod is likely to introduce a potential difference between it and its neighbours so that a noise current will flow between them. This noise current will therefore flow in the circuit shown in Fig. 10(b). If this were the only rod with a presynaptic source of noise, the reduction in voltage noise due to shunting could be calculated from the circuit resistances. In the turtle this would imply an 11-fold reduction. Since each rod in effect sums random currents generated by all rods in its receptive field, however, this reduction will be partially offset. The net effect in the turtle will be a reduction in presynaptic noise by a factor somewhat less than eleven.

Any signal to noise improvement depends strongly not only on the size of the stimulus but also on its intensity. When all rods absorb equal numbers of quanta, as in the case of large-field stimuli of moderate intensity, the photocurrent generated by each rod flows in the circuit of Fig. 10(a). Thus while presynaptic voltage noise will be reduced by shunting, responses to moderately intense, diffuse stimuli will not, and the signal to noise ratio will be improved. As the stimulus area is reduced so that more and more of the photocurrent is shunted, this improvement in the signal-to-noise ratio will also be reduced. When only a few isolated rods are excited, by a stimulus of any size, there is probably a net *reduction* in signal-to-noise ratio as a result of coupling.

Hecht, Shlaler and Pirenne showed that stimulus intensities at the human visual threshold are so low that only a few isolated photons are absorbed within the illuminated area. At such low intensities, therefore, even under large-field stimulation photocurrents will flow through the lateral pathways and responses will thereby be diminished. Hence, the electrotonic coupling of rods is likely to provide little or no improvement at these very low stimulus intensities.

A second major source of noise is the statistical fluctuation in the rate of transmitter release from the rod's synaptic terminal. The limitations imposed upon the reliability of signal transmission by this transmitter noise were recently discussed by Falk and Fatt,[3] as noted in the Introduction. Their analysis indicated that rod responses of only a few microvolts are much too small to be reliably distinguished from transmitter noise by the bipolar cell. They estimated that reliable detection would require a potential change of at least 1 mV in *every* rod presynaptic to the bipolar cell and that single quantal absorptions in a few widely separated rods, such as occur at the human visual threshold, would be undetectable. By these arguments, a diffuse stimulus that gives rise to an average absorption of one quantum in each rod may be sufficient to elicit a just-detectable response in the turtle bipolar cell since such a stimulus elicits a potential of more than 700 μV in the turtle rods. This value is more than two orders of magnitude larger than the value for rat rods obtained by Hagins *et al.*[2] and is close to the minimum required by Falk and Fatt. These arguments also suggest, however, that the turtle should be unable to detect either small-field stimuli of this intensity or diffuse stimuli of lower intensity.

These predictions may be checked, for example, by making direct intracellular recordings of bipolar cell and ganglion cell responses to such stimuli. If it should be found that turtles, like humans, can detect stimuli smaller and less intense than predicted, then the ability of the bipolar cell to distinguish weak signals from noise must be explained. Recent experiments by Fain imply that this will probably be the case (ref. 34 and also paper by Fain, p. 305, in this symposium).

As a possible explanation, Falk and Fatt suggested that rods might be coupled excitatorily, thereby increasing the total signal input to the bipolar cell. The results described in the present paper show that turtle rods are indeed coupled together, but in a dissipative network, not an excitatory one. We must conclude, therefore, that the electrotonic coupling of rods does nothing to resolve the question of how single quantal absorptions in a few widely separated rods might be signalled to the bipolar cell. The solution to this intriguing problem will involve further analysis and experimentation. Included in this effort must be a reevaluation of the transmitter noise limitations postulated by Falk and Fatt and an exploration of the characteristics of the synaptic transfer from rods to bipolar cells.

Acknowledgments

We sincerely thank Professor Kenneth T. Brown for his assistance and encouragement throughout this work, for providing us with the superior double-barrelled electrodes used in some of these experiments, and for his help in preparing this manuscript. We are especially grateful to Hal Leeper of UCLA for much helpful information and for generously allowing us to reproduce Fig. 8. Thanks also go to Professor Fred Crescitelli of UCLA for providing us with data used in drawing the difference spectrum of Fig. 2. This work was supported by grant number EY 00468 to Professor Brown from the National Eye Institute.

References

1. S. Hecht, S. Shlaer and M. M. Pirenne 1942, *J. Gen. Physiol.* **25**, 819–840.
2. W. A. Hagins, R. D. Penn and S. Yoshikami 1970, *Biophys. J.* **10**, 380–412.
3. G. Falk and P. Fatt 1972, *In* "Handbook of Sensory Physiology", vol. 7, part 1 (ed. H. J. A. Dartnall) pp. 220–244 Springer-Verlag, Heidelberg.
4. D. A. Baylor, M. G. F. Fuortes and P. O'Bryan 1971, *J. Physiol.* **214**, 265–294.
5. D. A. Baylor and A. L. Hodgkin 1973, *J. Physiol.* **234**, 163–198.
6. G. L. Fain 1975, *Science* (N.Y.) **187**, 838–841.
7. D. R. Copenhagen and W. G. Owen 1976, *J. Physiol.* **259**, 251–282.
8. D. R. Copenhagen and W. G. Owen 1976, *Nature* **260**, 57–59.
9. E. A. Schwartz 1973, *J. Physiol.* **232**, 503–514; 1975, *J. Physiol.* **246**, 617–638.
10. P. Liebman 1972, *In* "Handbook of Sensory Physiology", vol. 7, part 1 (ed. H. J. A. Dartnall) pp. 481–528. Springer-Verlag, Heidelberg.
11. E. A. Schwartz 1975, *J. Physiol.* **246**, 639–651.
12. J. N. Weakly 1973, *J. Physiol.* **234**, 597–612.
13. B. Katz and R. Miledi 1967, *Proc. R. Soc. Lond. B* **167**, 23–28.
14. G. M. Oliveira-Castro and W. R. Loewenstein 1971, *J. Membrane Biol.* **5**, 51–77.
15. E. A. Schwartz 1976, *J. Physiol.* **257**, 379–406.
16. F. S. Sjöstrand 1958, *J. Ultrastruct Res.* **2**, 122–170;
    L. Missotten, M. Appelmans and J. Michiels 1963, *Bull. Mem. Soc. Franc. Ophthal.* **76**, 59–82.

S. E. G. Nillsson 1964, *J. Ultrastruct. Res.* **11**, 147–165; A. I. Cohen 1964, *Invest. Ophthal.* **3**, 198–210; 1965, *J. Anat.* **99**, 595–610;
J. E. Dowling and D. B. Boycott 1966, *Proc. R. Soc. Lond. B* **116**, 80–111;
A. Lasansky 1972, *Invest. Ophthal.* **11**, 265–275;
E. Raviola and N. B. Gilula 1973, *Proc. Nat. Acad. Sci. USA* **70**, 1677–1681.
17. S. Uga, F. Nakao, M. Mimura and H. Ikui 1970, *J. Electr. Micr.* **19**, 71–84.
18. G. M. Gold, G. L. Fain and J. E. Dowling 1975, *Cold Spring Harbour Symp. Quant. Biol.* **40**, 547–561.
19. N. V. Custer 1973, *J. Comp. Neur.* **151**, 35–36.
20. Harold Leeper, personal communication.
21. G. Underwood 1970, *In* "Biology of the Reptilia" (ed. C. Gans and T. S. Parsons), Chapter 2, Academic Press, New York.
22. A. Lasansky 1971, *Phil. Trans. Roy. Soc. B.* **262**, 365–381.
23. P. Witkovsky, M. Shakib and H. Ripps 1974, *Invest. Ophthal.* **13**, 996–1009.
24. J. Korenbrot and R. A. Cone 1972, *J. Gen. Physiol.* **60**, 20–45.
25. G. Falk and P. Fatt 1973, *J. Physiol.* **229**, 185–239.
26. F. S. Werblin 1975, *J. Physiol.* **244**, 53–81.
27. A. J. Sillman, H. Ito and T. Tomita 1969, *Vision Res.* **9**, 1443–1451.
28. R. Zuckerman 1973, *J. Physiol.* **235**, 333–354.
29. A. Lasansky and P. L. Marchiafava 1974, *J. Physiol.* **236**, 171–191.
30. A. L. Hodgkin 1951, *Biol. Rev.* **26**, 339–409.
31. P. Fatt and B. Katz 1951, *J. Physiol.* **115**, 320–370.
32. E. J. Simon, T. D. Lamb and A. L. Hodgkin 1975, *Nature* **256**, 661–662.
33. L. H. Pinto and W. L. Pak 1974, *J. Gen. Physiol.* **64**, 26–48.
34. Gordon Fain, personal communication.

Discussion

G. L. Fain: In constructing your model, you assume that, when the retina is stimulated with diffuse light, the photoreceptors are equipotential and thus effectively isolated from one another. Although this is true for bright stimuli, it is not true for the dim ones you used to obtain the data for your model. When the retina is stimulated with dim light, there are large variations in the number of quanta caught by the receptors and consequently large non-uniformities in the photocurrents the cells generate. Under such conditions large currents will flow between the cells. The receptors will be nearly equipotential, it is true, but only *by virtue* of the currents flowing between them. Thus the assumption that the receptors are effectively isolated from one another under the conditions of your experiments is not true.

W. G. Owen: It must be conceded that, by opting for a conceptually simple model, we may have obscured the true nature of lateral current flow under dim, full-field illumination. We do not believe, however, that this simplification affects the reliability of the values obtained for the model. As you point out, under dim, full-field illumination, the statistical variations in quantal absorption from rod to rod are not reflected in the measured response amplitudes, due to the averaging effect of the lateral current flow. When dim small-field illumination is used, the response amplitude will vary from flash to flash as the number of absorbed quanta varies. It is important, therefore, when dim stimuli are used, that sensitivity be defined as the ratio of the *mean* response amplitude to the *mean* number of quanta absorbed by the illuminated rods. By taking the mean of an adequate series of measurements of response amplitude, a reliable value for the sensitivity can be calculated. All our values were obtained in this way.

Perhaps the most important step in developing our model was the estimation of the value of the shunt resistance, R_s, since many of the later conclusions depend upon the accuracy of that estimate. To do this, we used the ratio of the sensitivity measured under full-field illumination to that obtained when only the impaled rod was illuminated (a value extrapolated from our data). If the lateral flow of current between rods at low intensities had affected this ratio we would have obtained a different ratio in the presence of an illuminated background, where variations in photon capture between rods are not a significant factor. We carefully examined this possibility but found no significant difference in the ratio under the two sets of stimulating conditions.

K. Kirschfeld: The coupling between the rods you have worked out might be considered as a low pass spatial filter mechanism. It would be interesting to know the optical properties of the dioptric system within the retinal area investigated: if this also reduces high spatial frequencies to a similar degree, such a neural low pass filter would be an adequate device for matching optical and neural modulation transfer functions. The gain would be an improved signal to noise ratio at the individual sampling point.

W. G. Owen: I know of no estimate of the spatial modulation transfer function of the snapping turtle eye, so it is difficult to comment directly upon this point. In the region of the retina from which we record, however, the cones interact over a radius of only 60 μm or so, compared with 150 μm for the rods. The cones, therefore, should be more sensitive to higher spatial frequencies than the rods. If the dioptric apparatus operates as a low pass spatial filter it does not seem reasonable that it should attenuate spatial frequencies that the cones are capable of resolving. For this reason I doubt that the spatial filtering properties of the eye are matched to the spatial resolution of the rod network. Indeed, in view of the optical properties of the human eye, I suspect that the quality of the retinal image may be *more* than adequate for the neural network that interprets it.

P. B. Detwiler: My remark is in reference to your comparison of the voltage-current relation generated when both electrodes are in the same cell, with the V-I curve generated when the two electrodes are in different cells. I note that scaling allows only the linear portion of the two curves to be superimposed. Your conclusion that the coupling resistance is linear would be on firmer ground if the nonlinear parts of the V-I curves could be matched.

W. G. Owen: I agree that it would have been more satisfactory if the curves had been entirely superimposable by linear scaling. Taken in conjunction with our observation that the coupling pathway is reciprocal, however, we believe our suggestion that it is also linear to be reasonable. We found that it made no difference into which of the cells the current was injected, the V-I relation measured in the neighbouring cell was always the same. It is difficult to think of a nonlinear circuit that would yield this result.

Such arguments, however, tend to obscure the fact that over the normal operating range of the rods, the curves could be superimposed by linear scaling. This implies that under functional conditions the rods do respond as though connected by linear resistive pathways. Of course, it is important to bear in mind that these curves were measured in the steady state with long d.c. pulses of current and hence our suggestion only applies to slow changes in the rod's potential.

11

Transmission from Photoreceptors to Ganglion Cells in the Retina of the Turtle*

D. A. BAYLOR and R. FETTIPLACE†

Department of Neurobiology, Stanford Medical School, Stanford, California, U.S.A.

Introduction

This paper gives a brief description of recent experiments on signal transmission between photoreceptors and ganglion cells in the turtle retina. A principal technique was to inject electrical currents into a receptor while observing the impulse discharge in a ganglion cell. By substituting current for light and by observing their interaction one can test whether the responses to light of higher-order visual neurons are accounted for by the electrical changes recorded from the receptors with microelectrodes. The use of current also allows the form of the input signals to be simplified and extended. Thus the signals generated by weak lights are slow, probabilistic, and confined to the region negative to the resting potential in darkness. Currents can be used to generate deterministic inputs in the form of rectangular pulses of widely variable duration and either polarity. This strategy is useful in examining the statistical and kinetic properties of retinal transmission in the threshold region.

More complete accounts of the work described here can be found elsewhere.[1,2,3]

* Supported by Grant EY01543 from the National Eye Institute, USPHS.

† Present address: *Physiological Laboratory, Downing Street, Cambridge, England, U.K.*

Methods

Experiments were performed on eyecups isolated from freshwater turtles and maintained in a gassed moist chamber. *Pseudemys scripta elegans* was usually used, but sometimes *Chelydra serpentina* was selected because of its higher rod population density. Intracellular recordings were made from receptors with micropipettes filled with 4 M K acetate, and currents were passed through the recording electrode with a bridge circuit. Impulses in ganglion cells were led off with external microelectrodes consisting of sharpened tungsten wires insulated except at the tips. In some experiments antidromic shocks were delivered to the optic nerve to distinguish between ganglion and amacrine cells, both of which can give impulses. When this test was made, most of the cells gave antidromic spikes and were thus identified as ganglion cells. Certain units gave single impulses to light and to electrical stimulation of single receptors. No cell in this category, when tested, gave an antidromic spike, and it is possible that the entire class is composed of amacrine cells. The term "ganglion cell" will be used somewhat loosely whenever there was no evidence to the contrary.

Results and Discussion

Equivalence and interaction of light and electrical stimulation of a receptor

Impulses could usually be evoked in a ganglion cell by passing polarizing current in an underlying rod or cone. Much larger currents passed just outside a receptor failed to elicit a response. Current thus excited a ganglion cell by changing the membrane potential of a receptor. The impulses evoked by a step of current occurred in a transient burst after a delay of the order of 0·1 s. The discharge in successive trials was probabilistic, but the mean number of spikes per trial rose approximately linearly with the current intensity.

The pattern of a ganglion cell's response to electrical stimulation of a receptor correlated with its response to a small spot of light centered on the same receptor. Short pulses of hyperpolarizing current gave responses at the same phase as short pulses of light, as illustrated in Fig. 1. Above (left) is the response of an off-center cell to a weak pulse of hyperpolarizing current passed in a red-sensitive cone. On the right is the response of the same cell to a pulse of monochromatic red light applied over a 130 μm spot centered on the cone. The impulse in each case occurred at the termination of the stimulus. Below is a similar experiment with an on-center ganglion cell and red-sensitive cone. The response to hyperpolarizing currents (left) occurred at the make of the pulse and the response to light (right) was at "on". In many similar experiments the response to hyperpolarizing pulses never occurred at the opposite phase to the

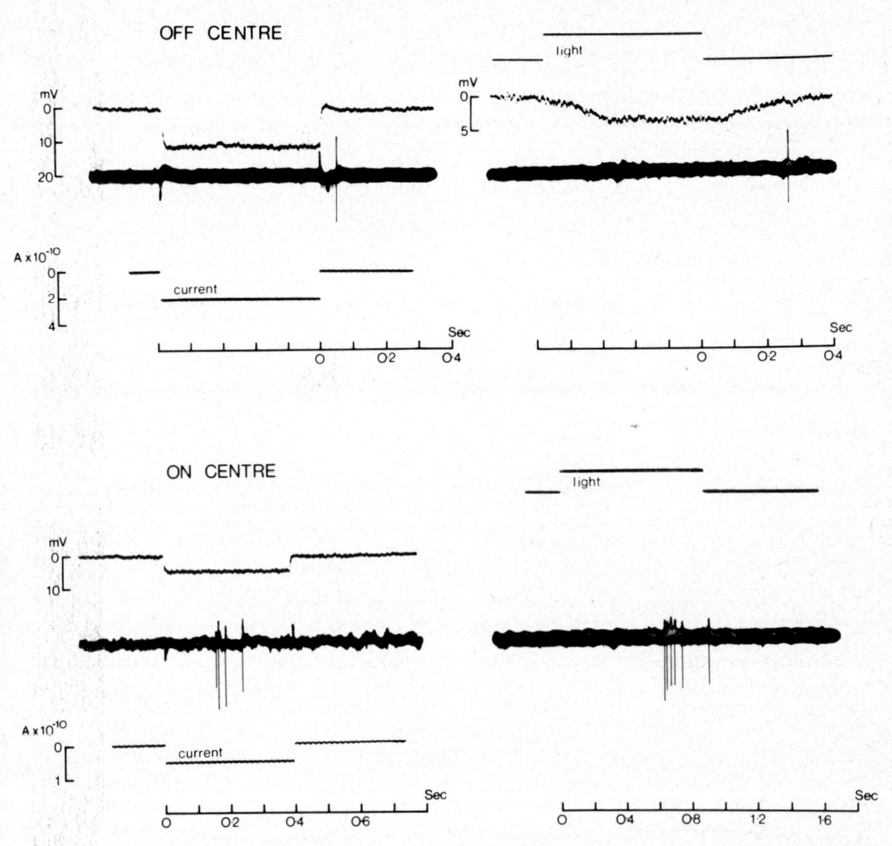

Fig. 1. Patterns of impulses evoked by a spot of red light and by hyperpolarizing current passed in a red-sensitive cone. Upper records—off center cell. On the left the trace above is the cone potential, hyperpolarization downwards, middle trace—external impulse recording, lower trace—current monitor, inward current downwards. On the right the upper trace monitors the light stimulus, a spot 130 μm in diameter, 640 nm, intensity $2{\cdot}7 \times 10^3$ photons $\mu m^{-2}\ s^{-1}$. Lower set of records—on center cell. Light stimulus on the right a 95 μm diameter spot, 640 nm, 26·9 photons $\mu m^{-2}\ s^{-1}$. Capacitative artifacts occur in the impulse-recording traces at the make and break of the current pulses.

light response. About half the ganglion cells which responded to both brightening and dimming of a light, however, responded only at the break of hyperpolarizing currents. The reason for the difference is not clear, but the other on-off cells behaved as expected and responded at both phases of hyperpolarizing pulses.

With light as the stimulus the membrane potential of a receptor is normally confined to the region negative to the dark level. Current allows one to examine the effects of changes in the region positive to the dark level. The off-center cell in Fig. 2 responded to positive-going changes in the potential of the cone, and the transient bursts of impulses triggered by the make of depolarizing pulses were similar in strength to the bursts triggered by the break

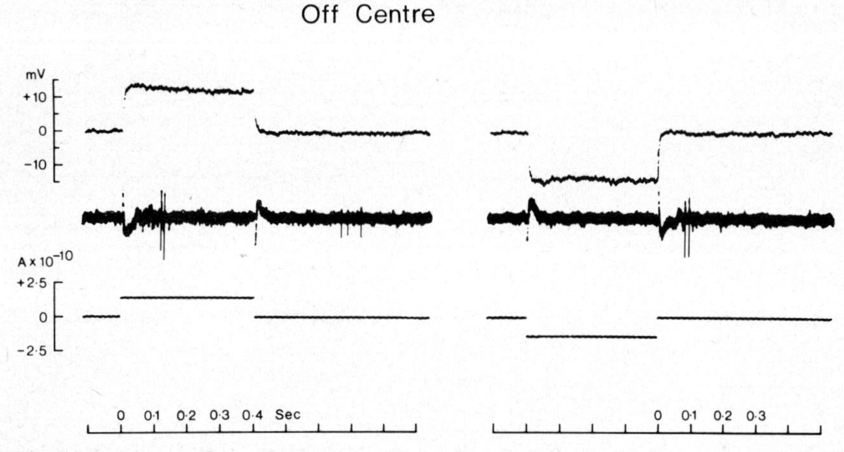

Fig. 2. Responses of an off-center ganglion cell to depolarizing (left) and hyperpolarizing (right) current passed in a red-sensitive cone. Upper traces—displacement of the cone potential from its resting value, middle—external impulse recording, lower—current monitor.

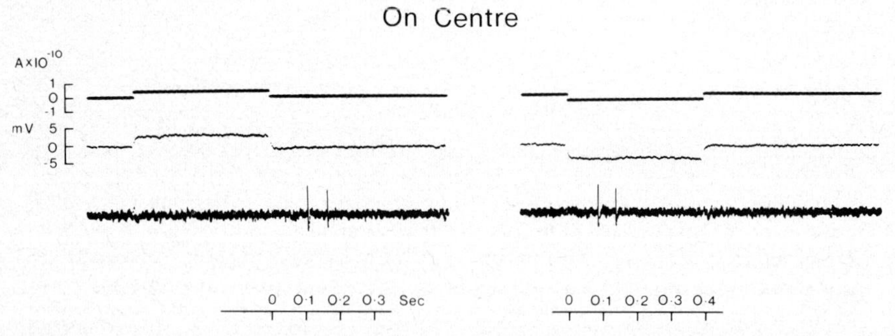

Fig. 3. Responses of an on-center ganglion cell to depolarizing and hyperpolarizing pulses of current in a red-sensitive cone. Upper traces—current monitor, middle—potential of the cone, lower—impulse recording. Current intensity 4×10^{-11} A.

of hyperpolarizing pulses. The analogous behavior of an on-center cell is illustrated in Fig. 3, where the responses to negative-going changes were similar whether they occurred in the region above or below the dark level. Experiments of this kind indicate that the chemical synapses at the receptor terminals can transmit signals in response to changes above and below the dark level of potential. This would be consistent with linear transmission across the synapse between receptor and bipolar cell.

The hyperpolarizations of the receptors generated by light and by extrinsic current had similar quantitative effectiveness in activating a ganglion cell. This comparison was made by determining the size of the hyperpolarizations occurring in a cone with a small (about 50 μm) spot of light of threshold intensity for the ganglion cell, and with a threshold hyperpolarizing current. Within a factor of two the threshold hyperpolarizations generated by light and current were equivalent, allowing for the differing spatial distributions of receptor excitation and assuming that all the receptors of the same kind (a) had equal access to the ganglion cell and (b) exerted additive effects in determining its excitation. The threshold current determined by polarizing a single red cone also agreed well with the total current estimated to be generated by all the red-sensitive cones when the ganglion cell's entire field center was illuminated with a threshold red light. When a small spot of red light was shone on the receptor layer and a current was simultaneously passed in a red-sensitive cone, the two stimuli interacted linearly in determining the ganglion cell's excitation. A subthreshold light and a subthreshold hyperpolarizing current exerted additive effects and brought the ganglion cell to threshold. A subthreshold depolarizing current antagonized the excitatory effect of a threshold light. These interactions of light and current were quantitatively consistent with assumptions (a) and (b) mentioned above.

The magnitude of the threshold current varied widely among different pairs of cells. In the most sensitive pairs, when the receptor was either a rod or a cone, the threshold current was about 2×10^{-11} A and changed the membrane potential of the receptor by 1–2 mV. Since the cone and rod receptors are usually electrically coupled in this preparation (see references 4, 5, 7, 10), a voltage change will have occurred not only in the impaled cell but also in the other receptors to which it is coupled. A current of 2×10^{-11} A can be estimated to correspond to that generated by about 130 photoisomerizations in a red-sensitive cone (taking the peak photocurrent as $1{\cdot}5 \times 10^{-13}$ A per isomerization) or perhaps 50 photoisomerizations in a rod. The large variations in the threshold current were satisfactorily explained by variations in the size of the central portion of the ganglion cell's receptive fields. Thus, there was a positive correlation between the threshold current for a pair and the area of the ganglion cell's receptive field center. This relation implies that the contributions of individual receptors are weighted inversely with the number of receptors which the ganglion cell surveys. The same organization could be inferred using light as the stimulus. Collected results from different ganglion cells showed the threshold intensities of light just covering the field centers to agree within a factor of two even though the areas of the field centers ranged over a factor of about twenty.

Conclusions from these experiments are that:

(1) the hyperpolarization generated by visual transduction is indeed responsible for regulating flow of visual signals to higher order cells;
(2) for small signals transfer between receptors and bipolar cells is linear;
(3) a ganglion cell responds to the sum of the signals issuing from all or nearly all of the receptors in its field center and weights the signals from a single receptor in inverse proportion to the area of its field center.

Statistical and kinetic properties of transmission between receptor and ganglion cell.

Deterministic electrical inputs, consisting of a series of constant current pulses passed in a receptor, yielded a probabilistic ganglion cell response in which the timing and total number of impulses per trial fluctuated considerably. The sequence of occurrence of successes (one or more total impulses per trial) and failures (no impulse) was found to satisfy run tests for a stationary random process when tested at intervals of 5 s for periods up to 6 min. This implies that the mean threshold remained constant and that there was no interaction between trials. Since fluctuation in the detection of very dim lights has sometimes been attributed solely to fluctuation in the stimulus itself,[8] it seemed interesting to try to quantify the magnitude of the "neural" fluctuations revealed in the current-passing experiments. This was done by determining "frequency of seeing" curves relating the percentage of successful responses to the applied current intensity. When the current axis was scaled logarithmically the results from different pairs were similar after normalizing to allow for differences in the threshold currents. The broad electrical threshold characteristic could be fitted by a cumulative Gaussian distribution with the ratio of the mean to the standard deviation about 2·5. These observations indicate that the retinal transmission lines contain considerable intrinsic noise and suggest that under some conditions noise in the retina could dominate the detection of light. For example, in the cone system of the turtle retina a sensitive ganglion cell might have an absolute threshold of about 150 photoisomerizations. The standard deviation in the stimulus would be $\sqrt{150}$ and the signal to noise ratio in the stimulus $150/\sqrt{150} = 12$. The signal to noise ratio of 2·5 from the electrical measurements is considerably smaller than this, pointing to the retina itself as the weaker link.

A dark noise has recently been discovered in turtle photoreceptors.[11] One would expect that the noise at the retinal input should be an important source of the output fluctuation revealed by a current-passing experiment. For a receptor noise of the magnitude reported,[9] and assuming that a ganglion cell integrates the voltage of all the red-sensitive cones in its receptive field center, calculations suggest that the cone noise would be enough to explain all of the fluctuation at the output. Since there are probably other sources of noise between input and output, it seems possible that the synaptic paths may filter the receptor noise; the experiments to be described point to the same conclusion.

By using electrical current the kinetics of retinal transmission can be examined without the constraints normally imposed by the slowness of the visual transduction mechanism. This allows one to explore the filtering effects of the pathway on small signals of variable duration and simple waveform. As indicated in Figs 1, 2 and 3, the ganglion cell discharge evoked by weak steps of current in a receptor reached peak about 0·1 s after the onset of the step and was transient. This suggests that the input was operated on by a delay process which slowed the beginning of the response and a differentiating process which blocked a steady state response. The mechanisms and sites of these operations are not yet clear, but some possibilities can tentatively be ruled out. Since the time constants of the membranes of both receptors and ganglion cells are in the region of 10 ms, the delay cannot be largely due to these factors. Intracellular recordings from ganglion cells show that the excitatory synaptic potentials evoked by steps of light or darkness undergo a pronounced decline which is not seen in the responses of the receptors under similar conditions. This suggests that the differentiation occurs in the synaptic linkage between receptor and ganglion cell, not in the ganglion cells' impulse-generating mechanism nor in the passive electrical properties of the receptors.

To further characterize the kinetics of synaptic transfer in the threshold region strength-duration and latency-duration relations were determined for the "on" and "off" pathways from red-sensitive cones and for the "off" pathway from rods. By interpolating from a series of trials, the size of the threshold current was determined as a function of pulse duration for currents of each polarity. The average latency to the ganglion cell impulses was also measured at each pulse duration.

Results from an experiment of this kind are shown in Fig. 4. The ganglion cell was excited by positive-going changes in the potential of the rod. Threshold current is plotted versus pulse duration on double log scales. The filled circles were obtained with depolarizing currents, which excited at make, open circles with hyperpolarizing currents, which excited at break. The threshold was lowest for relatively long pulses. It rose for depolarizing pulses shorter than about 100 ms and for hyperpolarizing pulses shorter than about 350 ms as indicated by the dotted lines whose intersections define these "utilization times". Latency-duration relations are not shown; with long pulses, however, the latency for both depolarizing and hyperpolarizing currents was about 140 ms. As the duration was shortened the latency with depolarizing pulses shortened and that with hyperpolarizing pulses lengthened. The continuous lines were drawn on the basis of a kinetic model of transmission between the rod and ganglion cell. It was assumed that the input in the rod was transformed by a linear band-pass filter before generating impulses. In the particular model chosen to fit this pair there were seven sequential stages of exponential delay and a single exponential differentiating stage; the time constants of the stages ranged from 10 to 200 ms. Multiple

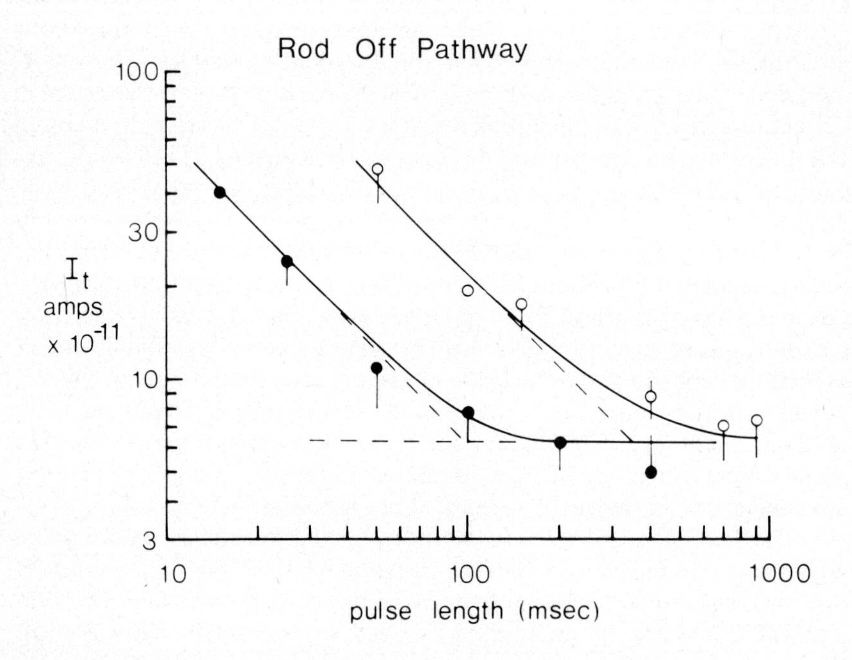

Fig. 4. Strength-duration relations for the "off" pathway from a rod to a ganglion cell. Double log scales. Closed symbols obtained with depolarizing currents which gave responses at make, open symbols with hyperpolarizing currents which excited at break. Dotted lines show method of estimating the utilization times for depolarizing pulses (100 ms) and hyperpolarizing pulses (350 ms). Continuous lines based on a linear filter model of synaptic transfer.

delay stages were needed to account for the small change in latency with changes in pulse duration. Probably many different schemes could be used for describing the kinetics of the pathway, but it would seem that any successful description will have to allow for multiple delays, which could arise from cell time constants, transmitter release and transmitter action kinetics.

The off pathway from red-sensitive cones was formally similar to the off path from rods, but the time scale in the cone path was shorter. This is illustrated in Fig. 5 which compares collected strength-duration relations from rod-ganglion cell pairs (above) and cone-ganglion cell pairs (below). The

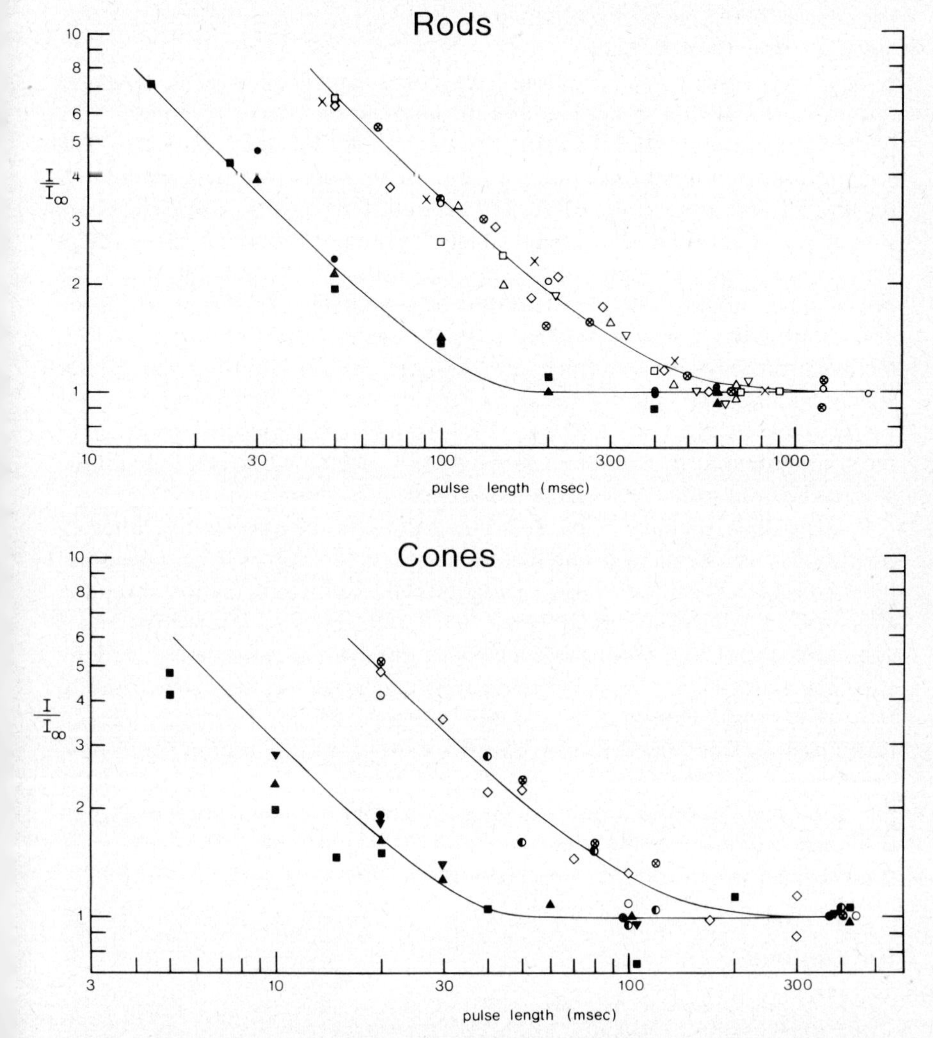

Fig. 5. Collected strength-duration relations for experiments on the off pathways from rods to ganglion cells (above) and red-sensitive cones to ganglion cells (below). Filled symbols obtained with depolarizing pulses exciting at make, open symbols with hyperpolarizing pulses which excited at break. Note the difference in the time scales. Continuous lines based on models of synaptic transfer having same form but with time scale in rod path 3·5 times slower than that in cone path.

curves are based on models of the same kind used in Fig. 4; the time scales differ by a factor of 3·5. This difference is similar to the difference observed in the speed of the linear responses to light in the rods and cones, suggesting that the kinetics of the synaptic paths may be scaled in relation to the kinetics of the signals normally processed.

The on pathway from red-sensitive cones to ganglion cells mediates responses to negative-going changes in the potential of the receptors. This pathway was slower than the off pathway from the red-sensitive cones. The strength-duration relations obtained from the on paths with hyperpolarizing pulses had utilization times of 90–180 ms in different pairs. These times are similar to the duration of the mean effect of a single photon in a red-sensitive cone, about 170 ms under the same conditions.[5] Signals of the form generated by single photons should thus be a relatively effective input for the "on" pathway. It seems possible that the faster "off" pathways, which mediate responses to dimming, are designed for transmitting the briefer signals which the receptors generate in background light.[6] No information is yet available for the "on" pathway from rods. By analogy with the design in the cone system, transmission in the rod "on" path might be extremely slow, with utilization times of perhaps half a second or more.

It is possible that the "on" synaptic pathways may perform a filtering operation which helps to sift light responses in the receptors from dark noise before impulses are initiated in the ganglion cells. From information about the cone noise kindly provided by Drs T. D. Lamb and E. J. Simon (personal communication) and from the kinetics of the on pathway determined by current-passing experiments, we tentatively calculate that the signal-to-noise ratio of a single photon effect would be improved by about 1·5 times in transmission. This value refers to the ratio of the peak amplitude of the photon response divided by the rms value of the receptor noise. A principal reason for the modest degree of the improvement is the similar frequency composition of the single photon effects and the noise. It will be interesting to study transmission in the rod "on" pathway under thorough dark adaptation.

References

1. D. A. Baylor and R. Fettiplace 1977, Transmission from photoreceptors to ganglion cells in turtle retina, *J. Physiol.* **271**, 391–424.
2. D. A. Baylor and R. Fettiplace 1977, Kinetics of synaptic transfer from receptors to ganglion cells in turtle retina, *J. Physiol.* **271**, 425–448.
3. D. A. Baylor and R. Fettiplace 1977, Synaptic drive and impulse initiation in ganglion cells of turtle retina, in preparation.
4. D. A. Baylor, M. G. F. Fuortes and P. M. O'Bryan 1971, Receptive fields of cones in the retina of the turtle, *J. Physiol.* **214**, 265–294.
5. D. A. Baylor and A. L. Hodgkin 1973, Detection and resolution of visual stimuli by turtle photoreceptors, *J. Physiol.* **234**, 163–198.
6. D. A. Baylor and A. L. Hodgkin 1974, Changes in time scale and sensitivity in turtle photoreceptors, *J. Physiol.* **242**, 729–758.
7. D. R. Copenhagen and W. G. Owen 1976, Functional characteristics of lateral

interactions between rods in the retina of the snapping turtle, *J. Physiol.* **259**, 251–282.

8. S. Hecht, S. Shlaer and M. H. Pirenne 1942, Energy, quanta and vision. *J. Gen. Physiol.* **25**, 819–840.
9. T. D. Lamb and E. J. Simon 1976, The relation between intercellular coupling and electrical noise in turtle photoreceptors, *J. Physiol.* **263**, 257–286.
10. E. A. Schwartz 1976, Electrical properties of the rod syncytium in the retina of the turtle, *J. Physiol.* **257**, 379–406.
11. E. J. Simon, T. D. Lamb and A. L. Hodgkin 1975, Spontaneous voltage fluctuations in retinal cones and bipolar cells, *Nature* **256**, 661–662.

12

Synaptic Interactions Mediating Bipolar Response in the Retina of the Tiger Salamander

F. S. WERBLIN

Department of Electrical Engineering and Electronics Research Laboratory, College of Engineering, University of California, Berkeley, California, U.S.A.

Introduction

Studies of visual cortical function show that the visual world is perceived through cues about the presence, movement and orientation of boundaries in the visual environment. The information about boundaries which reaches the cortex has been preprocessed in the retina where certain features in the visual world are abstracted and enhanced. This process begins with synaptic interactions at the outer plexiform layer which serve to optimize the bipolar cell response to small changes in intensity within local regions of receptive fields.[1] Under most ambient conditions, the bipolar cells are poised to respond with maximum sensitivity to small changes in intensity at boundaries.

This paper describes the response properties of the bipolar cells as functions of space, time and intensity. Then some of the synaptic mechanisms which mediate responses are inferred from electrical measurements in light, dark and, under certain pharmacological conditions, at bipolar and horizontal cells. Finally, an hypothesis to account for some of the bipolar response characteristics whose synaptic mechanisms are presently not directly measurable is presented. The results suggest that communication between the slow-potential cells at the distal retina is mediated by synaptic activity similar to that of spike-generating cells, and that the special features of response are generated by the specific interactions between the cells. Details of the experiments, and graphs showing specific data points are presented elsewhere

(Marshall and Werblin,[21] Wunk and Werblin, Skrzypek and Werblin, in preparation).

Graded Response of Rods, Horizontal Cells and Bipolar Cells

Recent studies of bipolar cells in a variety of animals have shown that the bipolar cell receptive field is concentrically organized into antagonistic zones such that illumination at the surround tends to decrease the response to illumination at the center of the receptive field.[2,3,4,5] This phenomenon can be seen in the bipolar cells of tiger salamander by measuring the response to spots of different diameter centered upon the receptive field of the cell. Figure 1 shows that as the diameter of the test spot is increased, up to about 400 μm, the response first increases in magnitude. As the test spot diameter is further

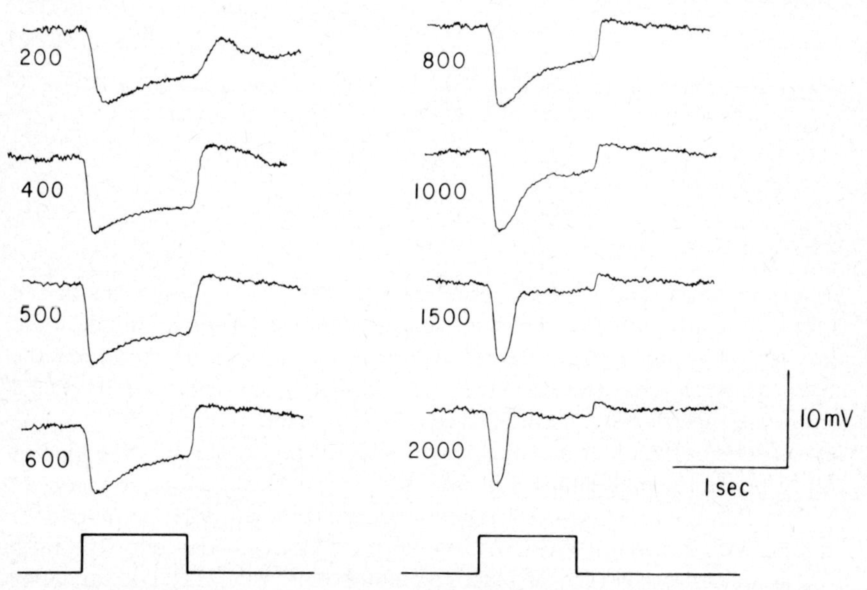

Fig. 1. Bipolar cell responses to spots of increasing diameter. Responses recorded from a hyperpolarizing bipolar cell in the eyecup preparation with single-barrel electrode.[1] The numbers to the left of the recordings specify the spot size. Intensity was fixed to elicit roughly a half-maximal response. The magnitude of the initial peak increases up to spot diameter of 400 μm, then decreases. The magnitude of the plateau following the peak continues to decrease as the spot diameter increases. From these measurements, the receptive field center is about 400 μm in diameter, while the antagonistic surround has no apparent upper bound. The 2000 μm spot, which simulates full field illumination, elicits a sizeable initial transient, but then the bipolar potential returns to a level somewhere between the extremes of the response limits. The specific final level varied from cell to cell. This and subsequent figures are tracings from figures to appear in other publications.

increased, the peak response decreases slightly but there is a dramatic decrease in the plateau response measured after about 150 ms. The plateau response continues to decrease for larger stimulus disks up to the maximum disk diameter of 2000 μm used here.

We have interpreted these data as indicating that the center of the receptive field for the bipolar cell is about 400 μm in diameter, because the response to center illumination increases as the test disk diameter increases up to about 400 μm. The decrease in response for larger diameters after about 150 ms is interpreted to be an expression of the antagonistic surround because it becomes more apparent as the outer margin of the text disk covers larger areas of the regions surrounding the excitatory center. In the experiments described below, we looked at the effects of surround illumination on the graded center response. The test disks stimulating the receptive field center were fixed at 400 μm and the receptive field surround was stimulated with annuli having about 700 μm inner diameter and 2000 μm outer diameter.

Graded intensity-response curves

The photoreceptors, horizontal and bipolar cells each respond with graded levels of polarization which span different log-intensity domains. Figure 2 shows the range of graded responses of the rods, horizontal and bipolar cells to 400 μm spots centered upon the receptive field. Spots of blue light (465 nm) were used to increase the sensitivity of the rods relative to the cones. The response of the bipolar cells, both the depolarizing and hyperpolarizing types, spans a much narrower log-intensity domain than that for either the rods or the horizontal cells. In terms of the hyperbolic tangent relation, these curves can be described by

$$\frac{R}{R_{\mathrm{max}}} = \frac{I^n}{I^n + k^n} \qquad (1)$$

where R/R_{max} is the fraction of the maximal response, I is the intensity of the test flash, and k is the intensity for the half-maximal response. The value of n, which is related to the steepness of the curves, is typically 1·0 for the rods,[6,7] about 0·8 for the horizontal cells, and about 1·4 for the bipolar cells.[1] The horizontal cell log-intensity domain is probably broader than that for the rods alone because horizontal cells receive some cone as well as rod input. The narrow domain of the bipolar cell curves shown here is consistent with some studies in turtle,[5,8] although other studies in turtle show very broad bipolar curves.[8]

The log-intensity domain for the bipolar response is dependent upon the size and position of the test flash. When test disks of diameter greater than the receptive field center are used, the intensity domain is extended as shown in

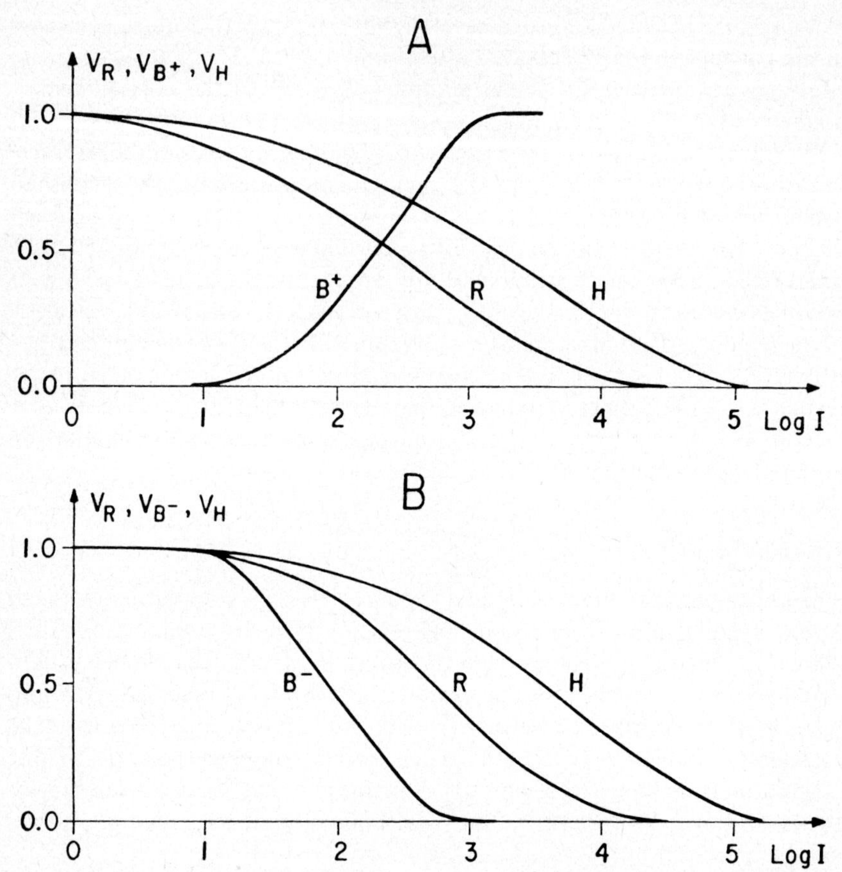

Fig. 2. Graded intensity-response curves for rod (*R*), horizontal cell (*H*) and bipolar cell (*B*). Upper figure shows a depolarizing bipolar; lower figure shows a hyperpolarizing bipolar cell in relation to the rod and horizontal cell responses. Test stimulus was 400 μm spot at the center of the receptive field at 450 nm to decrease the sensitivity of the cones relative to the rods. In all such experiments the domain spanned by the bipolar cells was less than that spanned by the rods or horizontal cells, but the graded responses began to appear at about the same intensities for all three cell types. In experiments using red light, the curves for horizontal and bipolar cells showed similar relations as above, and the intensity domain for the bipolars was considerably narrower than that for the horizontal cells.

Fig. 3. Here, the response to a 300 μm spot, presumably falling totally within the receptive field center, generated a relatively narrow log-intensity-response curve. The response to the larger 1500 μm spot, resembling that for the same size spot in Fig. 1, generated curves with different domains for peak and plateau. The plateau curve spanned nearly 2·5 log units and resembled the response curve for the horizontal cell in Fig. 2. The peak response curve was still rather narrow, like the response to the 300 μm spot, but shifted slightly to the right along the log-intensity axis.

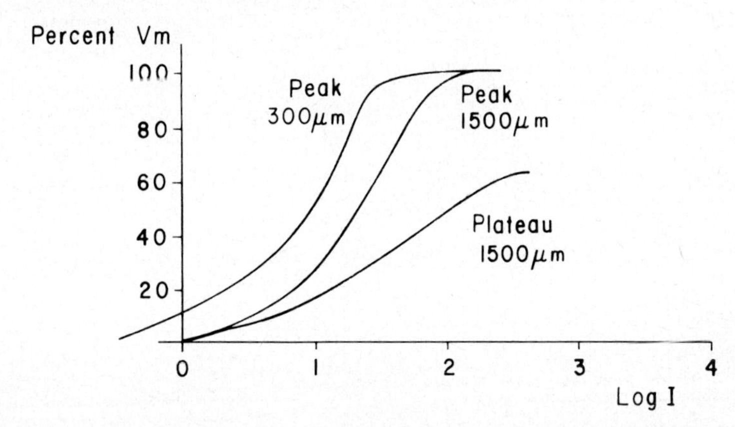

Fig. 3. Intensity-response curve for a hyperpolarizing bipolar with small (300 μm) and large (1500 μm) spots. The peak responses were measured about 100 ms after the stimulus at the obvious peak (see Fig. 1). Plateau was measured 0·5 s after the stimulus during the apparent steady-state. The intensity domain of the peak responses is characteristically narrower than the domain of the plateau responses.

To look more carefully at the effect of the surround illumination upon the center response, the graded center response was measured in the presence of 3 different fixed surround intensities, as shown in Fig. 4. These curves show that the position of the relatively narrow bipolar cell intensity-response curve can be aligned with different regions of the rod domain by changing the level of fixed surround illumination.

As the inner diameter of the annulus was decreased, so that the annulus approached a full field background, the intensity-response curves, like those in Fig. 4, shifted to a position such that the center point was at the background intensity. This suggests that one function of the antagonistic surround is to shift the response curves laterally to align the bipolar response curve with the background level.

In Fig. 5 the sensitivity of rod-to-bipolar transmission is measured by taking the ratio of the slopes of the bipolar response and the rod response verses log-intensity, and plotting it as a function of bipolar cell potential:

$$S = \frac{\mathrm{d}B/\mathrm{d}I}{\mathrm{d}R/\mathrm{d}I}. \qquad (2)$$

The curves in Fig. 5 show that rod-bipolar transmission is optimized when the bipolar cell potential is about half-way between the potential limits of its

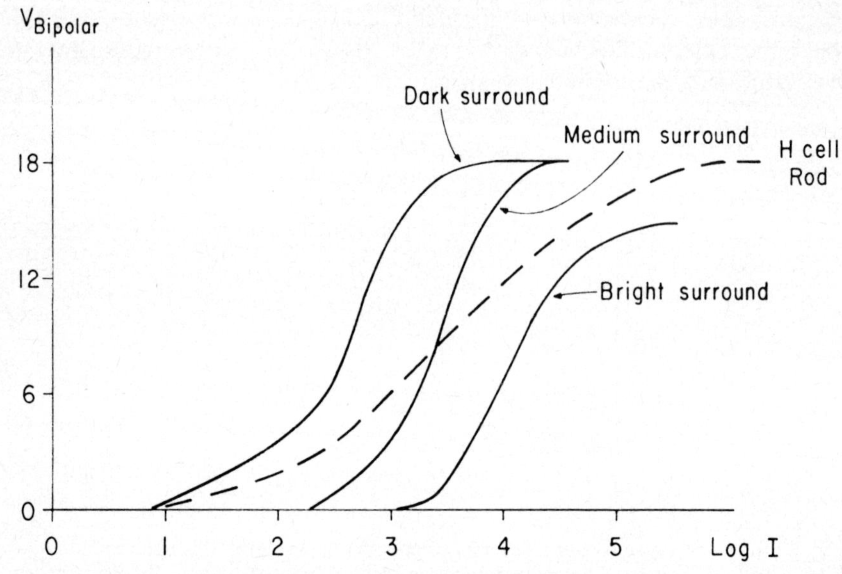

Fig. 4. Repositioning of the intensity-response curves in the presence of fixed intensity surround annulus. The solid curves are responses to 400 μm spot of increasing intensities presented in the presence of an annulus 700 μm ID, 2000 μm OD, at intensities: 0 (dark), 3·2 (medium) and 4·4 (bright) log units relative to the center test flash intensities plotted along the abscissa. Dashed curve shows the rod responses. Although the bipolar response domain spans only a fraction of that of the rods, it can be repositioned with surround illumination to different parts of the rod domain. In all cases the surround annulus intensity was brighter than the domain over which the center response is graded. But if the inner diameter of the surround annulus was decreased, to simulate full field illumination, the mid-point of the response curve approached the annular intensity. This suggests that each of these curves would be centered at the full-field background level.

response range. Under conditions of full-field illumination, one of the functions of the antagonistic surround seems to be to hold the bipolar cell potential at this optimum level.

The role of the antagonistic surround of the receptive field for bipolar cell can be expressed in at least three ways:

(1) The surround reduces the level of maximal light response in the bipolar cell after about 150 ms (Fig. 1).
(2) When included in the test flash, surround illumination extends the intensity domain of the bipolar response plateau (Fig. 3).
(3) Fixed levels of surround illumination align the narrow bipolar intensity response curve with different parts of the response domain of horizontal cells and rods (Fig. 4).

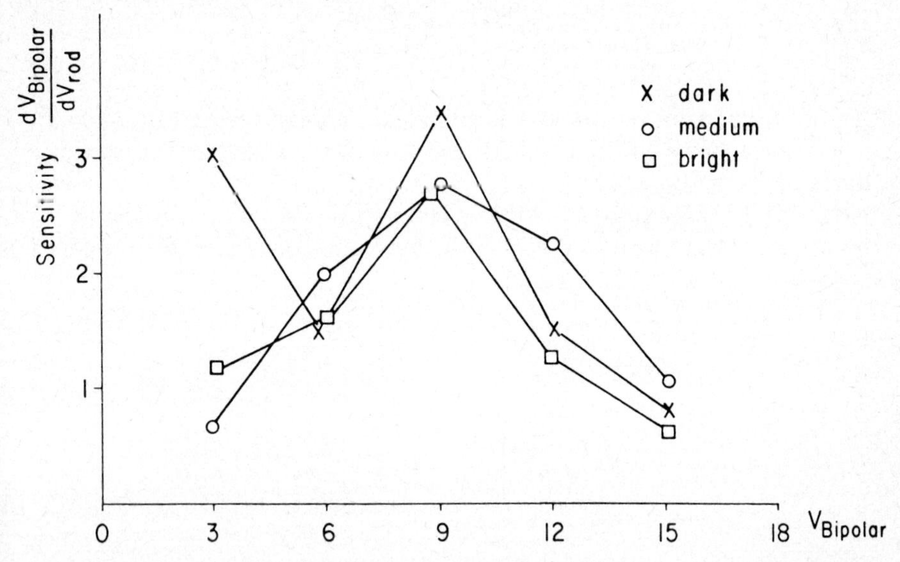

Fig. 5. Sensitivity of the rod-to-bipolar transfer curves. The curves were generated by taking the ratio of the slopes of the intensity-response curves for the rod and bipolar in Fig. 4 at different intensities, then plotting this ratio as a function of bipolar potential (see text). The sensitivity was always maximized for bipolar cell potentials about midway between the response limits.

One striking feature of the relationship between the curves in Fig. 2 is that the response domain of the bipolar cells is quite limited compared to that of the horizontal cells or rods. The near coincidence of the rod and horizontal cell domains in Fig. 2 suggests that rods release transmitter at graded levels over their full range of graded response. The bipolar cells, however, respond only over a fraction of the range of presumed graded release. This indicates that the bipolars are capable of responding to only a fraction of the range of photoreceptor transmission, or that the photoreceptors release graded levels of transmitter to bipolar cells only over a limited range of the photoreceptor response.

The specific region of the rod domain over which the narrower bipolar cell domain is limited can be altered by annular illumination as shown in Fig. 4. This suggests that graded release from rods is possible over the entire range of rod response, but that the *effect* of release at the bipolar cell is somehow limited to the narrower bipolar domain. A possible mechanism for the limited response domain for the bipolar cell will be outlined later. The synaptic mechanisms and pathways which mediate the centering and optimizing functions for the bipolar response, as inferred from electrical measurements at the horizontal and bipolar cells, is discussed first.

Electrical Properties of Horizontal Cells

Current-voltage curves and cobalt

Although we do not yet know the identity of the photoreceptor transmitter(s), it has been shown previously that release of the transmitter is apparently inhibited in the presence of Mg^{++}, Co^{++}, or low Ca^{++}.[9,10,11,12,13] We were able to show that the correct-voltage curve of the membrane in the presence of bright light is indistinguishable from the curve in the presumed

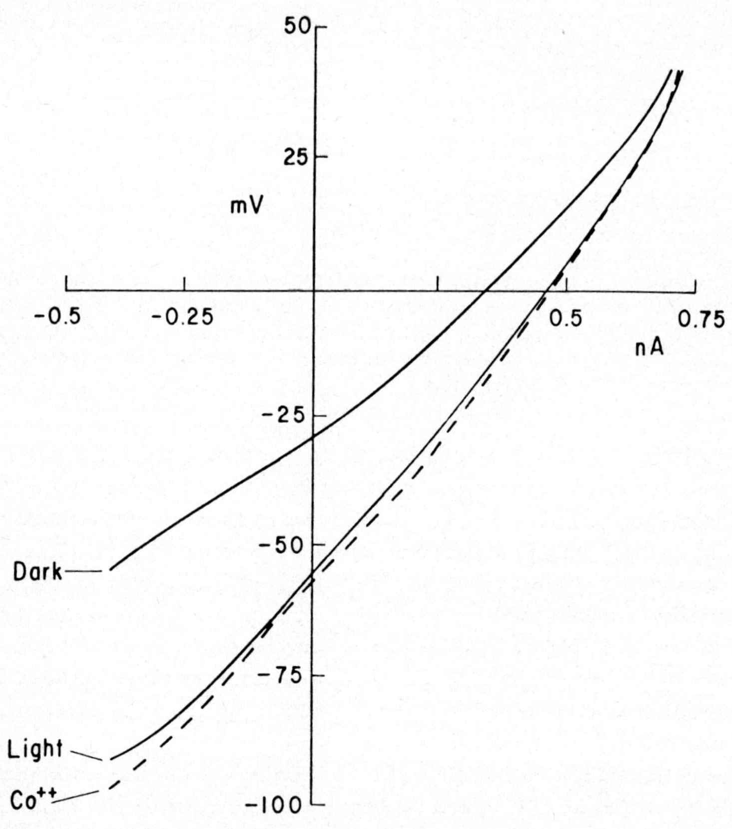

Fig. 6. Current-voltage curves for the horizontal cell in dark, light and in the presence of 2 mM Co^{++} bathed upon the retina. The measurements were made with double-barrel electrodes[1] in the eyecup preparation, superfused with oxygenated ringers containing 2 mM Co^{++}. All curves pivot about an apparent reversal potential of about +50 mV, but this must be taken as an upper limit, because there is probably significant electrotonic decrement between the recording site at the soma and the subsynaptic membrane. The electrical state of the membrane in the presence of bright light is indistinguishable from that in the presence of Co^{++}.

absence of the transmitter (presence of Co^{++}). These experiments further support the hypothesis of Trifonov[14] who suggested that the photoreceptors release synaptic transmitter in the dark, and that release is reduced by light which hyperpolarizes the photoreceptors.

The current voltage curves shown in Fig. 6 were taken from a horizontal cell in the dark, light and in the presence of 2 mM Co^{++}. The curves under all conditions are inward-rectifying, which tends to obscure conductance changes during responses involving permeability changes to ions with a positive reversal potential.[15] However, the slope resistance at any fixed potential level is greater in the light or in the presence of Co^{++} than it is in the dark. Furthermore, the light and Co^{++} curves are nearly identical and both pivot about a reversal potential for the response measured here at about +50 mV. These results also suggest that release of synaptic transmitter from the photoreceptors is reduced with illumination, and that the transmitter depolarizes the horizontal cells by increasing conductance to ions with a reversal potential more positive than the dark level.

Effects of mimetic substances

The amino acids glutamate and aspartate have been implicated as possible transmitter substances because they have been shown to depolarize the horizontal cells in a variety of preparations.[16,17,18,19] The results below show that glutamate, like the endogenous transmitter, depolarizes by increasing conductance to ions with a reversal potential more positive than the dark level. This measurement is an important verification of the earlier inferences about the mimetic action of the amino acids. In other experiments we have shown that acetylcholine, which also depolarizes horizontal cells and has been implicated as a photoreceptor transmitter,[20] acts by *decreasing* conductance[21] and is therefore less likely to be the photoreceptor transmitter. The curves in Fig. 7 show the current-voltage relations for a horizontal cell in the dark, in a normal ionic environment, in the presence of 2 mM Co^{++}, in the presence of glutamate alone, and in the presence of glutamate plus Co^{++} which presumably inhibits release of the endogenous transmitter. Similar results were obtained when light rather than Co^{++} was used to suppress release of transmitter from the photoreceptors. Two important observations can be made here: (1) Glutamate has the same effect on the horizontal cell potential and conductance whether or not the horizontal receives concurrent transmission from the photoreceptors. This indicates that glutamate acts directly upon the horizontal cells and not indirectly, for example, upon the receptors, which in turn affect the horizontal cells. (2) The results suggest that glutamate and the endogenous transmitter share the same limited number of postsynaptic receptor sites on the horizontal cell. If glutamate had access to

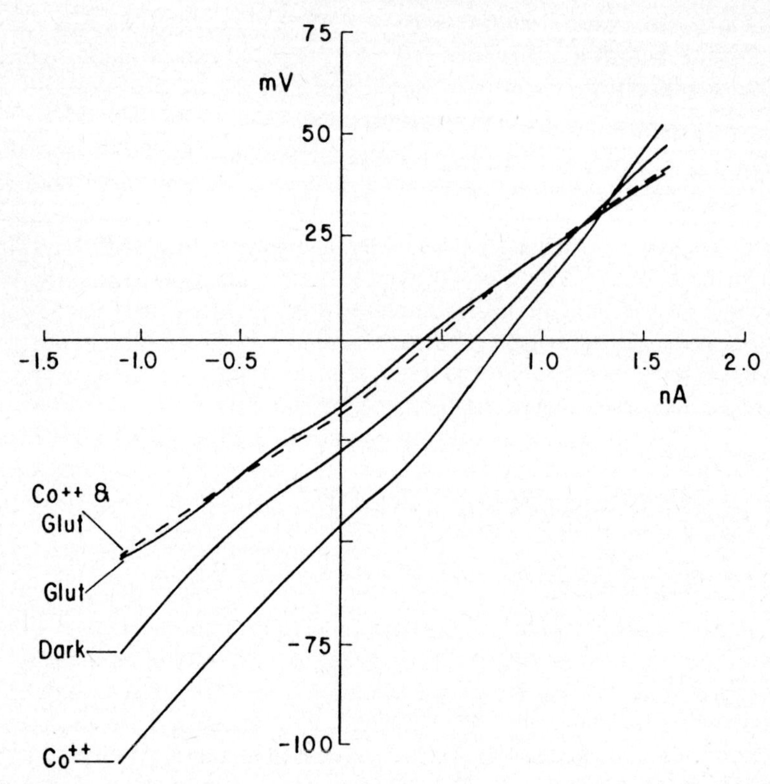

Fig. 7. Current-voltage curves for horizontal cell in the presence of glutamate. Measurements were made in dark, in the presence of 2 mM Co^{++}, in the presence of 10 mM glutamate, and with both Co^{++} and glutamate (dashed curve). Conductance is lowest in the presence of Co^{++} (or light). Conductance increased by 20% with darkness (endogenous transmitter) and by 30% in the presence of glutamate.

sites not available to the endogenous transmitter, then the effect of glutamate plus transmitter (without Co^{++}) would be greater than its effect alone in the absence of the photoreceptor transmitter.

The relatively positive reversal potential for the light response and for glutamate is consistent with recent studies in carp which show that in the presence of glutamate the horizontal cell potential can be changed by about 58 mV with a 10-fold change in sodium concentration.[13] However, other studies in the horizontal cells of salamanders show that ions other than sodium seem to be involved in the depolarizing response.[22]

Since the mimetic substances are added to the bathing medium and must diffuse through the entire depth of the retina before reaching the horizontal cell postsynaptic receptor sites, it is not possible to measure the absolute dose level required to activate the horizontal cells. However, the relative dose required to elicit given fractions of the total horizontal cell depolarization can be obtained. Although the absolute value of dose in the bathing medium required to elicit a response varied from one preparation to another, the ratio of dose eliciting, say a 10 and 90 % response, remained constant. A typical dose

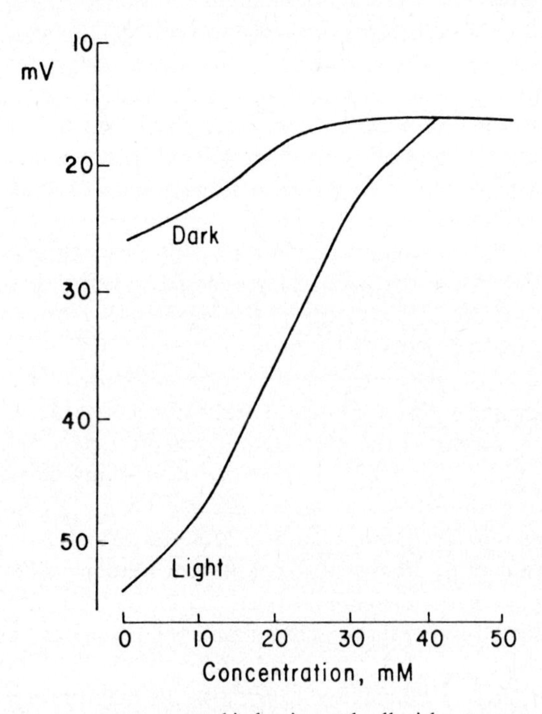

Fig. 8. Dose-response curve measured in horizontal cell with aspartate. Measurements made in the bathed eyecup preparation. Concentrations indicated correspond to levels in the bath, not at the synapse. The eyecup was "washed out" between application of successive doses. Responses were maximal for each dose after about 10 min. Response increases from 10–90 % of maximum over a concentration ratio of less than 10:1.

response curve for aspartate in dark and light is shown in Fig. 8. It is typical in the sense that the response rose from about 10–90 % of maximum over a dose ratio of less than 10:1. The curve is best fit by the hyperbolic relation

$$\frac{R}{R_{max}} = D^n/(D^n + K) \qquad (3)$$

where R/R_{max} is the fraction of maximum response and D is the mimetic concentration, when $n = 2{\cdot}8$. This suggests that more than two molecules of aspartate must be bound to a postsynaptic receptor site in order to open an ionic channel.[23]

The presence of a dose-response relationship at the postsynaptic receptor site requiring more than one transmitter molecule per ionic channel is not unprecedented. For example, two or more transmitter molecules appear to be required for glutamate at the molluscian ganglion cells:[24] for GABA at the crayfish muscle,[25] for acetylcholine at the vertebrate neuromuscular junction.[5(26, 27)] The apparent exponent in eqn (3) above is reduced in these preparations by desensitization to synaptic transmitter at the postsynaptic membrane. We could not measure desensitization in our bathing experiments because the concentration of the mimetic substance increased at the postsynaptic receptor sites only slowly, probably slower than any desensitizing effect. In the slow potential system at the photoreceptor synapse where transmitter release is tonic and maximum in the dark, the postsynaptic membrane is probably desensitized under normal physiological conditions. The exponent in eqn (3), before desensitization, say during recovery from a bright flash, could be even greater than 2·8.

The value of 2·8 for the exponent in the dose-response relation above must be qualified because we do not know if aspartate acts in the same way as the endogenous transmitter at the postsynaptic receptor sites. Assuming for the moment that the binding characteristics for opening ionic channels of aspartate and endogenous transmitter are the same, we can estimate the fraction of open postsynaptic channels and the relative concentration of endogenous transmitter at the postsynaptic membrane in the normal dark retina from Fig. 7. When all available postsynaptic receptor sites are bound by saturating doses of aspartate (or glutamate), the membrane conductance is increased by about 30% from the conductance in the totally unbound state (bright light or Co^{++}). Normally in the dark, the membrane conductance is about 20% greater than in the light. The same estimates can be made from potential levels in Fig. 8 because conductance is proportional to potential in these horizontal cells.[21] This suggests that in darkness about $\frac{2}{3}$ of the available postsynaptic channels are open. Therefore, the photoreceptors in the dark present a concentration of transmitter to the horizontal cells which is mid-way along the dose-response curve, at a position near but somewhat depolarized from maximum dose-response sensitivity in Fig. 8.

Electrical Properties of Bipolar Cells

Current-voltage curves

Studies in a variety of preparations have now shown two types of bipolar cells, with responses that are mirror images of each other with regard to membrane potential change.[2,3,4,5] One type hyperpolarizes in response to illumination at its receptive field center while the other type depolarizes. The current-voltage relations for both types, shown in Fig. 9, indicates that depolarization is always accompanied by a conductance increase with a reversal potential more positive than the dark level. Since the photoreceptor transmitter release is maximal in the dark and reduced by light, the transmitter must *increase* conductance in the hyperpolarizing or off-center bipolar, but *decrease* conductance in the depolarizing bipolar cell.[28] Although the phenomenon of a transmitter-mediated conductance decrease is unusual, it is not unprecedented in either vertebrate[29] or invertebrate[30] systems. These results are supported by recent pharmacological experiments in bipolar cells which show that Co^{++} causes center light-like polarizations in both types of bipolar cells in mudpuppy.[31] Also glutamate and aspartate have been shown to cause polarizations opposing center-light responses in both types of bipolar cell in carp.[32]

Antagonistic surround: feedback or feedforward

The current-voltage curves in Fig. 9 leave some question about how the surround antagonism is mediated. In both types of bipolar cell, illumination of the antagonistic surround seems to reverse the effect of center illumination. This reversal could be mediated in either or both of two ways: the antagonistic surround may activate horizontal cells which either feedback to the photoreceptors and depolarize them, reversing the light-elicited receptor hyperpolarization,[41,34] or the horizontal cells may feed forward to the bipolar cells and modulate conductance to ions with the same reversal potential as those mediating the center response, but in the opposite way.[40]

In order to distinguish between the feedback and feedforward alternatives we took advantage of an earlier, somewhat paradoxical observation, namely, that the surround response does not occur unless the center has been previously illuminated.[2] This phenomenon is referred to below as the "silent surround". Although it has always been found in the mudpuppy, it is common but not always the case in the tiger salamander, and has not been reported in some other preparations.[4,5,8,33] The current-voltage curves in Fig. 9 are inconsistent with the silent surround and feedforward because they show a surround-mediated conductance increase at all potentials with a reversal potential far from the dark level. This suggests that lateral antagonism, at least in silent-surround bipolar cells, is mediated only by feedback from horizontal cells to photoreceptors.

To test the silent-surround cells more directly, we looked for a conductance change in the bipolar cell response during presentation of the surround annulus as shown in Fig. 10. Here, an annular flash, which elicited a full

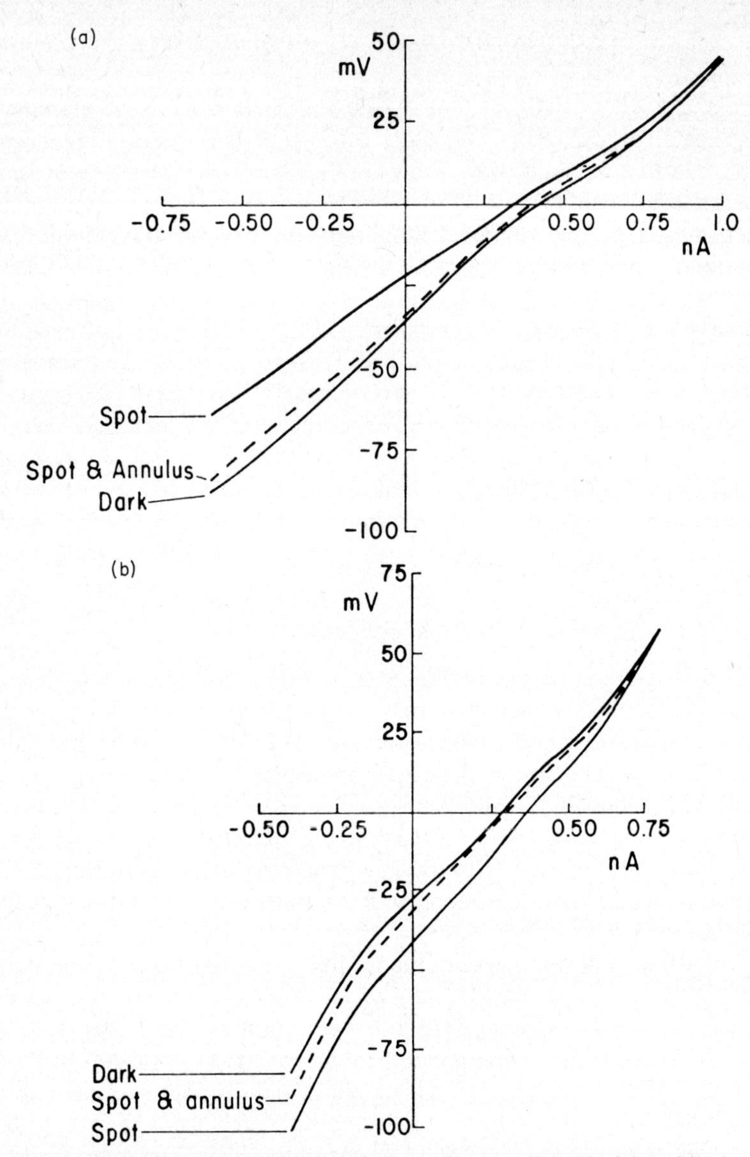

Fig. 9. Current-voltage curves for bipolar cells. (a) Depolarizing bipolar. (b) Hyperpolarizing bipolar. Dashed curve shows the relationship in the presence of center and surround illumination. In both cases, depolarization is associated with an increase in conductance to ions with an equilibrium potential shown here at about +50 mV. This is probably an upper limit because of electrotonic decrement between the recording site at the soma and the subsynaptic membrane. In both cases the effect of added surround illumination is to reverse the effect of center illumination. These curves suggest that the photoreceptor transmitter increases conductance in the hyperpolarizing bipolar cell, but decreases conductance in the depolarizing bipolar cell. The mechanism of surround antagonism is ambiguous in this measurement.

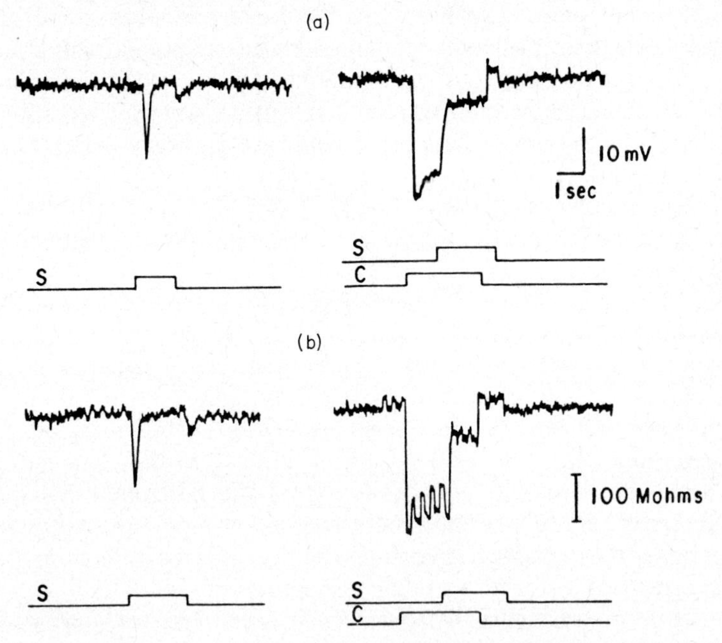

Fig. 10. The silent surround: with the center dark, the surround illumination elicits no antagonistic response, and there is no measurable conductance change. (a) The surround alone (S) elicits no maintained response, but the same surround (left) almost fully antagonizes the response to the center spot (C). (b) With bridge balanced, there is no measurable conductance change due to the surround alone (left), but the same surround completely reverses the center-elicited conductance change (right). This suggests that at least in the silent surround cells the effect of the antagonistic surround is not fed forward to the bipolar cells themselves.

antagonistic response when the center of the receptive field was illuminated, elicited no response when the receptive field was dark. Most important, there was no measurable conductance change in the bipolar cell during the presentation of the annulus alone, even though the same annulus caused a large conductance change when the center had been previously illuminated. These results were obtained for both the on-center and off-center bipolar cells, and are therefore somewhat different from the results of Toyoda[28] in carp. He found a conductance change in the off-center bipolars, but no conductance change associated with the response to surround alone in the on-center bipolars. The presence of a silent surround, and the observations from Fig. 2, showing that the intensity-response curves for center illumination can be shifted to various regions of the receptor response domain are both inconsistent with a feedforward pathway for the antagonistic surround of the bipolar cell receptive field. If the surround pathway were fed forward, the surround could never be silent, and the center response could not be unsaturated by further surround illumination. An exception to this argument might apply if the postsynaptic membrane were saturated electrically, but the reversal potentials shown in Figs 9(a) and 9(b), as well as Toyoda's (1973) estimates of reversal potential for the response in carp, argue against electrical saturation because the range of response potentials is displaced by at least 30 mV from the reversal potentials.

An alternative explanation for the curve shifting in Fig. 2 is that horizontal cells feedback to the photoreceptors, depolarize the terminals, and "reset" the relation between test intensity and photoreceptor graded hyperpolarization. One difficulty with this is that feedback to rods has never been measured, although the curve-shifting in Fig. 2 takes place over a range of intensities where only rods are responsive to the blue test light.

If the surround antagonism is mediated by a feedback synapse to the photoreceptors, then the existence of a silent surround has important implications for synaptic transmission from the horizontal cells to the photoreceptors. Since photoreceptors release synaptic transmitter at the highest rate in the dark, the release should be most affected by feedback, unless the reversal potential for the surround response is at the dark level. This is not inconsistent with the study of O'Bryan[34] who found the reversal potential for the surround component in turtle cones near the dark potential level.

An Hypothesis to Account for Differences in the Log-Intensity Domains of Bipolar and Horizontal Cells

The results presented above summarize some of the recent evidence related to synaptic transmission from the photoreceptors to horizontal and bipolar cells. They suggest that synaptic transmitter is released at high rate in dark, that the effect of the endogenous transmitter is to increase conductance at the postsynaptic membranes of horizontal and off-center bipolar cells, but to decrease conductance in the on-center bipolar cells. However, there is still no direct evidence to account for the difference in the intensity domains spanned by the horizontal and bipolar cell response curves shown in Fig. 2, or for the shifting of the domain of the bipolar curves in the presence of various levels of surround illumination (Fig. 3). The following hypothesis, based upon precedents from other systems of synaptic transmission, can account for the form of the bipolar and horizontal cell responses, and may provide some direction for further study. Some tests of the hypothesis are presented below.

The hypothesis is based upon the following four assumptions:

(1) Synaptic transmitter release from photoreceptors is an exponentially increasing function of membrane depolarization.

(2) Transmitter concentration at the postsynaptic sites is proportional to rate-of-release from the photoreceptors.
(3) The dose-response relation at the postsynaptic membranes of both bipolar and horizontal cells is complex but is similar to that measured experimentally in Fig. 8 and is described by eqn (3).
(4) Horizontal cells feedback to both cones and rods with a sign-inverting synapse to decrease inherent light-elicited changes in potential at the receptor terminals. The justification for these assumptions is given below.

We presently have no direct measure of rate-of-release of transmitter from the photoreceptors. However, in other systems where rate-of-release has been measured, it has been found to be exponentially related to presynaptic membrane depolarization. The rate of release appears to increase roughly tenfold for each 10–15 mV depolarization.[35,36] This exponential relationship has also been proposed for transmitter release from the photoreceptors.[37]

There is presently no way to affirm that transmitter concentration is proportional to the rate-of-release. However, a variety of mechanisms from passive diffusion of the transmitter to active uptake could account for the proportional relationship.

The dose-response curve for binding of photoreceptor transmitter to the postsynaptic receptor sites has not been measured in the bipolar cells because we have not identified the transmitter, and the required experimental time, even with mimetic substances, is prohibitively long. However, we have looked at the dose-response relation for aspartate at the horizontal cell. It is plotted in Fig. 7 and described in eqn (3). The curve suggests that binding of more than one transmitter molecule is required to open a postsynaptic ionic channel, but the results must be taken as conditional. Although aspartate causes the appropriate conductance increase, it may not bind to the postsynaptic receptors like the endogenous transmitter. Figure 7 shows a rather steep dose-response curve which rises from 10–90% of maximum conductance change with a change of transmitter concentration of less than 10 to 1. Similar dose-response relations have been measured at a variety of postsynaptic receptor sites including the neuromuscular junction,[26,27] the GABA receptors in crayfish muscle[25,42] and the glutamate receptors in mollusc.[24] In most other systems, the postsynaptic receptor sites appear to be desensitized by the extended presence of the transmitter. In the retina, where transmitter is always present, if it exists, desensitization must be a continuing physiological condition.

It is possible, with the above assumptions, to derive the appropriate form of the intensity-response relation for the bipolar cell both in the presence and absence of the antagonistic surround. The derivation is shown graphically in Fig. 11. Each quadrant of the graph represents one of the relationships, either measured or assumed, between the components in the chain of events leading

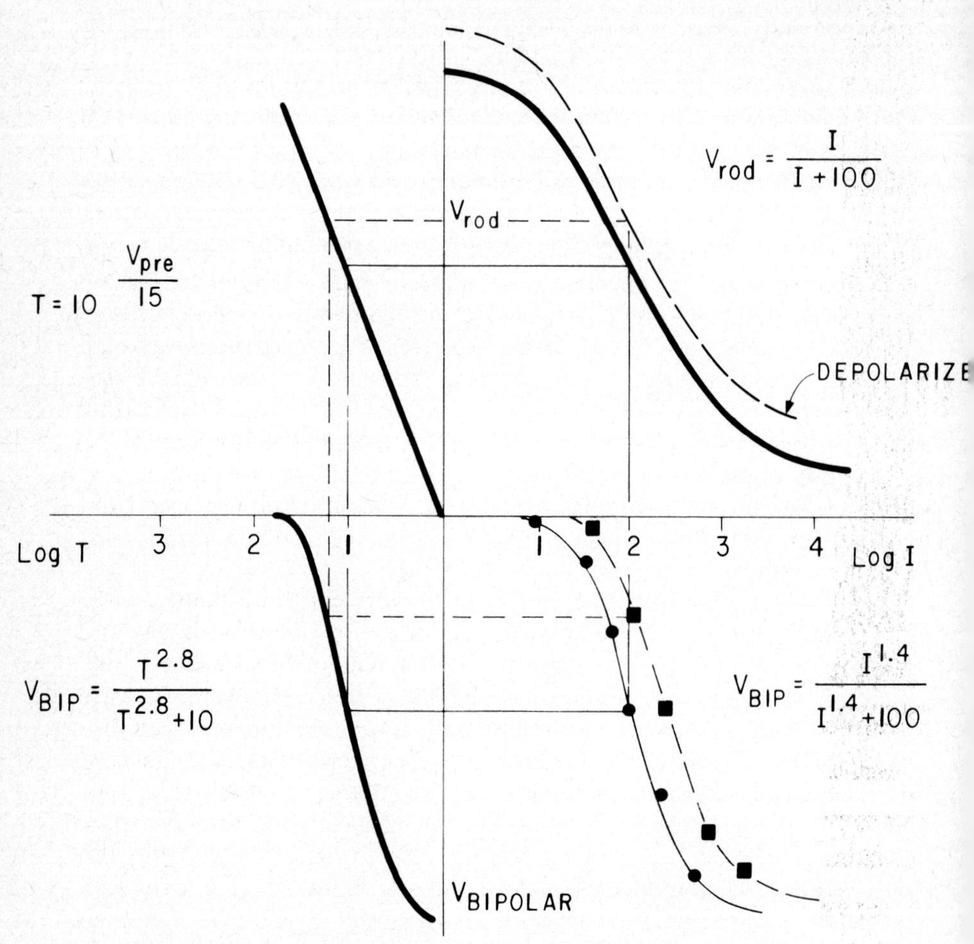

Fig. 11. Derivation of the bipolar cell response on the basis of the hypothesis. Each quadrant displays the relationship, either assumed or measured, between the quantities leading to the bipolar cell response. Proceeding anticlockwise from the upper right, quadrant 1 shows the measured intensity-response function for the rod; quadrant 2 shows the assumed rate of transmitter release as a function of rod polarization; quadrant 3 shows the dose-response curve assumed for the bipolar cell membrane; quadrant 4 shows the predicted intensity-response curves for the bipolar cell (circles). They have the same form as the measured intensity-response curve shown in Figs 2, 3 and 4. The dashed curves represent the system of transformations when the rod response has been shifted in the depolarized direction to simulate depolarizing synaptic feedback to the rod terminals from the horizontal cells. The upward shift of the rod curve, when carried around the release and binding curves, generates a bipolar intensity response curve shifted to the right (squares) like those shown in Fig. 4. These curves show that the four assumptions which comprise the hypothesis are sufficient to account for the form, steepness and position of the bipolar intensity-response curves.

from incident intensity to bipolar polarization. The graphs on the left are assumed as described above, while those on the right are measured intensity-response relations for the bipolar cells and rods taken from Figs 2 and 3.

The bipolar response can be derived from the graphs in the following way. For each intensity, the rod response potential corresponds to a particular rate of transmitter release specified by the curve in the upper-left quadrant, Log T, which in turn generates a specific value of V_{bip} on the dose-response curve at the lower-left. The intersection of this V_{bip} with the Log I defines one point on the derived bipolar intensity-response curve. In Fig. 11 the point corresponding to Log $I = 2$ is plotted. The complete bipolar intensity-response curve was constructed by carrying out this procedure for each intensity. The absolute value of Log T is not known, but only relative values are required here.

This four-quadrant graphical procedure shows that if release from the photoreceptors increases by ten-fold for each 15 mV depolarization, and if the exponent for the dose-response curve of the bipolar is like that for the horizontal cell in eqn (3) where n is 2·8, suggesting that more than two molecules of transmitter must be bound to the postsynaptic membrane for the opening of each ionic channel, then the measured intensity-response relation for the bipolar cell with $n = 1{\cdot}4$, as shown in Fig. 2 and described by eqn (1), will be generated.

The assumption that the horizontal cells feedback to the photoreceptors is sufficient to account for the shift in the position of the bipolar cell intensity-response curve along the intensity axis as shown in Fig. 4. The dashed curves for the rod response in Fig. 11 were generated by initially displacing the photoreceptor intensity response curve to a more depolarized level to simulate a depolarizing feedback to the photoreceptor terminal. This depolarized curve, when projected through the same release and binding curves on the left in Fig. 11, generated a dashed bipolar curve which is displaced to the *right* along the intensity axis, similar to the measured shift due to annular illumination in Fig. 4.

These curves suggest that both the form of the bipolar cell intensity-response curve and its shifting along the log intensity axis, can be accounted for by the four assumptions which are incorporated into the hypothesis.

Test of the assumptions

Since horizontal cells feed back to the photoreceptors with a sign-inverting synapse, their interaction with the photoreceptors tends to reduce the light-elicited activity in the photoreceptor terminals and therefore in all postsynaptic cells driven by the receptors. Small spots at the center of the bipolar cell receptive field tend to accentuate the feedforward synapse relative to the feedback, because the horizontal cells are relatively unaffected by the small spots. Thus, the intensity-response relation measured in bipolars with small spot stimulation simulates a steep, narrow "open loop" condition. On the other hand, when full field illumination is used, all presynaptic photoreceptor terminals are subject to feedback and the response versus intensity of all postsynaptic cells, both bipolar and horizontal, should be shallower. That this is the case is shown in Fig. 3. Intensity-response curves for a bipolar cell are shown with both a small spot and a full field stimulus. The bipolar responses to a small spot of increasing intensity generate a curve which is steep ($n = 1{\cdot}4$ in eqn. 1) reflecting the open loop condition. The responses of the bipolar cell to full field illumination following the initial peak generate a curve which is shallower and similar to that of horizontal cells ($n = 0{\cdot}8$), reflecting the closed loop condition.

The results above suggest that the center-surround antagonism at the distal retina is organized via feedback to the photoreceptor terminals themselves. Therefore, any local region of receptor terminals should show center-surround antagonism. This has been adequately demonstrated in the limiting case of a single cone in turtle,[40] and for the bipolar cells in a number of species, with dendrites which are driven by a local (100–400 μm) population of photoreceptors.[2,3,4,5] However, horizontal cells or their processes with small receptive fields should also possess antagonistic surrounds, and there may be some evidence for this in the work of Lasansky and Valerga,[38] who showed a depolarizing response to an annulus in horizontal cells of tiger salamander. Figure 12 shows the antagonistic surround properties of a narrow-field (500 μm) horizontal cell in the tiger salamander. When scatter to the center of the receptive field is reduced, annular illumination elicits a depolarizing response in horizontal cells. We have never been able to find a reversal potential for horizontal cell responses near the dark level, so this depolarizing response seems to be mediated by the same ionic channels which generate the hyperpolarizing response. These results are consistent with surround-elicited feedback to the photoreceptors. Horizontal cells with larger receptive fields do not show surround antagonism because they are hyperpolarized directly by the surround annulus so that the depolarizing antagonism is masked.

Techniques yet to be developed will be required to test the four assumptions which form the above hypothesis. We have no way at present to measure the level of release of the endogenous transmitter from the photoreceptors, nor do we have clear indication of its identity. Recent progress in measurement of postsynaptic noise may be useful.[39] We would need to identify the endogenous transmitter, and to develop a way of recording for extended periods from bipolar cells to verify the assumption about higher order binding ($n = 2{\cdot}8$, eqn. 3) to postsynaptic receptor sites. The hypothesis suggests that the intensity-response relation for the horizontal cells would be steeper, like that of the bipolar cells, in the absence of the synaptic feedback, but we presently have no way of selectively blocking release of transmitter from the horizontal cells, nor do we have a good indication of the identity of the horizontal cell transmitter. Although feedback has been measured directly in the cones,[40] feedback to rods has not been measured, even though there is good evidence from anatomical studies that horizontal cells feed back to the rods as well as cones in tiger salamander.[41]

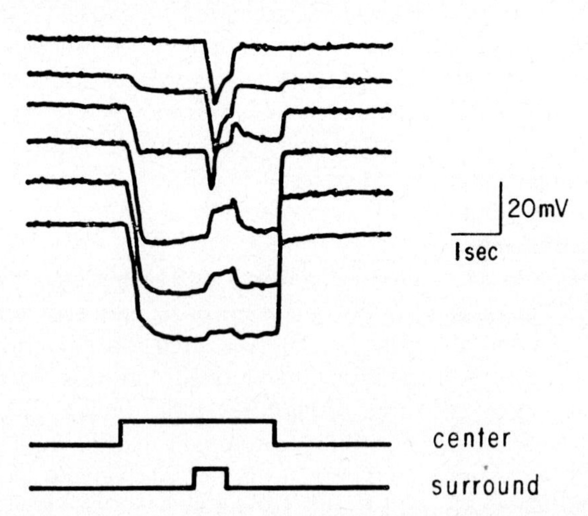

Fig. 12. Antagonistic surround measured in the horizontal cell. If lateral antagonism is fed back to the photoreceptors, then each local region of photoreceptor terminals should have antagonistic center-surround properties, and these should be "read out" by horizontal cells with narrow receptive fields. The responses shown in the figure, taken from one such horizontal cell, support this prediction. The cell was identified as a horizontal cell because its response time was relatively slow, and its intensity-response curve was relatively shallow like the horizontal cell in Fig. 2. Under appropriate conditions of surround and center illumination, the surround elicits a depolarizing response in the horizontal cell. For horizontal cells with broader receptive fields, the depolarizing surround response is obscured by direct hyperpolarizing input from neighboring horizontal cells (Skryzpek and Werblin, in preparation).

Conclusions

The hypothesis outlined above has been used as a bridge between two previously separate areas of retinal electrophysiology. The intensity-response curves for the rods, horizontal cells and bipolar cells were measured in studies viewing the retina as a visual machine.[1,6,44,45] The electrical measurements of cellular function were performed considering the retina as a slow-potential neuronal machine, without any presumptions about the visual function of the elements.[15,21] The linking hypothesis is used as an attempt to explain some aspects of visual function in terms of synaptic mechanisms.

The relatively steep bipolar response curves may be useful in optimizing retinal contrast sensitivity. Under optimal conditions, the contrast sensitivity for the human visual system is about 1 %. If the photoreceptors are operating near the center of their intensity-response functions, in which n in eqn (1) is about unity, then a 1 % change in intensity corresponds to a potential change of about 0·5 mV in the photoreceptors. If bipolar cells are held at the mid-point of their operating range under these conditions, and n in eqn (1) is 1·4, as suggested by Figs 1 and 2, then the potential change in the bipolars, assuming a maximum response range of about 30 mV, will be greater than 1 mV. Since the noise measured in the bipolar cells of the tiger salamander is normally less than 1 mV, synaptic transmission at the outer plexiform layer is probably capable of communicating changes of the order of 1 % from the photoreceptors via a single bipolar cell, but sensitivity may still be limited by noise at the postsynaptic membrane.[37]

If the dose-response curves for the endogenous transmitter at the bipolar cell membrane resemble those for aspartate at the horizontal cells (Fig. 8), and if transmitter is released at a rate that increases ten-fold for each 10–15 mV depolarization of the photoreceptors, then graded transmission can only take place over a very restricted range of photoreceptor terminal potential levels. The potential levels at the photoreceptor terminals must be tightly controlled by feedback from the horizontal cells, according to the hypothesis. Figure 2 shows that a bright surround annulus (which hyperpolarizes the horizontal cells by about 30 mV) can completely "reset" the bipolar cell potential to the dark level in the presence of a three log unit center test flash. The bipolar cell potential will be reset if the photoreceptor terminals are depolarized by the feedback to the same extent they are initially hyperpolarized by light. Perhaps the weakest point in the argument above is the fact that a clear feedback signal has never been measured in rods. The difficulty may be that the rod terminal is separated by too great an electronic distance from the cell body (the recording site) for a clear measurement of the feedback signal.

In systems where synaptic transmission is triggered by spike activity, the precise relationship between pre- and postsynaptic potentials is not as critical because the action potential displaces the presynaptic membrane potential from levels where the rate of release is quite low (near −70 mV) to levels where it becomes limited by the calcium equilibrium potential (above 0 mV). Even in those systems, when graded release is measured as a function of presynaptic potential, the major sensitivity of the synapse occurs for presynaptic potentials in the range of −50 to −30 mV.[36, 43] This corresponds closely to the response potentials over which the photoreceptors release rate is apparently graded.

The slow potential system at the outer plexiform layer seems to obey most of the rules of traditional neurophysiology established in spike-generating systems. The synaptic interactions at the outer plexiform layer, as outlined here, can account for the measured intensity-response functions and their "resetting" by surround input, provided the assumptions of the hypothesis are satisfied. A more detailed analysis of the synaptic mechanisms in the retina seems to require somewhat more refined experimental techniques. The answers to some of the questions raised by the hypothesis await the development of these techniques.

Acknowledgments

The research was supported by Grant EY00561-07 from the National Institutes of Health (Eye Institute). The author was a John Simon Guggenheim Memorial Fellow during the time this work was completed. The results are included in the doctoral theses of Dan Wunk, Larry Thibos, Josef Skryzpek and Larry Marshall, who shared in generating the thought included here. The initial hypotheses of Falk and Fatt[37] provided some of the direction for this research.

References

1. F. S. Werblin 1974, Control of reginal sensitivity: II. Lateral interactions at the outer plexiform layer, *J. Gen. Physiol.* **63**, 62–87.
2. F. S. Werblin and J. E. Dowling 1969, Organization of the retina of the mudpuppy, *Necturus maculosus*: II. Intracellular recording, *J. Neurophysiol.* **32**, 339–355.
3. A. Kaneko 1970, Physiological and morphological identification of horizontal bipolar and amacrine cells in goldfish retina, *J. Physiol.* **207**, 623–633.
4. N. Matsumoto and K. I. Naka 1972, Identification of intracellular responses in the frog retina, *Brain Res.* **42**, 59–71.
5. E. A. Schwartz 1974, Responses of bipolar cells in the retina of the turtle *J. Physiol.* **236**, 211–224.
6. R. A. Normann and F. S. Werblin 1974, Control of retinal sensitivity: I. Light and dark adaptation in rods and cones, *J. Gen. Physiol.* **63**, 37–61.
7. A. Lasansky and P. Marchiafava 1974, Light-induced resistance changes in retinal rods and cones of the tiger salamander, *J. Physiol.* **236**, 171–191.
8. S. Yazulla 1975, Cone input to bipolar cells in the turtle retina, *Vision Res.* **16**, 737–744.
9. J. E. Dowling and H. Ripps 1973, Effect of magnesium on horizontal cell activity in the skate retina, *Nature* (*Lond.*) **242**, 101–103.
10. L. Cervetto and M. Piccolino 1974, Synaptic transmission between photoreceptors and horizontal cells in the tutle retina, *Science* **183**, 417–419.
11. J. N. Weakly 1973, The action of cobalt ions on neuromuscular transmission in the frog, *J. Physiol.* **234**, 597–612.
12. Yu. A. Trifoniv, A. L. Byzov and L. M. Chailahian 1974, Electrical properties of subsynaptic membranes of horizontal cells in fish retina, *Vision Res.* **14**, 229–241.
13. A. Kaneko and H. Shimazaki 1975, Effects of external ions on the synaptic

transmission from photoreceptors to horizontal cells in the carp retina, *J. Physiol.* **252**, 509–522.
14. Yu. A. Trifonov 1968, Study of synaptic transmission between photoreceptors and horizontal cells by means of electric stimulation of the retina, *Biofizika* **13**, 809–817.
15. F. S. Werblin 1975a, Anomalous rectification in horizontal cells, *J. Physiol.* **244**, 639–657.
16. L. Cervetto and E. F. MacNichol 1972, Inactivation of horizontal cells in turtle retina by glutamate and aspartate, *Science* **178**, 767–768.
17. A. Kaneko and H. Shimazaki 1975, Synaptic transmission from photoreceptors to bipolar and horizontal cells in the carp retina, *Cold Spring Harbor Symp. Quant. Biol.* **40**, 537–546.
18. M. Marakami, K. Ohtsu and T. Ohtsuka 1972, Effects of chemicals on receptors and horizontal cells in the retina, *J. Physiol.* **227**, 899–913.
19. J. E. Brown and L. H. Pinto 1974, Ionic mechanisms for the photoreceptor potential of the retina of *Bufo Marinus*, *J. Physiol.* **236**, 575–591.
20. D. M. K. Lam 1972, Biosynthesis of acetylcholine in turtle photoreceptors *Proc. Nat. Acad. Sci.* **69**, 1897–1991.
21. L. M. Marshall and F. S. Werblin, in preparation.
22. G. Waloga and W. L. Pak 1976, Horizontal cell potentials: Dependence on external sodium ion concentration, *Science* **191**, 964–966.
23. J. I. Hubbard, R. Llinas and D. M. J. Quastel 1969, "Electrophysiological Analysis of Synaptic Transmission", Edward Arnold, London.
24. Y. Oomura, H. Ooyama and M. Sawada 1974, Analysis of hyperpolarizations induced by glutamate and acetylcholine on onchidium neurons, *J. Physiol.* **243**, 321– 341.
25. A. Feltz 1971, Competitive interaction of g-guanidino propionic acid and g-aminobutyric acid on the muscle fibre of the crayfish, *J. Physiol.* **216**, 391–401.
26. F. Dreyer and K. Peper 1975, Density and dose-response curve of acetylcholine receptors in frog neuromuscular junction, *Nature* **253**, 641–643.
27. H. C. Hartzell, S. W. Kuffler and D. Yoshikami 1975, Postsynaptic potentiation: Interaction between quanta of acetylcholine at the skeletal neuromuscular synapse, *J. Physiol.* **251**, 427–463.
28. T. Toyoda 1973, Membrane resistance changes underlying the bipolar cell response in the carp retina, *Vision Res.* **13**, 283–294.
29. F. S. Weight and J. Votava 1970, Slow synaptic excitation in sympathetic ganglion cells: Evidence for synaptic inactivation of potassium conductance, *Science* **170**, 755– 758.
30. H. M. Gerschenfeld and D. Paupardin-Tritsch 1974, Ionic mechanisms and receptor properties underlying the responses of molluscan neurons to 5-hydroxytryptamine, *J. Physiol.* **243**, 427–456.
31. R. F. Dacheux and R. F. Miller 1976, Photoreceptor-bipolar cell transmission in the perfused retina eyecup of the mudpuppy, *Science* **191**, 963–964.
32. M. Murakami, T. Ohtsuka and H. Shimazaki 1975, Effects of aspartate and glutamate on the bipolar cells of the carp retina, *Vision Res.* **15**, 456–458.
33. A. Kaneko 1973, Receptive field organization of bipolar and amacrine cells in the goldfish retina, *J. Physiol.* **235**, 133–153.
34. P. M. O'Bryan 1973, Properties of the depolarizing potential evoked by peripheral illumination in cones of the turtle retina, *J. Physiol.* **235**, 207–223.

35. A. W. Liley 1956, The effects of presynaptic polarization on the spontaneous activity of the mammalian neuromuscular junction, *J. Physiol.* **134**, 427–443.
36. B. Katz and R. Miledi 1967, A study of synaptic transmission in the absence of nerve impulses, *J. Physiol.* **192**, 407–436.
37. G. Falk and P. Fatt 1973, Limitations to single-photon sensitivity in vision. *In* "Lecture Notes in Biomathematics", vol. 4, Physics and Mathematics of the Nervous System (eds. M. Conrad, W. Guttinger and M. Dal Cin) pp. 171–204. Springer-Verlag, Berlin.
38. A. Lasansky and S. Vallerga 1975, Horizontal cell responses in the retina of the largal tiger salamander, *J. Physiol.* **251**, 145–165.
39. E. J. Simon, T. D. Lamb and A. L. Hodgkin 1975, Spontaneous voltage fluctuations in retinal cones and bipolar cells, *Nature* **256**, 661–662.
40. D. A. Baylor, M. G. F. Fuortes and P. M. O'Bryan 1971, Receptive fields of cones in the retina of the turtle, *J. Physiol.* **214**, 265–294.
41. A. Lasansky 1973, Organization of the outer plexiform layer in the retina of the tiger salamander, *Phil. Trans. R. Soc. B* **265**, 471–489.
42. A. Takeuchi and N. Takeuchi 1967, Anion permeability of the inhibitory postsynaptic membrane of the crayfish neuromuscular junction, *J. Physiol.* **191**, 575–590.
43. A. R. Martin and G. L. Ringham 1975, Synaptic transfer at a vertebrate central nervous system synapse, *J. Physiol.* **251**, 409–426.
44. F. S. Werblin 1971, Adaptation in a vertebrate retina: Intracellular recording in *Necturus*, *J. Neurophysiol.* **34**, 228–241.
45. L. Thibos and F. S. Werblin, Receptive field properties of the steady antagonistic surround in the mudpuppy retina. In preparation.

Discussion

Peter B. Detwiler: Your current voltage curves indicate that the input resistance of horizontal cells in darkness is about 100 MΩ. This is rather surprising when one considers the fact that horizontal cells are members of an electrically coupled two-dimensional network. Is it possible that the horizontal cells you have chosen to study are not electrically coupled? If this is the case I have the following additional questions. (1) How do you identify an electrically isolated horizontal cell? (2) What proportion of the horizontal cells in Tiger Salamander is isolated? (3) Are isolated horizontal cells noisy and does glutamate influence their noise?

F. S. Werblin: The horizontal cells reported here were abstracted from a more complete study of horizontal cells to be published shortly (Marshall and Werblin, in preparation). We selected those with the narrowest receptive fields, about 500 μm in diameter, because these had also the lowest extrapolated reversal potentials for the response (near +50 mV). The finding that the horizontal cells with broader receptive fields (up to 2000 μm in diameter) had reversal potentials extrapolating to nearly +200 mV, suggests that part of the response in these cells is associated with a synaptic input which does not alter membrane conductance. It is on the basis of this kind of reasoning that we think the narrow-field horizontal cells are not coupled and that the broad-field horizontal cells are electrically coupled. Against this argument are the finding that all horizontal cells in tiger salamander have extremely high input resistances (averaging about 80 MΩ), and that there is no clear correlation between input resistance and receptive field size. It may be that in tiger salamander the horizontal cells are coupled by chemical as well as electrical synapses, and Lasansky has shown evidence for both in a recent anatomical study. The narrow field horizontal cells reported here comprise about 25 per cent of all horizontal cells studied. We have not looked carefully at the effect of glutamate on noise in horizontal cells.

A. Kaneko: Since the same transmitter substance is supposed to work on the two types of postsynaptic cells, namely the horizontal cell and the bipolar cell, one should expect a similar intensity-amplitude relationship in both types of cells. In your presentation, the intensity-amplitude curve for the bipolar cell was steeper than that for the horizontal cell. How do you interpret this difference?

F. S. Werblin: The hypothesis outlined in this paper suggests that the relationship between photoreceptor and either bipolar- or horizontal-cell potential is controlled by both the feedforward synapse from the photoreceptors and the feedback synapse from the horizontal cells to the photoreceptors. The observation that the intensity-amplitude relation is steeper for the bipolar-cell response under some conditions than for the horizontal-cell response supports the hypothesis. There are two conditions under which the feedback synapse from horizontal cells to photoreceptors is not operating: one is the condition of small-spot illumination, which is relatively ineffective in driving horizontal cells, the other is during the first 100 ms of the bipolar response, when feedback from the horizontal cells has not yet developed (see Fig. 1). Under these conditions the "gain" from photoreceptors to bipolars is high, and the amplitude-response curves for the bipolar cell are steep, as you point out (Fig. 3). However, when the stimulus spot is large, encompassing the surround of the bipolar-cell receptive field, and one waits until the antagonistic surround has been activated, the intensity-response curve for the bipolar is shallow, resembling that for the horizontal cell (Fig. 3). The crucial test of the hypothesis will come when we learn how to block the effect of the horizontal-cell transmitter itself, because then the amplitude-response curve for the horizontal cell should become steeper.

13

Responses of Second-order Neurons to Photic and Electric Stimulation of the Retina

J. TOYODA, M. FUJIMOTO and T. SAITO

Department of Physiology, St. Marianna University, School of Medicine, Takatsu-ku, Kawasaki, Japan

Introduction

Photoreceptors in the vertebrate retina respond to light with hyperpolarization. They differ in this respect from most sensory receptors that respond to stimulus with depolarization. The question is how such an electrical signal is transmitted to second-order neurons, namely bipolar cells and horizontal cells. Histological studies on the synaptic structure at photoreceptor terminals suggest that the transmission is chemically mediated.[1] Generally in the nervous system, transmitter substances are liberated by depolarization of presynaptic terminals. If the same rule is applied to the retina, the release of a transmitter from photoreceptors must be facilitated in the dark when they are relatively depolarized and be suppressed in the light when they are hyperpolarized. Trifonov[2] first proposed the hypothesis that horizontal cells are kept depolarized in the dark by the excitatory transmitter released from receptors and become hyperpolarized by illumination which hyperpolarizes receptors and consequently suppresses the release of transmitter. There are several studies on the properties of horizontal cells that support this hypothesis. For example, transretinal current in the direction to depolarize receptor terminals caused a transient depolarization of horizontal cells.[2,3] The membrane conductance of horizontal cells in the mudpuppy was higher in the dark than in the light, suggesting an action of a transmitter in the dark.[4] Application of chemical agents, which is known to block synaptic transmission, hyperpolarized the horizontal cell and abolished the response to light.[5,6] The membrane potential of horizontal cells is dependent on the extracellular sodium ions.[7,8]

Histologically, both bipolar and horizontal cells are postsynaptic to receptors and their dendritic processes constitute a triad structure within the same terminals. The simplest assumption, therefore, is that the transmitter released from receptors acts simultaneously on both horizontal and bipolar cells. However, the electrical properties of bipolar cells are known to be rather complex. There are two major types of bipolar cells. One of them, the off-center cell, responds to light spot on the recording site with hyperpolarization and the other, the on-center cell, responds to the same stimulus with depolarization. It is reported that the hyperpolarization response of off-center cells accompanies an increase in input resistance whereas the depolarization response of on-center cells a decrease in input resistance. Both responses are enhanced by hyperpolarizing extrinsic current and suppressed by depolarizing current, and their equilibrium potential has been estimated to be near zero membrane potential.[9] These results indicate that the hypothesis on the synaptic transmission from receptors to horizontal cells is applicable to off-center cells but it needs some modifications to explain the response properties of on-center cells. Admitting that the transmitter is released in the dark, explanation of the two types of bipolar cell responses is possible only if we assume that the same transmitter opens excitatory subsynaptic channels of off-center cells as well as of horizontal cells but closes those of on-center cells.[9,10] The present experiments were attempted to test this working hypothesis in more detail by methods affecting the synaptic transmission such as transretinal electric stimulation and application of synaptic blocking agents.

Methods

Preparations used in the present experiments were either the opened eye-cup or the isolated retina of the carp, *Cyprinus carpio*. After carefully draining off the vitreous humor, the eye-cup preparation was placed on a chlorided silver plate serving as an indifferent electrode. Moist oxygen was supplied continuously into a lucite box which was used to cover the preparation. Another chlorided silver ring was placed on the vitreous side of the retina for transretinal current stimulation. A pair of two recording micropipettes were lowered vertically from above to the center of the ring electrode. One of them, which serves as a reference electrode, was left in the vitreous humor on the surface of the retina while another micropipette was inserted into the retina for intracellular recording. The resistance of the micropipette was between 40–80 MΩ in Ringer solution. The potential recorded between the two microelectrode was differentially amplified to reduce the stimulus artifact.

The isolated retina was used mainly in perfusing experiments. The retina was placed with its receptor side up in a shallow chamber about 12 mm in diameter and was perfused with a Ringer solution, covering 1 mm in thickness over its surface at a flow rate of about 1 ml/min. The normal solution contained NaCl 100 mM, KCl 2·5 mM, $CaCl_2$ 2·2 mM, $NaHCO_3$ 20 mM and Glucose 20 mM and was equilibrated with 98% O_2 and 2% CO_2. A test solution contained 1–2 mM $CoCl_2$, but was otherwise identical to the normal solution. The settings of both stimulating and recording electrodes were similar to those for the eye-cup preparation. The transretinal current was 0·5 ms in duration and its intensity was adjusted to 2·5 mA which is sufficient to give an almost maximal response in horizontal cells.

The change in the membrane resistance was tested by passing current through the recording electrode. The current pulses used were in the order of 1 nA and were about 100 ms in duration. They were triggered by the same pulses which triggered the transretinal current to compare the membrane resistance before and during illumination. The voltage drop across the electrode and the membrane resistance was brought close to a balance by a bridge circuit built in the preamplifier. Pulses of both polarity were used. But the results shown later in this paper were selected from the experiments using depolarizing current. In this case, the upward deflection of the pulse indicates an increase in resistance. The effect of polarization of the membrane was tested by passing d.c. current instead of pulses through the same bridge circuit.

The light stimulus was mostly a white light from a glow modulator tube (Sylvania 1131C) or from a tungsten-halogen lamp operated at slightly below the rated supplying voltage. In preliminary experiments the center-surround organization of the response was tested by a light spot of 300 μm in diameter and an annulus of 400 μm inner diameter. In later experiments, the retina was illuminated diffusely from above.

Results

Transretinal current stimulation

Byzov and Trifonov[3] reported that a brief transretinal current flowing from receptor side to vitreous side, i.e. in the direction to depolarize receptor terminals, elicited a depolarization response in the horizontal cell of the carp retina. A current of the opposite direction was mostly ineffective. These results were confirmed in the present experiments in two types of horizontal cells; Cajal's external horizontal cells which receive inputs from cones and Cajal's intermediate horizontal cells which receive inputs exclusively from rods.[11] Sample records are shown in Fig. 1. The shape of each electric response was dependent on the type of horizontal cells recorded, on the intensity of the background light and on the condition of the preparation. However, in a given unit the profile of initial hyperpolarization to light resembled that of the falling phase of the electric response in the presence of the steady illumination. For example, in units responding to light with a prominent initial transient, the electric response also showed a prominent hyperpolarizing rebound. When the electric stimulation was repeated at a sufficiently high rate, a steady

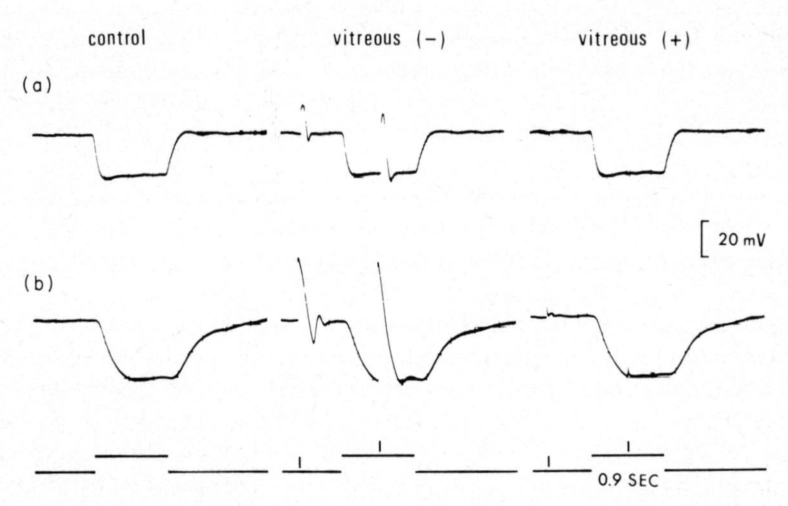

Fig. 1. Responses of two types of horizontal cells to photic and electric stimulation of the retina. (a) Responses of an external horizontal cell. (b) Responses of an intermediate horizontal cell. Records in the left column marked as control are responses to photic stimulation of 0·9 s duration as indicated by upward displacement of the stimulus trace at the bottom. The membrane potential in the dark was about −20 mV in the external horizontal cell and −40 mV in the intermediate horizontal cell. Records of the middle column marked as vitreous (−) are responses of the same unit to both photic and electric stimulation. A transretinal current pulse of 2·5 mA and 0·5 ms duration with vitreous side as an anode was applied once before and once during light stimulus at points indicated by a vertical bar on the stimulus trace. The electric stimulus evoked a transient depolarizing response, which is larger during the hyperpolarization response to light, followed by a slight rebound. Responses shown in the right column marked as vitreous (+) were recorded at the same condition as the middle records but with the transretinal electric stimulus of the opposite polarity which was hardly effective in evoking a response.

depolarization level was attained by a fusion of each electric response as shown in Fig. 2. These observations are consistent with the idea that horizontal cells kept depolarized in the dark by a continuous release of a transmitter substance are polarized when the transmitter release is suppressed by illumination.

Figure 3 shows the effect of transretinal current on bipolar cells. A current in the direction to depolarize receptor terminals usually elicited a diphasic electric response. It was positive-negative in off-center cells, but was negative-positive in on-center cells. Often, however, the late positive phase of the latter was lacking especially when the depolarization response to light was not large. A current of the opposite direction was usually ineffective. Occasionally, however, a small response of opposite polarity to those described above was

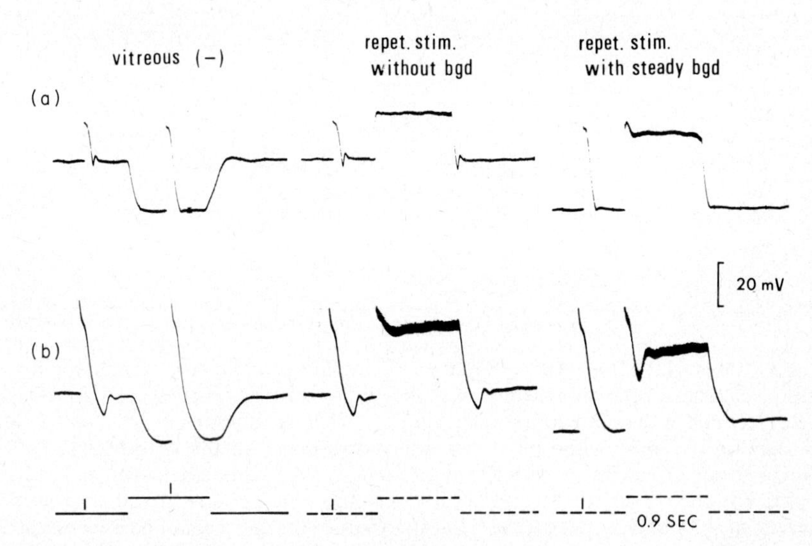

Fig. 2. Effect of repetitive transretinal electric stimulation on horizontal cells. (a) Responses of an external horizontal cell. (b) Responses of an intermediate horizontal cell. Records on the left column marked as vitreous (−) were obtained at the same experimental conditions as those of the middle column of Fig. 1. In records shown in the middle column, a single electric stimulus (vitreous side anodal) was followed by repetitive application of the same current pulse at a rate of 100 pulses/s in the absence of photic stimulation for the period indicated by an upward displacement of a dashed line at the bottom. The repetitive stimulation elicited a steady depolarization due to a fusion of each response to a single electric stimulus. In records shown in the right column, the same sequence of electrical stimulation as in the previous records was repeated but in the presence of a steady background illumination which hyperpolarized the membrane to a level indicated by the shift of the response trace. The background illumination usually augmented the depolarization elicited by electric stimulation. But, occasionally there was a slight suppressing after-effect in the electric response depending on the stimulus condition as seen in the lower record of this column.

elicited as seen in the figure. These results are to be expected if it is assumed that the current flowing from vitreous side to receptor side hyperpolarizes the receptor terminals and consequently suppresses the release of a transmitter. The polarization of the membrane by extrinsic current is one of the simplest methods to estimate the passive ionic movement involved in the synaptic transmission. The effect of polarizing extrinsic current on two types of bipolar

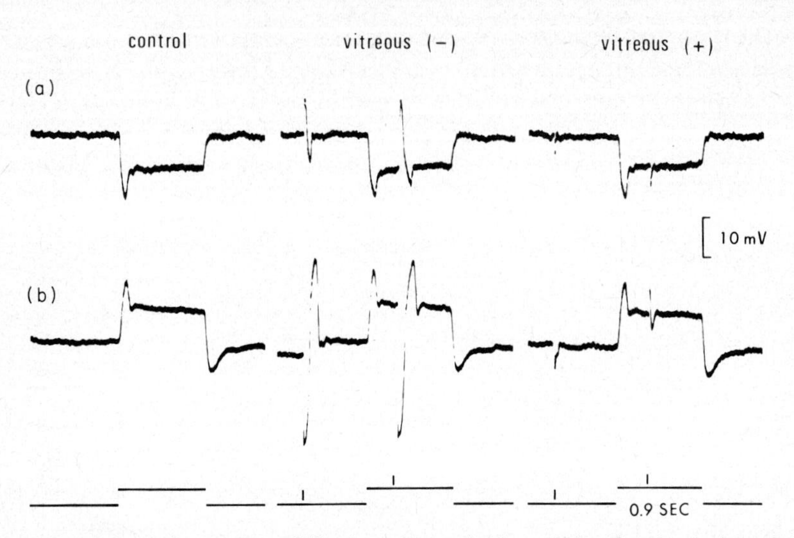

Fig. 3. Responses of two types of bipolar cells to photic and electric stimulation of the retina. (a) Responses of an off-center cell. (b) Responses of an on-center cell. Responses were recorded at the same conditions as in Fig. 1. The membrane potential in the dark was about −25 mV in the off-center cell and −30 mV in the on-center cell. The electric stimulation with vitreous side as an anode evoked a diphasic response; positive-negative in the off-center cell and negative-positive in the on-center cell. The electric current of the opposite polarity was hardly effective though a small positive-negative response was evoked in the on-center cell shown in this figure.

cells is shown in Fig. 4. In off-center cells, both photic and electric responses were augmented by hyperpolarizing extrinsic current and were depressed by depolarizing current. But a closer observation revealed a change in the proportion of the positive and negative phases of the electric response; the early positive phase was enhanced by hyperpolarizing extrinsic current more than the late negative phase. In most on-center cells, both photic and electric responses were also augmented by hyperpolarizing extrinsic current and were depressed by depolarizing current. Generally, however, the effect of extrinsic current on the photic response and on the electric response was not parallel. For instance, the hyperpolarizing extrinsic current augmented the photic response and the late positive phase of the electric response, but its effect on the early negative phase was slight. Often the early negative phase was preceded by another positive component if the membrane was hyperpolarized by extrinsic current. Such an effect, although not prominent, is observed in the records of Fig. 4. If the membrane was depolarized by extrinsic current, the early negative phase became sharper than the control as if the component preceding this phase reversed its polarity. Occasionally, the hyperpolarizing electric response of on-center cells became even smaller than the control by hyperpolarizing

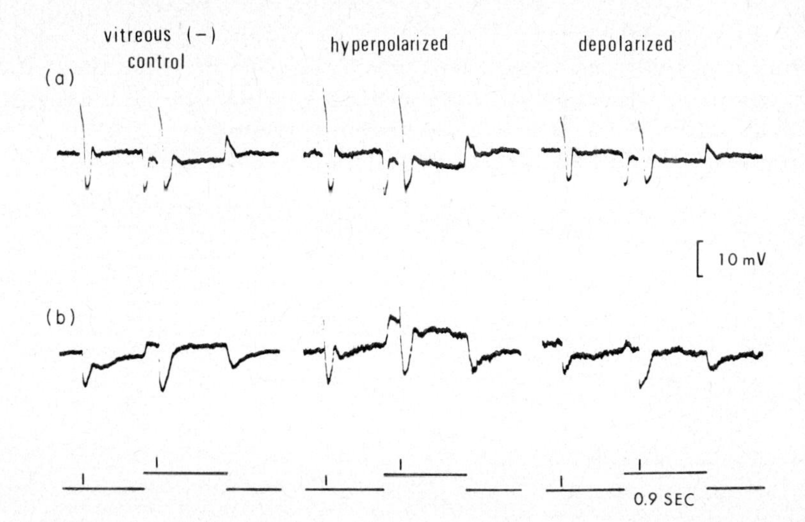

Fig. 4. Effect of polarizing extrinsic current on bipolar cell responses. (a) Responses of an off-center cell. (b) Responses of an on-center cell. Records of the left column are control responses to photic and vitreous-side anodal electric stimulation. The condition and sequence of stimuli were the same as in the previous experiments. Responses shown in the middle column were recorded at the same stimulus condition as the control but while the membrane was kept hyperpolarized by polarizing extrinsic current of 1 nA through the recording electrode. A shift in the membrane potential was balanced by a bridge circuit and therefore is not indicated in this figure. Responses shown in the right column were recorded during steady depolarization of the membrane by extrinsic current. In these bipolar cells, responses to both photic and electric stimulation were enhanced by hyperpolarization and were depressed by depolarization of the membrane by current.

extrinsic current. It is suggested from these results that at least two different mechanisms are involved in the negative phase of the electric response. As will be discussed later, one of these mechanisms appears to be a disfacilitation of the EPSP and the other most likely the IPSP.

As in horizontal cells, repetitive application of the transretinal current elicited a sustained polarization of the membrane in bipolar cells. It was usually a sustained depolarization in off-center cells and a hyperpolarization in on-center cells.

Effect of cobalt ions

Certain divalent cations, including cobalt ions, are known to block synaptic transmission by antagonizing the action of calcium ions which are essential in the release of transmitter substances from presynaptic terminals. The effect of a synaptic blocking agent was studied in the present experiments using a perfusing solution containing cobalt ions to see whether or not the electric response as well as the photic response of second-order neurons is abolished during its application, and to see whether their membrane is depolarized or hyperpolarized when all of the synaptic inputs on them are suppressed.

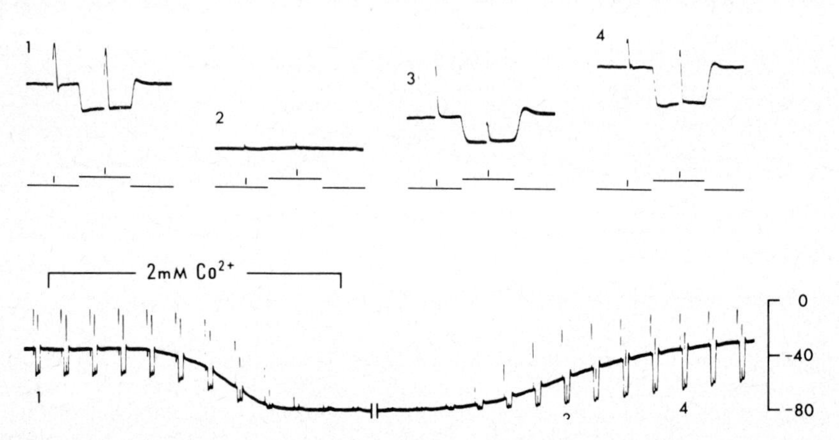

Fig. 5. Effect of cobalt ions on the membrane potential and response to photic and electric stimulation in an external horizontal cell. Responses of an external horizontal cell were recorded continuously in the retina perfused with a normal Ringer solution. After recording control responses the perfusing solution was switched to a Ringer solution containing 2 mM Co^{++} for the duration indicated in the figure. Application of Co^{++} resulted in hyperpolarization of the membrane with concomitant abolition of responses to both photic and electric stimulation. Responses sampled at different times (points 1–4 of the lower continuous record) are shown above with expanded time scale. Each sequence of photic and vitreous-side anodal electric stimulation is the same as in previous experiments. The photic stimulus indicated by upward displacement of the stimulus trace at the bottom is 0·9 s in duration and is repeated every 7 s. Five stimulus sequences are abridged between vertical bars. Responses are slightly retouched.

When the normal perfusing solution was switched to a test solution containing cobalt ions, horizontal cells, both external and intermediate, were hyperpolarized to a final potential level of about −80 mV and their responses to both photic and electric stimulation were abolished. A sample record of an external horizontal cell is shown in Fig. 5. After switching back to a normal solution, the membrane potential of horizontal cells recovered gradually and often became slightly more depolarized than before. The photic response recovering from the effect of cobalt ions was often slower but larger than the control response.

Figure 6 shows the effect of cobalt ions on an off-center bipolar cell. Off-center cells were hyperpolarized by cobalt ions and their photic and electric responses were abolished. The level of the membrane potential finally attained was variable but was usually between −30 and −45 mV. Although the intracellular concentration of potassium ions is not known, this value seems to be rather small as compared with −80 mV for horizontal cells. Hence, it is possible that the bipolar cell membrane is more or less permeable to sodium

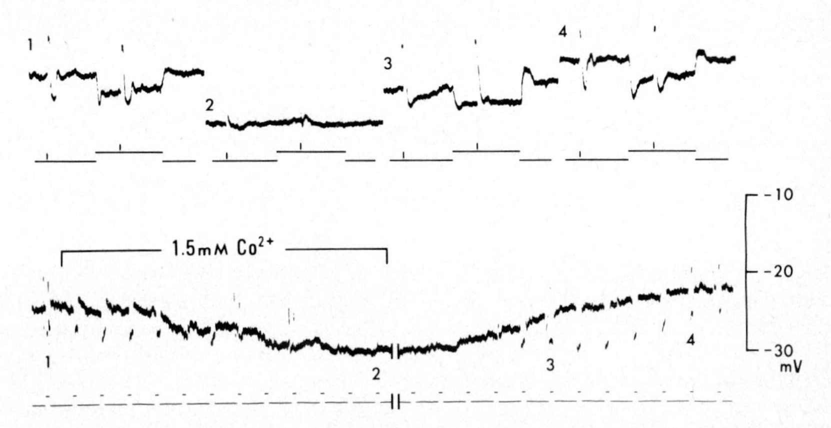

Fig. 6. Effect of cobalt ions on the membrane potential and response to photic and electric stimulation in an off-center bipolar cell. Responses of an off-center cell were recorded in the isolated perfused retina. Switching of the perfusing solution from a normal Ringer to a Ringer solution containing 1·5 mM Co^{++} for the duration indicated in the figure resulted in hyperpolarization of the membrane with concomitant abolition of the responses. After switching back to a normal solution, the membrane potential and the responses recovered gradually. Although transretinal electric stimulation (vitreous side anodal) was applied only occasionally in this record, sample records (1–4) shown above with expanded time scale were selected from the corresponding points (1–4) of the continuous record to show responses to both photic and electric stimulation. Timing of the electrical stimulation is indicated in these sampled records by a vertical bar on the lower stimulus trace. Photic stimulus repeated every 6·4 s is 0·9 s in duration and is indicated by upward displacement of the stimulus trace. Six stimulus sequences are abridged between vertical bars. Responses are slightly retouched.

ions in the absence of synaptic input. The effect of cobalt ions on an on-center cell is shown in Fig. 7. On-center cells were depolarized by cobalt ions and their photic and electric responses were abolished. The membrane potential level finally attained was variable but it never exceeded −15 mV. In a few units, the responses were abolished without a prominent change in the membrane potential.

Occasionally, the electric response of both horizontal and bipolar cells was more resistant to cobalt ions than their photic response. Thus the electric response was detectable for more than one minute after the complete

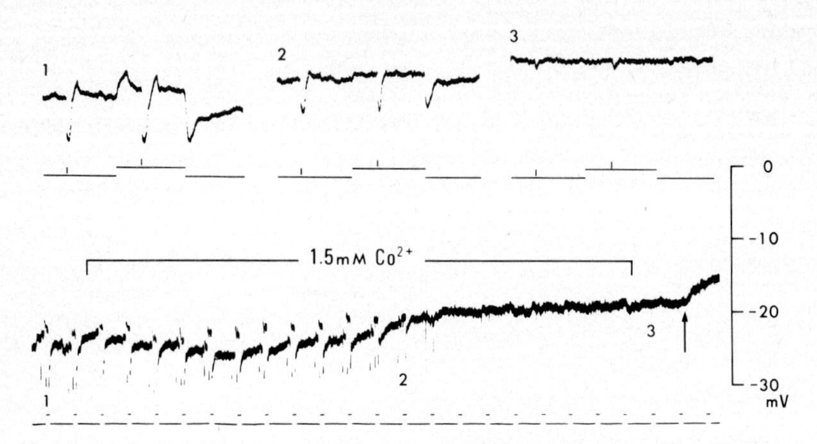

Fig. 7. Effect of cobalt ions on the membrane potential and response to photic and electric stimulation in an on-center bipolar cell. Intracellular responses of an on-center cell were recorded under the same experimental condition as in Fig. 6. Application of 1·5 mM Co^{++} resulted in a depolarization of the membrane with concomitant abolition of the responses in the on-center cell. The recording from this unit was not stable enough to follow the recovery from the effect of Co^{++}. It began to deteriorate at the point indicated by an arrow due to a slight dislocation of the electrode resulting from solution exchange.

disappearance of the photic response though their response became smaller and smaller. The reason for such variations in the effect of cobalt ions is not clear.

Rectifying membrane properties

If the membrane potential change so far described is the result of a conductance change of the subsynaptic membrane, such a change must be reflected in a change in the input resistance. In most horizontal cells, however, the change in input resistance was hardly detectable. It is possible that the conductance change of the subsynaptic membrane of horizontal cells is masked by the rectifying property of their non-synaptic membrane.[12] But even without any rectifying behaviour of the membrane, a change of input resistance would be very difficult to detect in horizontal cells owing to their extensive electrical coupling. The input resistance of these cells has been shown to vary nearly in proportion to the internal resistance of the cell and the cell-to-cell coupling resistance, but to be only weakly dependent on the membrane resistance itself.[13] In bipolar cells, the photic response always accompanied a change in input resistance. It is necessary, however, to study the electrical properties of the bipolar cell membrane before analysing its synaptic activity.

The current-voltage relation was studied in both horizontal and bipolar cells using a double-barrelled electrode. A current changing at a rate of about 2·5 nA/s was passed through one barrel. The magnitude of the current was monitored through an electronic galvanometer, and its output was fed to the x-axis while the membrane potential recorded through the other barrel was amplified and fed to the y-axis of the plotter. Sample records from an intermediate horizontal cell and from an on-center bipolar cell are shown in Fig. 8. The falling phase of current was used to plot these curves and the light stimulus with a spot of 600 μm diameter was applied at a constant rate during

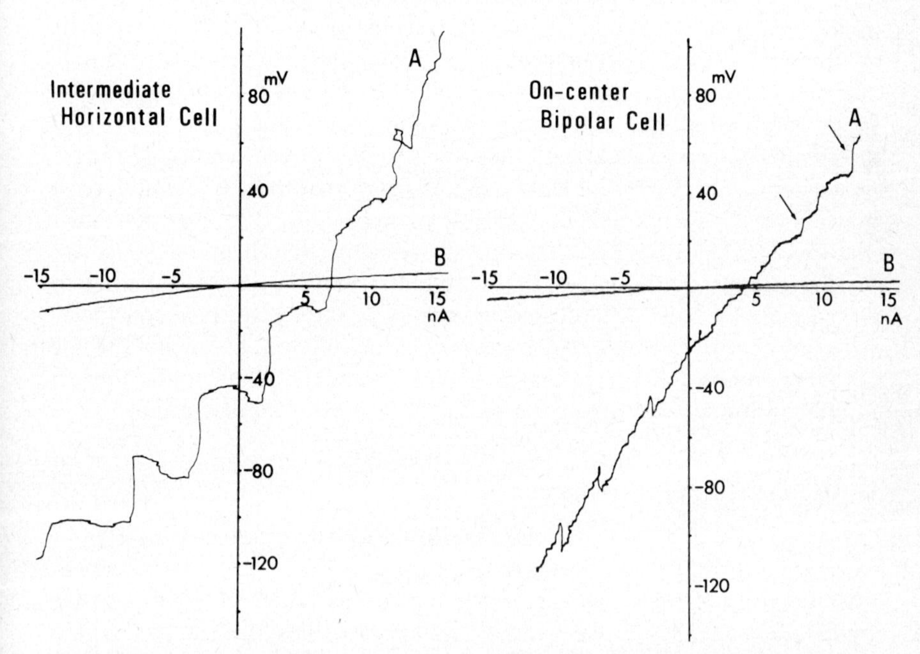

Fig. 8. Current-voltage relations of an intermediate horizontal cell and an on-center bipolar cell. Curves were plotted from right to left during polarization of the membrane by current changing almost linearly from positive to negative at a rate of about 2·5 nA/s. A light spot of 600 μm in diameter and 300 ms in duration was presented repetitively every 2 s to observe the response to light at the same time. In both records, curve A was obtained with the double-barrelled electrode inside the cell and curve B with the same electrode immersed in a Ringer solution to show the coupling resistance of the electrode. The horizontal cell membrane was anomalously rectifying. Its response to light was enhanced slightly by hyperpolarization and depressed by depolarization of the membrane by current. The membrane of the on-center bipolar cell was slightly rectifying. The depolarizing response to light of the on-center cell was enhanced by hyperpolarization and depressed or inverted by depolarization of the membrane by current. The reversal of the response is indicated by an arrow. Note that the response is plotted from right to left in this record.

these polarization experiments. In both records, the curve B shows a coupling resistance in a Ringer solution and the curve A the current-voltage relation recorded within the cell. As reported on other preparations, the horizontal cell membrane of the carp was anomalously rectifying. In some horizontal cells, the photic response was slightly augmented by hyperpolarization and depressed by depolarization of the membrane. The effect of polarization on the photic response was more prominent in intermediate horizontal cells.

The membrane of on-center cells, on the other hand, showed a rectifying property common to the nerve membrane. Namely, the membrane resistance was higher during hyperpolarization and was lower during depolarization. A similar current-voltage relation has been reported on the bipolar cell of the mudpuppy.[14] The rectifying property of the bipolar cell membrane raised a question as to whether the change in the input resistance accompanying the photic response as well as the augmenting or depressing effect of extrinsic current on the response is attributable solely to this property or not. However, it is evident from the record shown in this figure that there is a conductance increase of the membrane in response to light, since the current-voltage relation in the dark is steeper than the curve connecting the photic response. Another support on this view is the fact that the photic response is inverted by depolarizing current, as indicated by arrows in the figure. It is concluded, therefore, that both conductance changes in the subsynaptic membrane and rectifying properties of the non-synaptic membrane contribute to the changes in input resistance accompanying photic responses of bipolar cells.

Membrane resistance changes

The application of cobalt ions and the repetitive stimulation of the retina by transretinal current elicit a sustained polarization of the membrane in second-order neurons. If these changes in the membrane potential are due to the conductance changes in the synaptic membrane, they must accompany a change in input resistance comparable to that accompanying the photic response. It has been shown in a previous report[9] that the resistance change in on-center cells was proportional to the amplitude of the response for a fairly wide range of light intensities. Measurements of membrane resistance changes were performed from this point of view with special attention on the proportionality of changes in a potential and a resistance.

Figure 9 shows the effect of cobalt ions on the membrane resistance of two types of bipolar cells. Before the application of cobalt ions, the hyper-polarization response to light of off-center cells accompanied a resistance increase. On application of cobalt ions, the off-center cell was hyperpolarized and its photic response was abolished. Hyperpolarization induced by cobalt ions also accompanied a resistance increase. The change in the membrane resistance was almost proportional to the change in the membrane potential irrespective of whether it was elicited by illumination or by application of cobalt ions. In on-center cells, depolarization of the membrane elicited by light and by cobalt ions accompanied a decrease in input resistance. The change in the resistance and the potential was also proportional in both cases. It is, therefore, probable that both illumination and cobalt ions affect the same

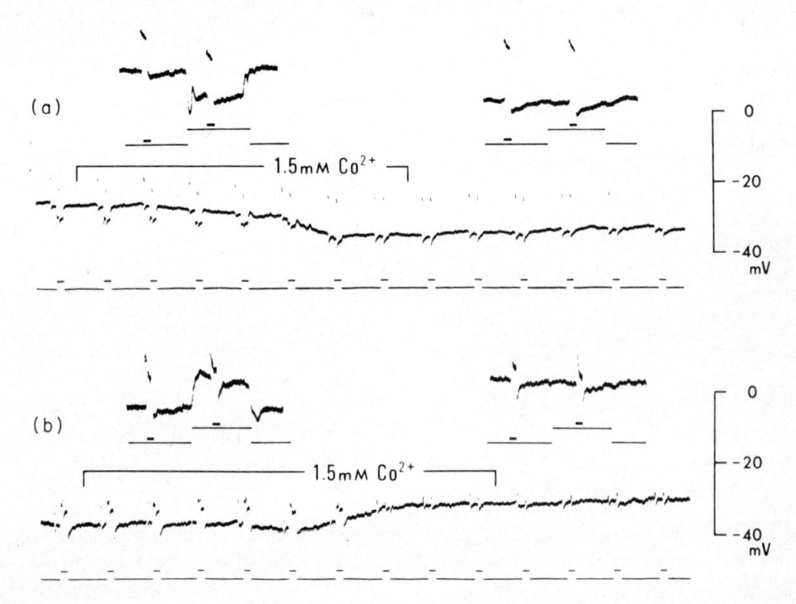

Fig. 9. Effect of cobalt ions on the membrane resistance of bipolar cells. (a) Off-center cell. (b) On-center cell. Changes in the input resistance were measured by passing a depolarizing current pulse of 1 nA and of 100 ms duration through the recording electrode. The pulse was applied once before and once during photic stimulation in place of transretinal electric stimulation used in previous experiments. The potential drop across the electrode and membrane resistance was partly compensated by a bridge circuit so that an increase in the positive pulse height in these records indicated an increase in the input resistance. Responses sampled from the continuous records are shown above with expanded time scale and with 1·5-time magnification in voltage scale. Timing of the pulse for the resistance measurement is indicated by a horizontal bar in the lower stimulus trace of these sampled records. In the off-center cell, the hyperpolarizing response to light was accompanied by an increase in the input resistance as judged from the change in the pulse height, and hyperpolarization of the membrane induced by application of 1·5 mM Co^{++} for the duration indicated in the figure was also accompanied by an increase in the resistance although the response and resistance change to light were abolished. In the on-center cell, the depolarizing response to light and the depolarization induced by Co^{++} were accompanied by a decrease in the input resistance. A change in the pulse height of 1 mV corresponds to a resistance change of 1 MΩ. The photic stimulus of 0·9 s duration repeated every 6·4 s is indicated by upward displacement of the lower stimulus trace.

membrane site to produce similar changes in the membrane potential. If cobalt ions act to suppress the synaptic activity, the illumination of the retina must also act to suppress the synaptic activity to these second-order neurons.

Figure 10 shows the effect of repetitive transretinal stimulation on the membrane resistance of bipolar cells. The retina was stimulated by a train of

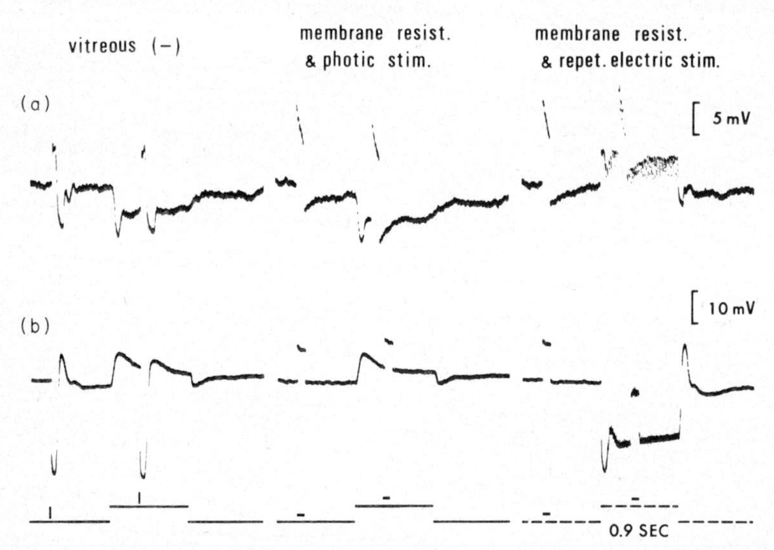

Fig. 10. Membrane resistance changes of two types of bipolar cells accompanying the potential change elicited by repetitive electric stimulation of the retina. (a) Off-center cell. (b) On-center cell. Records in the left column are control responses of bipolar cells to both photic and electric stimulation (vitreous side anodal) of the retina. The stimulus and recording conditions are the same as in Fig. 3. Records in the middle column show the input resistance changes accompanying the responses to light measured by the same method as described in the legend of Fig. 9. The measurement of the input resistance change of these cells in response to light is essential for later estimation of the proportionality between the potential and resistance change. As in the previous experiment shown in Fig. 9, the hyperpolarizing response of the off-center cell accompanied an increase and the depolarizing response of the on-center cell a decrease in the input resistance. In record of the right column, transretinal electrical stimulation was applied repetitively at a rate of 100 pulses/s in place of photic stimulation for the duration indicated by upward displacement of the dashed line at the bottom. The repetitive electric stimulation elicited a steady depolarization in the off-center cell and a steady hyperpolarization in the on-center cell. The input resistance changes accompanying these potential changes were measured by a current pulse through the recording electrode applied before and during repetitive transretinal stimulation as indicated by a horizontal bar above the stimulus trace at the bottom. The depolarizing response of the off-center cell elicited by the transretinal stimulation was accompanied by a decrease and the hyperpolarizing response of the on-center cell by an increase in the input resistance. A change in the pulse height of 1 mV corresponds to a change of 1 MΩ in the input resistance.

transretinal current at a frequency of about 100 pulses/s. Changes in the membrane resistance were studied by passing a pair of trans-membrane current pulses through the recording electrode before and during repetitive stimulation of the retina. As seen in these records, the depolarization of an off-center cell accompanied a resistance decrease and the hyperpolarization of an on-center cell accompanied a resistance increase. However, the proportionality between the potential change and the resistance change was not always held strictly in on-center cells. Often the hyperpolarization elicited in on-center cells by repetitive electric stimulation accompanied a resistance decrease. Since both photic and electric responses in these latter cells were small, we regarded this hyperpolarization response accompanied by a resistance decrease as one of the components involved in the synaptic process.

Discussion

Effect of transretinal current

The present experiments confirmed the previous observation on the carp retina that transretinal current in the direction to depolarize receptor terminals elicited a depolarization in horizontal cells.[2,3] The effect was similar on both external and intermediate horizontal cells, suggesting that there is no qualitative difference in the output of rods and cones. The current of the same polarity was also effective in eliciting a response in bipolar cells. The electric responses of bipolar cells were usually diphasic, and their polarity was opposite in two types of bipolar cells. Similar results have been reported by Kaneko and Shimazaki.[15] A direct action of current on the second-order neurons can be excluded from an observation that electric responses disappear after the application of cobalt ions. These results suggest that the transmitter acting on bipolar cells is also released when the receptor terminals are depolarized as is assumed for horizontal cells, but ionic mechanisms activated by the transmitter is different in two types of bipolar cells.

The early depolarizing phase of the electric response in off-center cells is similar in its characteristics to the EPSP observed generally in the nervous system. It is enhanced by hyperpolarizing extrinsic current and depressed by depolarizing extrinsic current. It accompanies a decrease in input resistance. The early hyperpolarizing phase of on-center cell, on the other hand, is rather complex. Its main component is enhanced by hyperpolarizing current, depressed by depolarizing current and is accompanied by an increase in the input resistance. It resembles in this respect the hyperpolarizing response to light of photoreceptors associated with a decrease in inward sodium current. A possible mechanism underlying this component is that a transmitter released by electric stimulation acts on the subsynaptic membrane to decrease inward current by closing excitatory channels which remain open in the absence of transmitter action. The other component appears to have characteristics common to the IPSP. Namely, some of the hyperpolarizing responses are augmented by depolarizing extrinsic current and depressed or inverted by hyperpolarizing extrinsic current. The equilibrium potential therefore is expected to be slightly below the membrane potential of −20 to −40 mV in the dark. Furthermore, they accompanied a decrease in input resistance. Most of the electric responses of on-center cells appear to be a combination of these two components.

The late phase of the responses to electric stimulation will not be discussed here, since there will be a contribution of a surround effect which is not the main topic of this report.

Effect of cobalt ions

Cobalt ions were used in the present experiments to suppress the synaptic transmission because of its potent inhibitory action on the transmitter release.[16] Unfortunately, no reliable intracellular record was obtained from photoreceptors in the present experiments, but it has been reported that cobalt ions effectively abolish activities of horizontal cells without affecting the receptor potential.[6,15,17,18]

As reported already, horizontal cells were hyperpolarized by cobalt ions. The membrane potential attained was about −80 mV and agrees with the results of Kaneko and Shimazaki[7] using a low calcium and high magnesium medium to block synaptic activity.

Cobalt ions hyperpolarized off-center bipolar cells and depolarized on-center cells. These results agree with those of Kaneko and Shimazaki[15] on the carp and those of Dacheux and Miller[17] on the mudpuppy. The membrane potential attained was between −30 and −45 mV for off-center cells and between −20 and −35 mV for on-center cells. If this membrane potential level is due solely to the concentration difference of potassium ions across the membrane, one has to assume the intracellular concentration of potassium in bipolar cells to be one quarter to one tenth of the concentration in horizontal cells. It seems more reasonable to assume that the bipolar cell membrane is more or less permeable to sodium ions in the absence of synaptic inputs. In order to test this possibility, it may be necessary to examine whether the membrane potential after application of cobalt ions is dependent on the external sodium concentration or not. Although this type of experiment was not successful in the present study, Kaneko and Shimazaki[15] concluded from their observations on the hyperpolarizing effect of a sodium-free medium that the membrane of on-center cells has a relatively high permeability to sodium ions in the absence of the transmitter action.

The hyperpolarization induced by cobalt ions in off-center cells accompanies an increase in input resistance. We assume that this change in input resistance indicates a conductance change of subsynaptic membrane beside the rectifying property of the membrane, since the change in the input resistance was almost proportional to the potential change irrespective of whether it was elicited by illumination or by cobalt ions. Similarly, the depolarization induced by cobalt ions in on-center cells suggests a conductance increase of the subsynaptic membrane.

Applicability of the hypothesis

The hypothesis proposed to explain two types of bipolar cell responses assumes that the transmitter released in the dark from photoreceptor terminals opens subsynaptic excitatory channels of off-center cells as in horizontal cells, but closes those of on-center cells. According to this hypothesis the following results are expected. The transretinal current which acts to depolarize receptor terminals will be effective in increasing the transmitter release, and consequently elicit a depolarization of off-center cells accompanied by a decrease in input resistance but a hyperpolarization of on-center cells accompanied by an increase in input resistance. The suppression of synaptic transmission by cobalt ions will result in changes of both membrane potential and resistance to the opposite direction, namely a hyperpolarization accompanied by a resistance increase in off-center cells, and a depolarization accompanied by a resistance decrease in on-center cells. Since the ionic channels involved are excitatory having an equilibrium potential at a depolarized level, both hyperpolarization and depolarization elicited by photic as well as electric stimulation must be augmented by hyperpolarizing extrinsic current and be depressed or inverted by depolarizing extrinsic current. These predictions were substantiated qualitatively in the present experiments. However, as far as the on-center cells are concerned, a certain modification of the hypothesis is necessary to explain all of the results obtained in the present experiments. The effect of polarization of the membrane on the electric response and the measurements of its accompanying resistance changes suggest two different ionic mechanisms in on-center cells. One of them is the closure of the excitatory subsynaptic channels by the transmitter and the other is the opening of the inhibitory subsynaptic channels. A possibility one has to take into account in this regard is that one of these ionic mechanisms might be activated by the horizontal-bipolar synapses found in some vertebrate,[19] although such synapses have not been reported in the carp. There is, however, another piece of evidence that indicates a participation of two different ionic mechanisms in the receptor-bipolar transmission. The depolarization response of on-center cells elicited by a small light spot not to

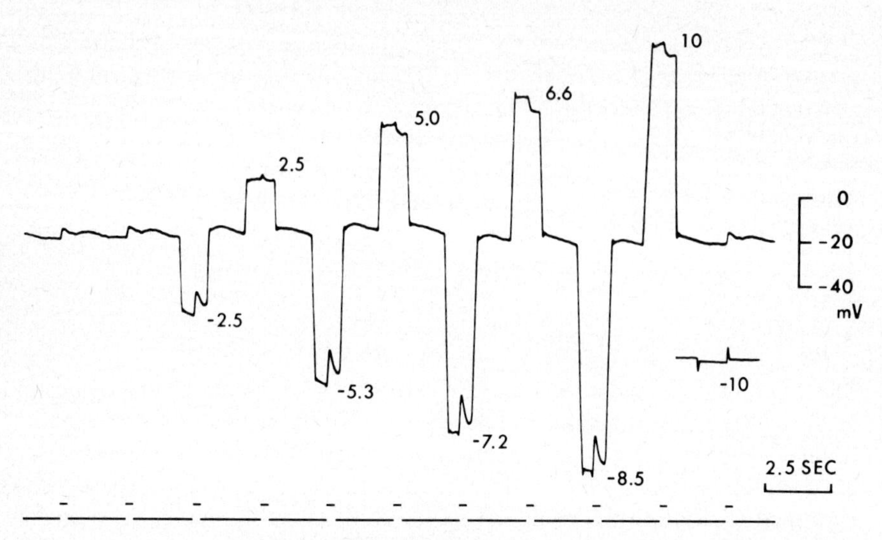

Fig. 11. Effect of polarization of the membrane on the response to light of an on-center bipolar cell. A light spot of 300 μm in diameter and 300 ms in duration was presented at a constant rate during continuous recording from an on-center cell as indicated by upward displacement of the stimulus trace at the bottom. Extrinsic current was applied through one barrel of the double-barrelled electrode about 400 ms before the photic stimulus for 1 s to polarize the membrane to different levels. Intensities of the current are indicated in nA. On the lower right, the coupling resistance of the electrode is shown by a potential shift to the current of -10 nA with its tip immersed in a Ringer solution. Note that the response to light shows an initial depolarizing transient followed by a hyperpolarization at 5·0 nA and an initial hyperpolarizing transient followed by a depolarization at $-5{\cdot}3$, $-7{\cdot}2$ and $-8{\cdot}5$ nA.

activate horizontal-bipolar synapses can be inverted by depolarizing extrinsic current, but usually an initial transient remains uninverted so that the response becomes diphasic. An example is shown in Fig. 11, where the effect of polarizing current was tested on the photic response of an on-center cell elicited by a light spot of 300 μm diameter. The current of various intensities was passed from one barrel of a double-barrelled electrode for 1 s. The photic stimuli of 300 ms duration were presented in the middle of the polarizing current pulse. Kaneko and Tachibana (personal communication) first observed in the light-adapted carp retina that the depolarization response of on-center cells to a small light spot could be inverted by hyperpolarizing extrinsic current. A similar effect of polarizing current was also observed in the mesopic preparation used in the present experiments. As seen in this figure, the response first augmented by hyperpolarizing extrinsic current is inverted in its initial phase by further increase in current so that the response becomes diphasic. Thus two components with a different time course are involved. The fast component is hyperpolarizing to the transretinal current which facilitates the release of a transmitter and is depolarizing to light which suppresses its release, but is inverted by hyperpolarization of the membrane. It is suggested that its ionic mechanism is similar to the IPSP.

Acknowledgments

This work was supported in part by grants from the Ministry of Education of Japan and by U.S. Public Health Service Grant (No. 5 Rol EY00017-09: Principal Investigator, T. Tomita).

References

1. E. De Robertis and C. M. Franchi 1956, *J. Biophys. Biochem. Cyt.* **2**, 307; W. K. Stell 1972, *In* "Handbook of Sensory Physiology", vol. 7, part 2 (ed. M. G. F. Fuortes) p. 111, Springer-Verlag, Heidelberg.
2. Yu. A. Trifonov 1968, *Biofizika* **13**, 809.
3. A. L. Byzov and Yu. A. Trifonov 1968, *Vision Res.* **8**, 817.
4. J. Toyoda, H. Nosaki and T. Tomita 1969, *Vision Res.* **9**, 453.
5. J. E. Dowling and H. Ripps 1973, *Nature* (*London*) **242**, 101.
6. L. Cervetto and M. Piccolino 1974, *Science* **183**, 417.
7. A. Kaneko and H. Shimazaki 1975, *J. Physiol.* (*London*) **252**, 509.
8. G. Waloga and W. L. Pak 1976, *Science* **191**, 964.
9. J. Toyoda 1973, *Vision Res.* **13**, 283.
10. A. Kaneko 1971, *Vision Res. Suppl. No.* 3, p. 17.
11. W. K. Stell 1967, *Amer. J. Anat.* **121**, 401; A. Kaneko and M. Yamada 1972, *J. Physiol.* (*London*) **227**, 261.
12. Yu. A. Trifonov, A. L. Byzov and L. M. Chailahian 1974, *Vision Res.* **14**, 229; F. S. Werblin 1975, *J. Physiol.* (*London*) **244**, 639.
13. A. V. Minor and V. V. Maksimov 1969, *Biofizika* **14**, 328.
14. R. Nelson 1973, *J. Neurophysiol.* **36**, 519.
15. A. Kaneko and H. Shimazaki 1976, *Cold Spring Harbor Symp. Quant. Biol.* **40**, 537.
16. J. N. Weakly 1973, *J. Physiol.* (*London*) **234**, 597.
17. R. F. Dacheux and R. F. Miller 1976, *Science* **191**, 963.
18. E. A. Schwartz 1976, *J. Physiol.* (*London*) **257**, 379.
19. J. E. Dowling and F. S. Werblin 1969, *J. Neurophysiol.* **32**, 315; A. Lasansky 1971, *Phil. Trans. Roy. Soc. Lond. B* **262**, 365.

Discussion

A. Kaneko: Recently, Mr Tachibana and I have made similar experiments on the depolarizing (on-centre) bipolar cells in the carp retina with results close to those Dr Toyoda has presented. I would like to add a small piece of information on input resistance changes in the depolarizing bipolar cells. In the photoptic retina in which bipolar cells receive a dominant input from cones, the input resistance showed a biphasic change accompanying the photic response, there being an initial increase followed by a delayed decrease (the same as Dr Toyoda's finding). But in the scotopic retina, the depolarizing bipolar cell showed a monophasic resistance decrease, the time course of which was parallel to the time course of the potential change. It is very puzzling that the synaptic mechanism from cones to bipolar cells is different from that between rods and bipolar cells, and one should be careful about the state of adaptation in this type of experiment.

J. Toyoda: Since the transretinal current was considered to be effective in stimulating both rod and cone inputs to bipolar cells, the light stimulus used in the present experiments was adjusted to excite both of these inputs. Judging from our preliminary experiments, there is certainly a piece of evidence suggesting that the relative contribution of the two ionic components, although it is variable in units even under the same experimental conditions, is dependent on the level of adaptation. Thus, it is possible that the synaptic mechanisms involved in the rod system are different from those of the cone system. However, it will be difficult at present to exclude another possibility, namely that the synaptic mechanisms of the two systems are not qualitatively different, consisting of two ionic components but in different proportions.

P. B. Detwiler: Dr Toyoda you state that the applied transretinal current is in the direction to depolarize the photoreceptor terminals. Have you established directly, by recording from photoreceptors, that the applied transretinal current does in fact depolarize the receptors?

J. Toyoda: The effect of transretinal current on photoreceptors was estimated on the basis of the neurophysiological principle that the extrinsic current applied through extracellular electrodes hyperpolarizes the membrane at the point where it crosses from outside to inside, but depolarizes the part of the membrane where the current crosses in the opposite direction. Thus, the transretinal current flowing from receptor side to vitreous side hyperpolarizes the photoreceptor membrane at the outer segment but depolarizes the membrane at the synaptic terminal. We studied the effect of transretinal current on photoreceptors, mainly to test the feedback effect from horizontal cells, but without convincing results because of the small amplitude of photoreceptor responses recorded.

14a

The Interaction in Photoreceptor Synapses Revealed in Experiments with Polarization of Horizontal Cells

YU. A. TRIFONOV and A. L. BYZOV

Institute for Problems of Information Transmission, The Academy of Sciences, Moscow, U.S.S.R.

In experiments on the turtle retina it was shown that hyperpolarization of horizontal cells evokes a depolarization in cones.[1] In fish cones such an effect is absent.[2] However in studying the inversion of horizontal cell responses during depolarization of its membrane we obtained results which indicate that the feedback (from horizontal cells to cones) exists in fish retina too. But although it is effective at the subsynaptic level, this feedback is not accompanied by changes of cone potential.

The horizontal cell membrane was polarized by means of current passed through extracellular low resistance silver electrodes. The idea of the method is illustrated in Fig. 1b. The layer of horizontal cells is represented as a cable, because the horizontal cells are coupled electrically.[3,4] The extracellular current is passed along the cable. Some part of this current crosses the membrane changing the membrane potential. The distribution of extracellular resistances can be adjusted in such a way that this displacement of membrane potential would be uniform in a sufficiently large segment of the cable. To achieve the uniform polarization one has to make the extracellular longitudinal gradient of potential proportional to distance from one of the edges of the cable (left edge in Fig. 1b).

The length of segment along which the potential gradient is proportional to distance must be large in comparison with the space constant of the cable. In this case, firstly, the polarization of the membrane is uniform and, secondly, the density of current crossing the membrane does not depend on the membrane resistance. The density of membrane current is proportional to the intensity of longitudinal extracellular current. The method has been described in detail earlier.[5, 6, 7]

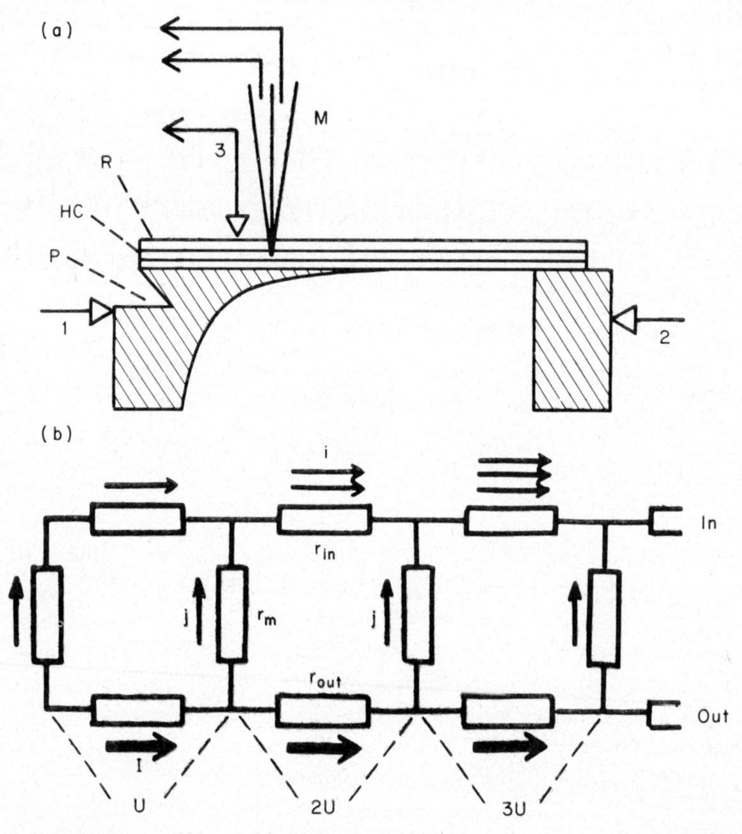

Fig. 1. Illustration of the method of polarization of the horizontal cell layer with extracellular current. (a) The scheme of chamber. *R*—retina; *HC*—the layer of horizontal cells; *P*—the agar plate; 1 and 2—the electrodes for passing the current; 3—reference recording electrode; *M*—two microelectrodes for differential recording of membrane potentials. The represented profile of the agar plate is near to the real one, but the retina is drawn thicker. (b) The equivalent circuit explaining the principle of uniform polarization of membrane of horizontal cell layer with longitudinal current applied extracellularly. in and out—inner and outer sides of membrane respectively; r_m—resistance of the membrane; r_{in} and r_{out}—inner and outer longitudinal resistances; *j*—density of current crossing the membrane; *i*—inner longitudinal current; *I*—external longitudinal current. If the density of membrane current is uniform, the inner and outer longitudinal gradients of potential increase linearly with distance from the left edge. Therefore the linearity of increase of outer potential gradient is the necessary condition of uniform polarization.

This idea was realized in the following way. A strip of carp or pike retina was placed receptors up on an agar plate. The width of the strip was about 3 mm. The thickness of the plate was maximal at one edge and decreased toward the other edge (Fig. 1a). The profile was chosen so that the extracellular longitudinal gradient of potential (f) was proportional to distance from the left edge in Fig. 1a (x) for the space of about 6 mm. The potential gradient is proportional also to the intensity of longitudinal current (I). Thus $f = qIx$, where q is a coefficient characterizing the particular profile of the agar plate including the strip of the retina on it. The value of q can be calculated from this equation because f, I and x can be measured.

The density of membrane current (j) is $j = Iq/r$, where r is the longitudinal inner resistance of the horizontal cell layer per unit of length of the strip.[5] We were unable to calculate the absolute value of j because r is unknown. Instead of j the values of I and q are indicated in figures.

Differential recording was used. Two microelectrodes stuck together were inserted into the retina. The vertical distance between tips was in various cases from 5–40 μm. Only the lower microelectrode penetrated the cells. This method permits one to record the membrane potential without adding the extracellular longitudinal potential difference which can be very large. The values of vertical (radial) potential difference outside the cells were negligible: not more than 2–3 mV with the strongest currents and with a distance between the tips of 20–40 μm.

The horizontal cells respond with changes of membrane potential both to light and to radial current shocks (anode on receptor side, cathode on vitreal side of the retina).[8] The inversion of these responses takes place when the horizontal cell membrane is depolarized and displaced beyond the zero level of membrane potential (Fig. 2a and b). The inversion of horizontal cell responses to radial current shocks can be demonstrated easier than the inversion of light responses. The light response disappeared often without inverting. In Fig. 2a, for instance, illumination influenced only the size of inverted responses to radial current shocks, but there was no inverted light response.

Thus it is not always possible to invert the light response of horizontal cells. The voltage-current curves of horizontal cell membrane in Fig. 3 show how the light responses change with depolarization. The curves were obtained with slow (about 20 s for each curve) increases of longitudinal current. The intensity of current is shown on the abscissa. The dark voltage-current curve in Fig. 3a is practically a straight line. In the saturating light the curve is nonlinear because of nonlinearity of non-synaptic membrane.[4, 9] The dark curve and the bright light curve intersect near zero level of membrane potential indicating the position of equilibrium potential.

In light of intermediate intensity the curve (curve "dim light" in Fig. 3a) occupies the intermediate position until the potential approaches the equilib-

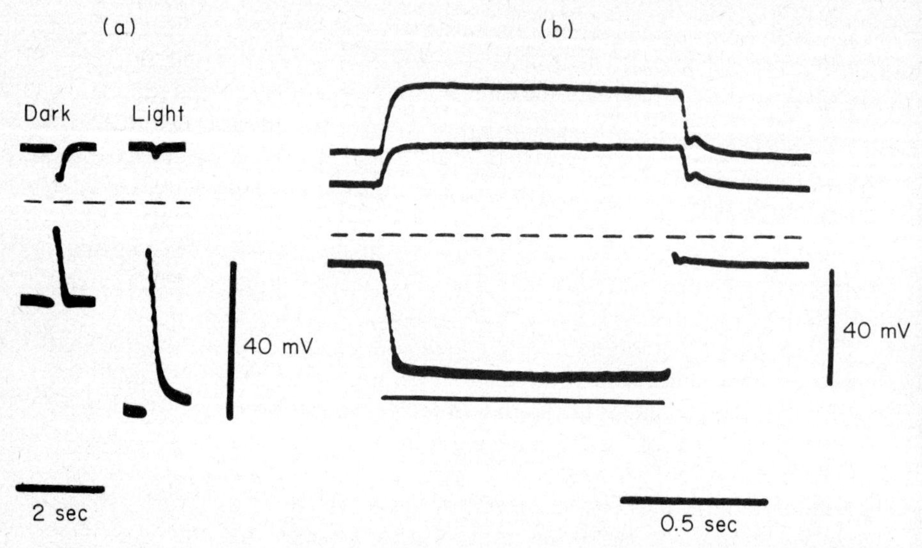

Fig. 2. Inversion of the responses of horizontal cells with depolarization. (a) Inversion of responses to radial current shocks in darkness and in light. Responses are evoked by short (about 2 millisecond) current shocks passed across the retina (anode on the receptor side, cathode on the vitreal side). Depolarization over the zero level of membrane potential (interrupted line) was evoked by longitudinal current of 0·5 mA, $q = 90\ \Omega \cdot mm^{-2}$. (b) Inversion of light response. Intensity of depolarizing longitudinal currents is 0·5 mA and 0·75 mA, $q = 90\ \Omega \cdot mm^{-2}$. Interrupted line shows approximately the zero level of potential. Line under the records shows the light stimulation. Records (a) and (b) were obtained on the pike retina.

rium level, but above this level it follows close to the curve "bright light", practically coinciding with it in this particular case. Sometimes the curves obtained in a dim light, which evokes a very small response, coincide after inversion with the curve recorded in the saturating light (Fig. 3b).

In Fig. 3c another case is represented. In this retina it was impossible to get the inverted light response: responses disappeared without their inversion when the potential crossed the equilibrium level, and the voltage-current curves for all intensities coincided. This case is similar to that illustrated by Fig. 2a.

The possibility of obtaining the inverted light response depends to some extent on the state of adaptation. In Fig. 4a the inversion occurs in a light adapted retina: it is seen from the fact that the dark curve and the light curve intersect. After dark adaptation (5 min) the curves changed: depolarization diminished and abolished the light response without inversion (Fig. 4b). Light adaptation restored the inversion again. These changes could be reproduced many times in one and the same cell.

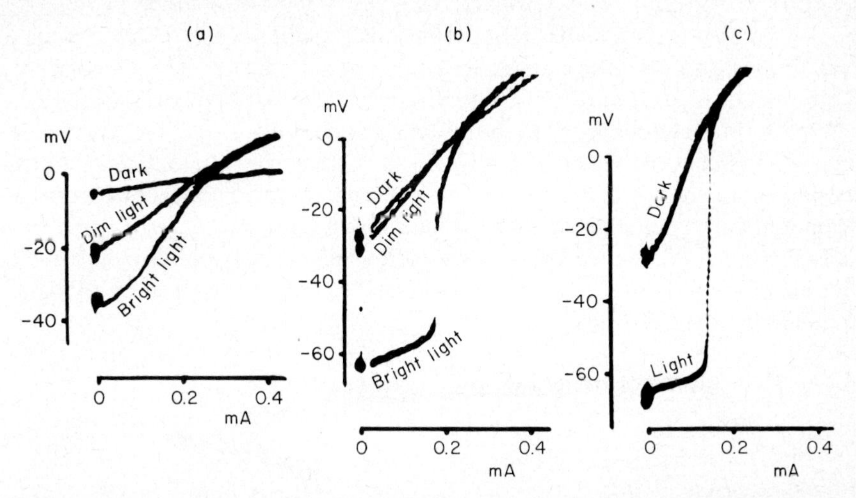

Fig. 3. Voltage-current curves of the membrane of horizontal cells in the light and in darkness. Abscissae—longitudinal current, depolarizing the horizontal cell membrane, $q = 90\ \Omega{\cdot}\mathrm{mm}^{-2}$. Ordinate—membrane potential of horizontal cells. The pictures were taken directly from an oscilloscope in which the horizontal shift of the beam is proportional to intensity of current. The intensity of current was changed very slowly (about 20 s for each curve). a, b and c were obtained in different experiments on the pike retina.

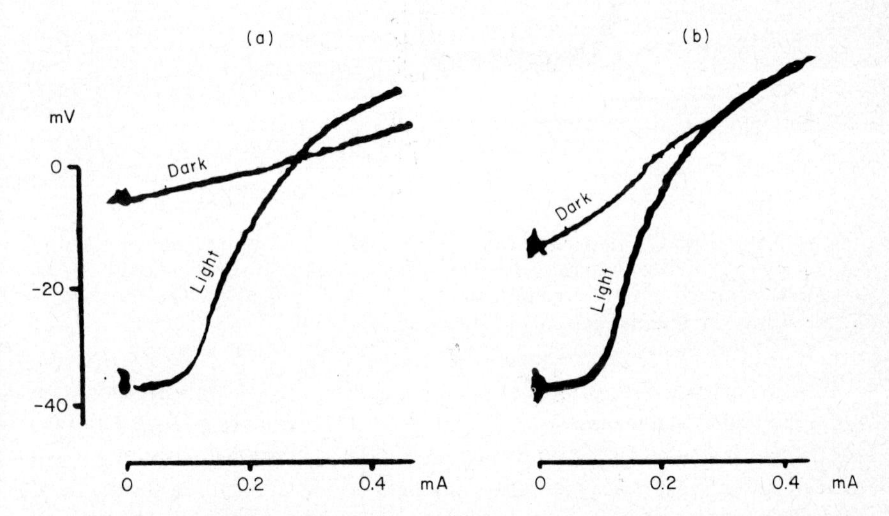

Fig. 4. Disappearance of inversion of the light response after dark adaptation. Voltage-current curves of horizontal cell membrane in light adapted retina (a) and after 5 min dark adaptation (b). Records are obtained in one and the same cell on the pike retina, $q = 90\ \Omega{\cdot}\mathrm{mm}^{-2}$. The axes are the same as in Fig. 3.

As a rule the inversion of a light response can be obtained in retinas in which the horizontal cells have a low dark potential and accordingly a large light response. If the state of the retina became worse, the dark potential increased, and this usually coincided with the disappearance of inversion. This resembles the effect of background light. In Fig. 3(a and b) it can be seen that, if the initial conditions were not darkness but a weak background light, depolarization would abolish the response without inverting it. It has to be noted that the disappearance of the inverted responses after dark adaptation is accompanied by increase of the dark potential (compare in Fig. 4 the position of dark potentials in a and b).

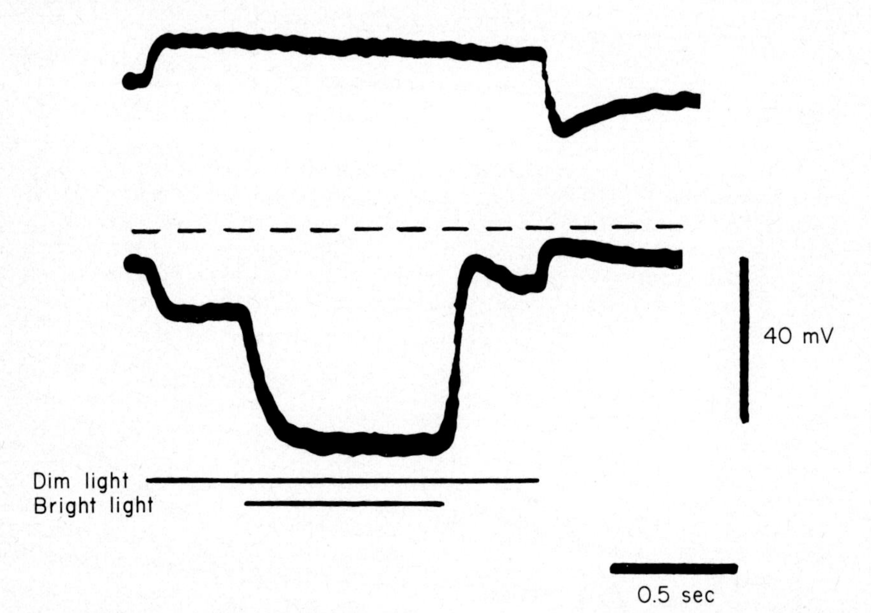

Fig. 5. Responses of a horizontal cell of carp retina to diffuse dim and superposed saturating lights under normal conditions (lower trace) and with depolarization (upper trace). Interrupted line is approximate zero level of potential. The intensity of depolarizing longitudinal current is 1·5 mA, $q = 25\ \Omega \cdot mm^{-2}$.

In Fig. 3(a and b) it can be seen also that the inverted responses reach their maximal size at lower intensities of light than the normal responses. In Fig. 5 this is demonstrated in the other way. A dim light if applied without current evoked a small response. The saturating light added to it caused the maximal hyperpolarizing response. During depolarization above the equilibrium level the response to dim light became a maximal one as can be seen from the fact that the additional bright light did not evoke an additional deflexion of potential.

Evidently the maximal response should arise if the light stopped liberation of transmitter completely. The response to the dim light was small under normal conditions. Therefore the dim light by itself failed to stop the liberation of transmitter completely under normal conditions, but was able to do so if the transmembrane potential of horizontal cells was inverted. One can conclude from this that such a depolarization of horizontal cells, if applied during the dim light, stops the liberation of transmitter from the cone endings because of some mechanism of feedback.

In Fig. 6b calculated voltage-current curves of horizontal cell membrane are represented which show how these curves should look without feedback between horizontal cells and cones. The calculation was based on the model shown in Fig. 6a[6] where R_n is the resistance of nonsynaptic membrane which is voltage dependent; R_s is the resistance of subsynaptic membrane; E is the

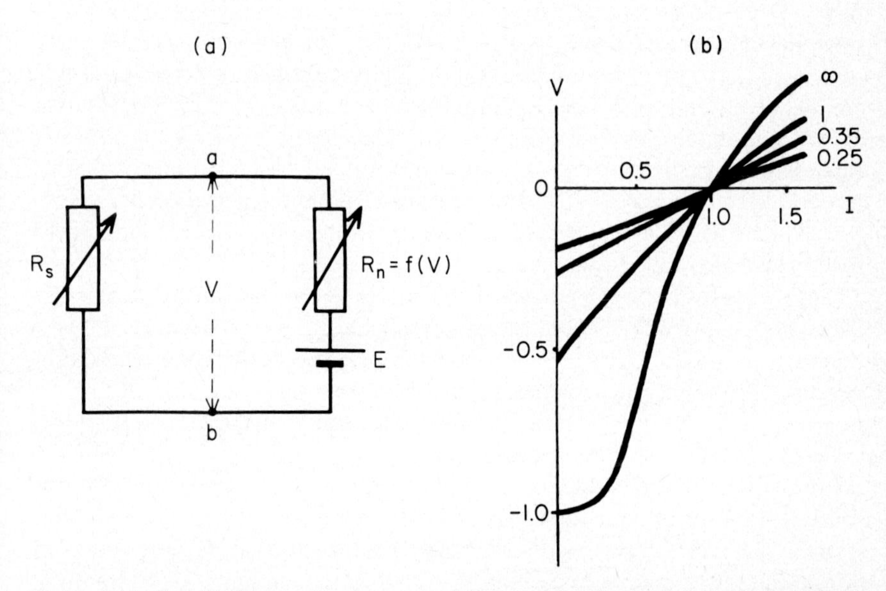

Fig. 6. Calculated voltage-current curves of horizontal cell membrane obtained from the model. Calculations are done without taking into account the feedback of horizontal cells with cones. (a) The model of horizontal cell membrane,[6] E—e.m.f. of non-synaptic membrane; R_n—non-synaptic membrane, the resistance of which is a function of membrane potential; R_s—subsynaptic membrane, the resistance of which is low in darkness and high in the light; V—membrane potential; a and b—inner and outer sides of membrane. The function $R_n = f(V)$ represented graphically can be evaluated from voltage-current curve in (b) at $R_s = \infty$. (b) Voltage-current curves at various values of R_s (see numbers near the right ends of curves). The units in the model: the unit of voltage is e.m.f. of E, the unit of current is the current depolarizing the membrane up to zero potential, the unit of resistance is the resistance of nonsynaptic membrane at zero potential.

battery of the resting potential. It was assumed that the voltage-current curve of R_n coincides with the voltage-current curve of the horizontal cell membrane in the supersaturating light. The equilibrium potential of subsynaptic membrane was assumed to be zero, accordingly there is no battery in series with R_s. The curves were calculated for several different values of R_s, corresponding to various intensities of light. The curves calculated for intermediate intensities of light occupy, in contrast with some real curves, intermediate positions at any membrane potential, including parts of the curves above zero level. The difference between the calculated and the real curves demonstrates the effect of supposed feedback between horizontal cells and cones.

Concerning the mechanism of feedback, one possibility would be to suppose that the horizontal cells have inhibitory synapses on cones. If this were the case the depolarization of horizontal cells would lead to hyper-polarization of cones,[1] and as a result to diminution or cessation of liberation of transmitter. In order to abolish the light response of horizontal cells completely (as in Fig. 3c) the supposed IPSP in cones should have at least the same value as the cone light response. In the course of experiments about 30 records of cone responses were obtained. In all cases depolarization of horizontal cells failed to evoke any measurable potential deflexion in cones of fish retina (Fig. 7). The hyperpolarization of horizontal cells and illumination by annulus also failed to evoke any response in cones.[2]

Thus the simplest explanation of the feedback mechanism by means of a chemical synapse was not confirmed experimentally, at least in fishes. It has to be noted, however, that the microelectrode inserted in the cone records the potential difference across the somatic membrane, but not across its presynaptic membrane, because the space between presynaptic membrane of the cone and subsynaptic membrane of the horizontal cell is separated from outer media by the resistance of the synaptic gap. But only the potential of the presynaptic membrane controls synaptic transmission.

To explain the above results we have to postulate that depolarization of horizontal cells evokes a hyperpolarization in cones confined to the region of the presynaptic membrane only. Such a local polarization is made possible by an hypothesis according to which the current generated by horizontal cells, or passed through their membrane, can flow in some part through the cones and displace the potential in their presynaptic membrane.[7,10] During artificial depolarization of horizontal cells the current should flow out of them. Current crossing the subsynaptic membrane of horizontal cells flows partly through the presynaptic membrane of cones and evokes local hyperpolarization there. Certainly this current also has to cross the somatic membrane of cones and has to evoke there a potential drop of opposite polarity. However the density of the current in the somatic membrane of cones should be much less than that in

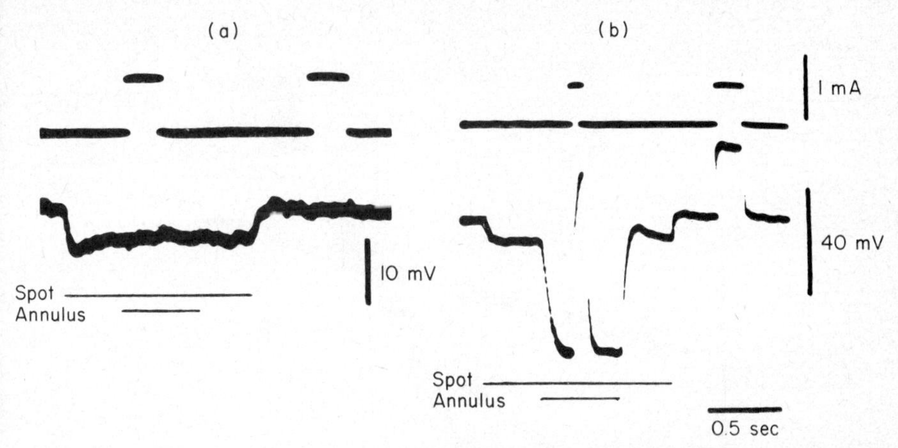

Fig. 7. Absence of change of potential in the somatic membrane of the cone with depolarization of horizontal cells. Upper trace—intensity of longitudinal current, upward deflexion corresponds to current which depolarizes horizontal cells (see b). Lower traces—responses of the cone (a) and of horizontal cell (b). Diameter of light spot is 250 μm, inner and outer diameters of annulus are respectively 250 μm and 750 μm. Records are obtained in one and same penetration on the carp retina, $q = 25\ \Omega\cdot\text{mm}^{-2}$.

presynaptic membrane because of the difference in their areas. Accordingly the potential drop in somatic membrane can be very small. Thus the hypothesis permits one to suppose the significant changes of potential to be localized in the presynaptic membrane.

It has to be noted that voltage-current curves similar to those in Fig. 3 were reproduced in an automatic model which embodies the latest version of the above hypothesis developed by Byzov[7] (Fig. 8). The essential assumption of this model is a supposition that the resistance of the cone presynaptic membrane decreases with depolarization. This assumption was made to explain the nonlinearity of cone membrane observed experimentally by Byzov and Cervetto[11] and the effect of an annulus described in turtles.[1] Reproduction of voltage-current curves of horizontal cells was an accessory result which did not require any additional changes of the model.

Polarization of horizontal cells also produces changes of membrane potential in hyperpolarizing bipolars. These cells were penetrated immediately before the horizontal cells. They responded with hyperpolarization to a spot centered on the microelectrode and with depolarization to an annulus (Fig. 9a). Toyoda has shown that the cells with such properties in carp retina are hyperpolarizing bipolars.[12]

In darkness the depolarization of horizontal cells evoked in bipolars a hyperpolarization which disappeared during hyperpolarizing responses to a

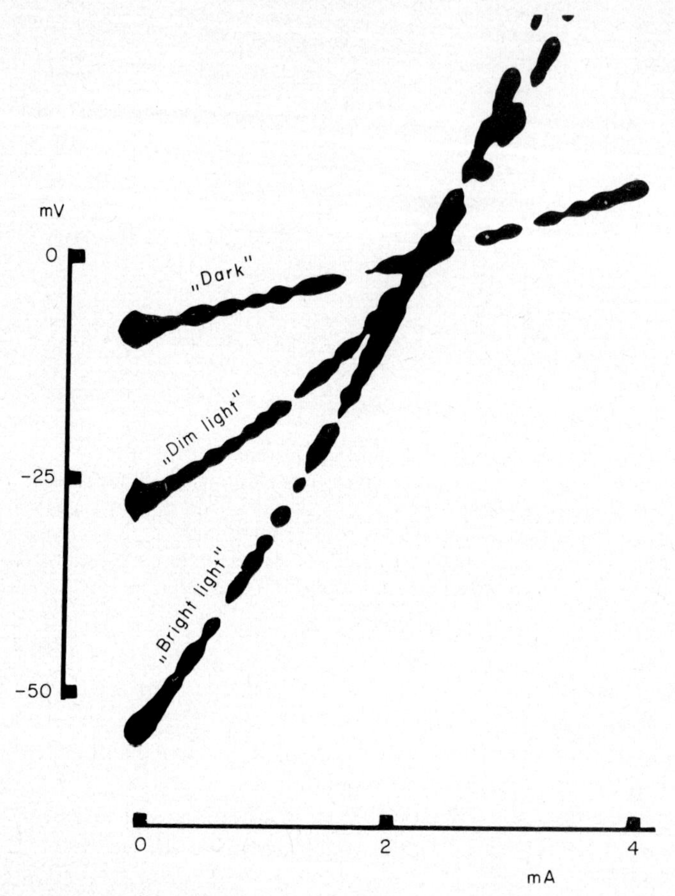

Fig. 8. Voltage-current curves of "horizontal cell membrane" obtained on the automatic model of synaptic transmission in which the feedback to cones was taken into account[7]—see text.

bright light spot and recovered during depolarizing responses to an annulus (Fig. 9b and c). Hyperpolarization of horizontal cells produced depolarization in these bipolars (Fig. 9d) resembling the effect of an annulus.

What is the mechanism by which polarization of horizontal cells influences the potential in the hyperpolarizing bipolars? In view of the absence of changes of potential in the cone somatic membrane during polarization of horizontal cells, we may discuss only two alternative explanations of this interaction. The first is direct synaptic action of horizontal cells upon the hyperpolarizing bipolars, the second is the influence on synaptic transmission between cones and hyperpolarizing bipolars.

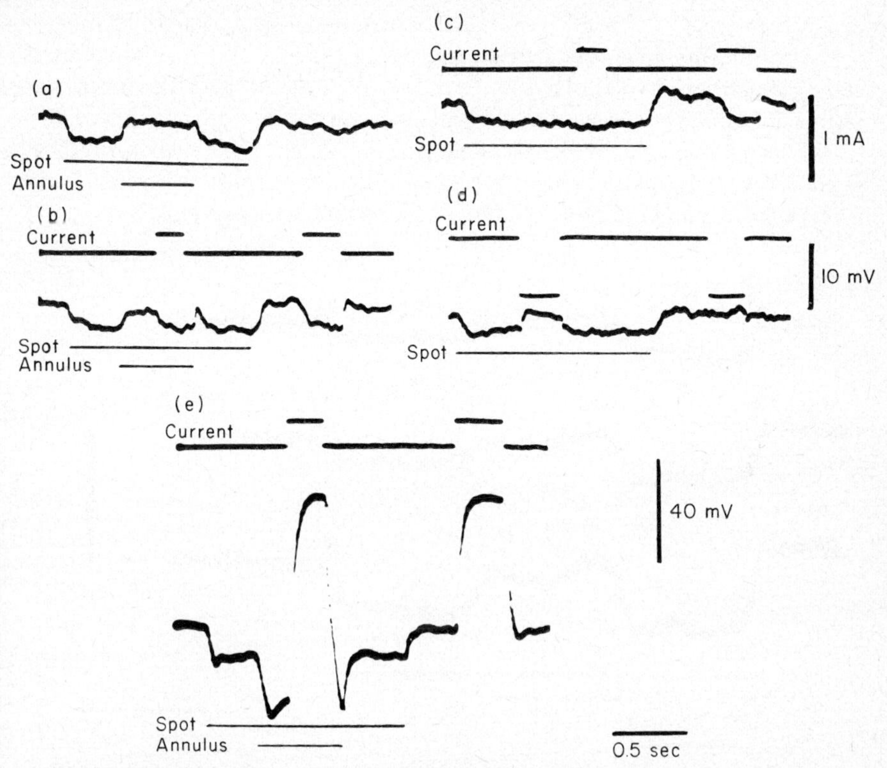

Fig. 9. Responses of a hyperpolarizing bipolar to polarization of horizontal cells in the carp retina. (a) Response of bipolar to spot (diameter 250 μm) and annulus (inner and outer diameters 250 μm and 750 μm). (b), (c) and (d). Changes of potential in the same bipolar with current polarizing the horizontal cells. Upper trace—longitudinal current, upward deflexions (b and c) and downward deflexions (d) correspond respectively to current depolarizing and hyperpolarizing the horizontal cells. (e) Response of horizontal cell to spot and annulus and to depolarizing longitudinal current in the same penetration, $q = 90\ \Omega\cdot\text{mm}^{-2}$.

The first supposition can be checked experimentally. The hypothetical synapses between horizontal cells and hyperpolarizing bipolars must be an inhibitory one. Depolarization of horizontal cells should provoke the liberation of hyperpolarizing transmitter and give rise to i.p.s.p.'s in hyperpolarizing bipolars. The equilibrium potential of this IPSP is expected to be near the level maintained during a bright spot because this level corresponds to the maximal hyperpolarizing responses (Fig. 9c). Therefore the responses of bipolars to polarization of horizontal cells are expected to decrease and even to invert if the bipolar is hyperpolarized with current injected through the microelectrode.

The experiments with intracellular polarization of hyperpolarizing bipolars in the carp retina have been carried out by Toyoda,[12] and we obtained the same results on hyperpolarizing bipolars of turtle (Fig. 10). With the intracellular hyperpolarization of a bipolar its responses both to the spot and to the annulus increased. Depolarization diminished both responses. This result shows that the hyperpolarization of a bipolar by injected current shifts its membrane potential away from the equilibrium levels for responses to the

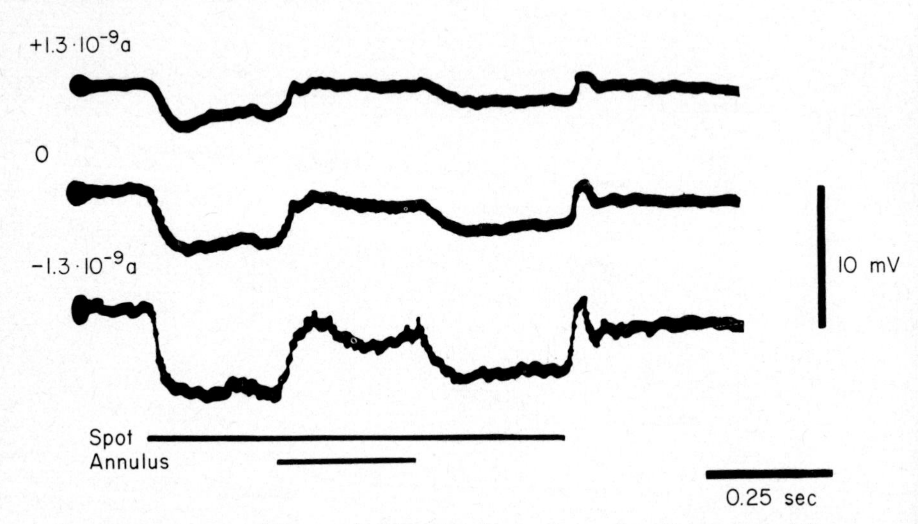

Fig. 10. Responses of hyperpolarizing bipolar in the turtle retina to spot (diameter 210 μm) and annulus (diameters 220 μm and 1200 μm) with polarization by the current injected through recording microelectrode. The intensity of current is indicated near the traces. (−) and (+)—hyperpolarizing and depolarizing currents respectively. In this figure vertical shift does not correspond exactly to real changes of membrane potential with current.

annulus as well as to the spot. This is in contradiction with the hypothesis about an inhibitory synapse of horizontal cells on the hyperpolarizing bipolars. Therefore the second alternative about the influence of polarization of horizontal cells upon synaptic transmission from cones to bipolars is preferable. Such a supposition was made earlier from experiments which have demonstrated the strong effect of polarization of horizontal cell on local ERG in turtles.[13]

The supposition used to explain feedback between horizontal cells and cones is able to account for the effects observed in hyperpolarizing bipolars. By analogy with horizontal cells it seems to be most probable that diminution of liberation of transmitter leads to hyperpolarization in hyperpolarizing bipolars. Therefore current that hyperpolarizes the presynaptic membrane in cones must also evoke hyperpolarization in these bipolars. If the current depolarizing the horizontal cells evokes hyperpolarization of the cone presynaptic membrane and decreases liberation of transmitter, the potential in bipolars should be changed as well. That is, hyperpolarization should occur in hyperpolarizing bipolars, but such a change of potential should be absent if the liberation of transmitter has already been stopped by a bright light spot. Current hyperpolarizing the horizontal cells should evoke depolarization of the cone presynaptic membrane and, as a result, depolarization in hyperpolarizing bipolars. Such changes of potential were really observed in hyperpolarizing bipolars. The suggested explanation implies that one and the same presynaptic membrane subserves the synaptic transmission both to horizontal and bipolar cells.

Thus it is supposed that due to the resistance of the synaptic gap the current flowing out of the subsynaptic membrane of horizontal cells can polarize presynaptic membrane of cones. Such a polarization should be effective only inside the invaginations. Therefore it seems to be probable that bipolars, which have superficial contacts with cones, should be less sensitive to polarization of horizontal cells than bipolars sending their processes into invaginations.

References

1. D. A. Baylor, M. G. F. Fuortes and P. M. O'Bryan 1971, *J. Physiol.* **214**, 265.
2. T. Tomita 1965, *Cold Spring Harb. Symp. quant. Biol.* **30**, 559.
3. V. V. Maksimov 1968, *In* "Synaptic Processes" p. 247, Naukova Dumka, Kiev, in Russian;
   K. I. Naka and W. A. Rushton 1967, *J. Physiol.* **193**, 437;
   A. Kaneko 1971, *J. Physiol.* **213**, 95.
4. Yu. A. Trifonov, L. M. Chailakhyan and A. L. Byzov 1971, *Neurophysiol* (*Kiev*) **3**, 89, in Russian;
   Yu. A. Trifonov, A. L. Byzov and L. M. Chailakhyan 1974, *Vision Res.* **14**, 229.
5. Yu. A. Trifonov and L. M. Chailakhyan 1975, *Biophysica* **20**, 107, in Russian.
6. A. L. Byzov, Yu. A. Trifonov and L. M. Chailakhyan 1975, *Neurophysiol.* (*Kiev*) **7**, 74, in Russian.
7. A. L. Byzov, K. V. Golubtzov and Yu. A. Trifonov 1976, Symposium on "Processing of Information in Visual System," Leningrad, in Russian;
   A. L. Byzov, K. V. Golubtzov and Yu. A. Trifonov 1977, This volume, p. 265.
8. Yu. A. Trifonov 1968, *Biophysica* **13**, 809, in Russian;
   A. L. Byzov and Yu. A. Trifonov 1968, *Vivision Res.* **8**, 817.
9. F. S. Werblin 1975, *J. Physiol.* **244**, 639.
10. A. L. Byzov and Yu. A. Trifonov 1968, *In* "Synaptic Processes" p. 231, Naukova Dumka, Kiev, in Russian.
11. A. L. Byzov and L. Cervetto, 1977, *J. Physiol.* **265**, 85.
12. J. I. Toyoda 1973, *Vision Res.* **13**, 283.
13. A. L. Byzov 1967, *Physiol. J. USSR* **53**, 1115, in Russian.

14b

The Model of Mechanism of Feedback between Horizontal Cells and Photoreceptors in Vertebrate Retina*

A. L. BYZOV, K. V. GOLUBTZOV and JU. A. TRIFONOV

Institute for Problems of Information Transmission, The Academy of Sciences of the U.S.S.R., Moscow, U.S.S.R.

Synaptic transmission between photoreceptors and horizontal cells is carried out by means of a depolarizing transmitter, which is liberated in darkness and ceases to be liberated in light.[1] Feedback is also known to exist between horizontal cells and cones: as shown by Baylor, Fuortes and O'Bryan,[2] hyperpolarization of horizontal cells by current or by illumination with an annulus evokes a depolarizing response in cones.

What is the mechanism of this feedback? The fact that the response of the cone is of opposite sign with respect to the change of potential in horizontal cells seems to indicate the chemical nature of the feedback synapse. But this assumption[2] has until now had no experimental verification: O'Bryan[3] failed to find the expected inversion of this response with polarization of the cone by extrinsic current. We propose here a new explanation, which is based on the assumption that the current generated by subsynaptic membrane of horizontal cells flows, in some part, through the receptors and, when crossing their presynaptic membrane, evokes here a potential drop and changes its resistance. The magnitude of electric coupling between horizontal cells and cones necessary for this mechanism to work was shown to be very small ($\frac{1}{10}$–$\frac{1}{40}$). This corresponds to what can be calculated taking into account the real structure of the synapse.

The model seems to be of wider interest and probably is also applicable to

* This paper was not presented at the symposium. It is included here as it provides a basis for the analysis of the preceding paper.

other synapses, in the first place to those controlled by gradual changes of membrane potential.

Figure 1 shows schematically the structure of the synapse between photoreceptor and horizontal cell (a) and its equivalent circuit (b). The upper part of the circuit is the cone with "resting potential" battery (E_c), its internal resistance (R_c) and presynaptic membrane (R_p). The lower part is the horizontal cell with nonsynaptic (E_{hc} and R_{hc}) and subsynaptic (R_s) membranes; the resistance R_s is low in darkness and high in light (i.e. EPSP's in

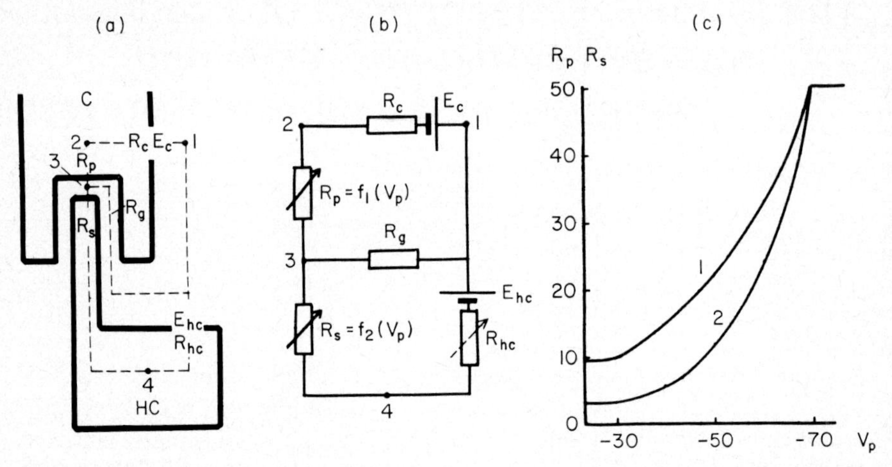

Fig. 1. The model of the synapse between photoreceptor and horizontal cell. (a) schematic representation of synapse, equivalent circuit of which is shown in (b); the external segment is absent in this model. *C*—the cone, *HC*—the horizontal cell, R_c is the resistance of cone somatic membrane (3 kΩ), E_c is the battery of cone membrane (−75 mV), R_p is the resistance of presynaptic membrane of the cone, R_s is the resistance of the subsynaptic membrane of horizontal cell; both R_p and R_s are functions of the potential difference across presynaptic membrane V_p (see (c)), R_g is the longitudinal resistance of synaptic gap (1·8 – 2·6 kΩ), E_{hc} is the battery (−75 mV) and R_{hc} is the resistance of nonsynaptic membrane of horizontal cell (the values of the last resistance are different in measurements with "local" and "diffuse" illumination—see later pp. 269–272); (c) R_p (1) and R_s (2) in kΩ as functions of V_p (in mV); outside the range of −30– 70 mV R_p and R_s are constant in the model.

darkness). R_g is the longitudinal resistance of the synaptic gap. Figure 1c shows R_p and R_s (curves 1 and 2, respectively) as functions of the potential difference across the presynaptic membrane (V_p). The nonlinearity of the presynaptic membrane is accepted in the model in accordance with recent interpretation of the mechanism of liberation of transmitter as a result of Ca^{++} influx in presynaptic endings because of increased permeability.[4] The function $R_s = f_2(V_p)$ reproduces the mechanism of chemical transmission between cone and horizontal cell. Only steady states without transients are reproduced by the model at present. In its last realization the model is made automatic. The role of the presynaptic membrane is played by the tunnel diode, which has characteristic ($R = f(V)$) similar to that shown in Fig. 1c (curve 1). The subsynaptic membrane was mimicked by a field effect transistor, the resistance of which was controlled through an operational amplifier by the potential difference across the presynaptic membrane. This function was similar to that shown by curve 2 in Fig. 1c. The absolute values of resistances in the automatic model were different from those of the "manual" model. However the ratios of resistances, as well as the functions $R_p = f_1(V_p)$ and $R_s = f_2(V_p)$, were almost the same. That is why the results obtained on both models practically coincide.

In the automatic model the action of light is reproduced, not by polarization of the cone from a separate source (as in the "manual" model), but by an increase of membrane resistance of the external segment (R_{es}), put in parallel with the somatic membrane of the cone. Some other details of the model are described elsewhere.[5]

The qualitative operation of the model can be illustrated by three examples which show the manifestation of feedback at pre- and postsynaptic levels.

Modelling of Depolarizing Effect of Annulus

Figure 2a shows the effect of illumination of the retina of the turtle with an annulus. When switched on after the spot, it evokes an additional hyperpolarization in horizontal cells and depolarization in the cones. Similar depolarization in the cones can be evoked by injection of hyperpolarizing current in horizontal cells.[2] In Fig. 2b this effect is reproduced in the automatic model. The upper curve shows the response to the "light" (e.g. to the increase of R_{es}) in a cone, the lower curve is the same in a horizontal cell. One can see that hyperpolarization of a horizontal cell by extrinsic current during the "light period" is accompanied by depolarization in the cone. Qualitatively, this effect can be explained as follows: the current leaking into the cone during polarization of the horizontal cell slightly hyperpolarizes the somatic membrane of the cone and strongly depolarizes its presynaptic membrane. As a result, the resistance of the latter drops and the somatic membrane of the cone is shunted, i.e. depolarized.

Voltage-Current Relationships of the Cone Membrane

In some experiments the shape of the V-I curves of the cones indicates the nonlinear properties of their membrane. In particular, the depolarizing part of the V-I curve measured in the light with current pulses of moderate intensity is

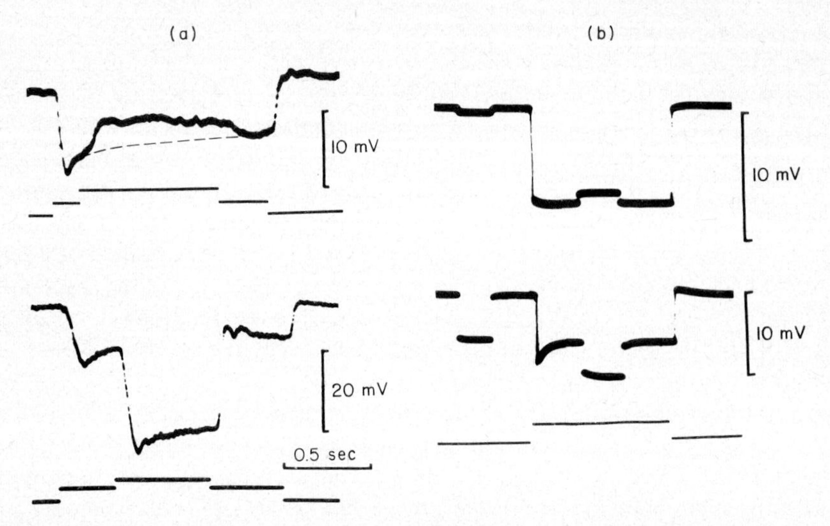

Fig. 2. Modelling of depolarizing effect of annulus in the cone. (a) an experiment on the turtle retina. The lower line indicates the onset of the light spot (diam. 125 μm) and of annulus (inner diam. 150 μm, outer diam. 1·2 mm) superposed. The upper curve is the response of the cone, the lower curve the response of horizontal cell. (b) reproduction of the effect of annulus in automatic model (about details of this model see text). Hyperpolarizing current pulse passed through "horizontal cell" (the lower curve) during the "light" (indicated below) evokes depolarizing deflection in the "cone" (upper curve). $R_s^{min} < R_{hc} < R_s^{max}$.

steeper than the hyperpolarizing one (Fig. 3a).[6] Nonlinearity of the rod membrane was also reported by Werblin.[7] The model reproduces this nonlinearity (Fig. 3b). It can be also seen in the model that in darkness the hyperpolarizing part of the curve is somewhat more steep than the depolarizing one (Fig. 3b). The same was observed experimentally in the retina (Fig. 3a; see also reference 7). As shown by the measurements in the model, the feedback from horizontal cells plays a significant role in the nonlinearity of *V-I* curve in diffuse light, but not in darkness.

Voltage-Current Relationships of Horizontal Cell Membrane

The properties of the horizontal cell membrane, in particular its *V-I* relationships in fish retina, were described in detail earlier.[1] Figure 4a illustrates one example of such a *V-I* curve taken in darkness (the upper curve) and in light of three different intensities. One can see the inversion of light response near zero level of membrane potential (the intersection of the curves) and the nonlinearity of nonsynaptic membrane of horizontal cells revealed in brightest light. These features of *V-I* curves were shown to reflect the

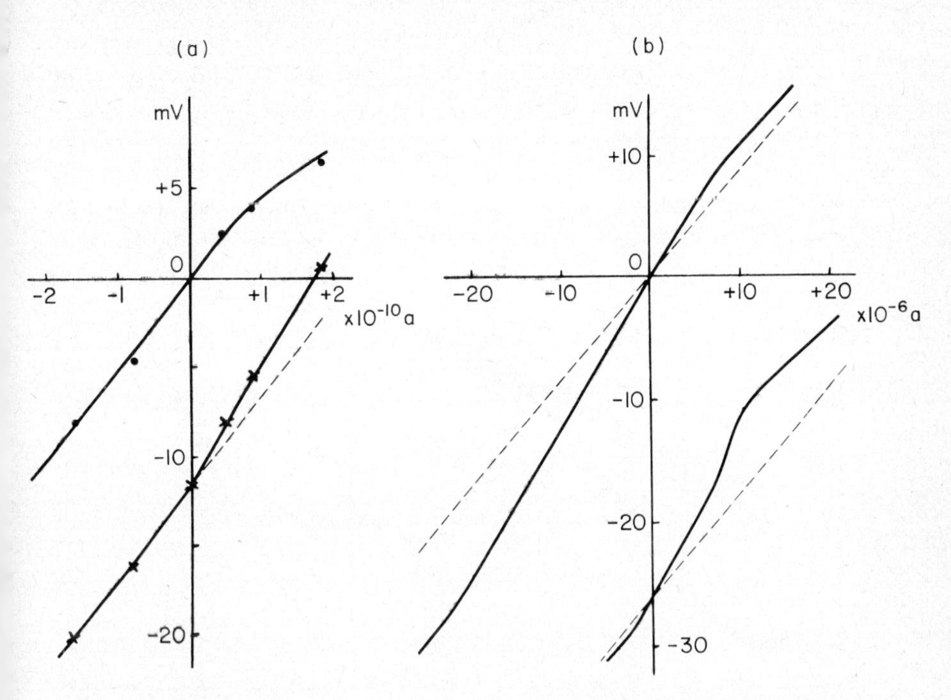

Fig. 3. The voltage-current relationships of the cone membrane in turtle retina (a) and in the model (b). (a) *V-I* relationships in darkness (points) and in diffuse light (crosses). The dashed line is the continuation of the hyperpolarizing part of the curve in light.[6] (b) *V-I* relationships in the model in "darkness" ($R_{hc} = 200$ kΩ, so that $R_{hc} \gg R_s^{max}$) and in "diffuse light" ($R_{hc} = 0{\cdot}2$ kΩ, so that $R_{hc} \ll R_s^{min}$). The dashed lines show the *V-I* curves in the case when R_p = const. = 44 kΩ and R_s = const. = 40 kΩ. The slopes of the lines for "light" and "darkness" in this model are the same because there is no external segment membrane, the resistance of which is normally controlled by the light.

properties of horizontal cells themselves. However there is one feature which cannot be explained this way: with weak lights the curves starting from the intersection point are fused with the curve for the brightest light. To explain this phenomenon, the hypothesis on the mechanism of feedback, as developed here, was applied. The current leaking through the cone during depolarization of horizontal cell should hyperpolarize the presynaptic membrane of the cone and as a result decrease the liberation of transmitter from it. Accordingly, the subsynaptic membrane of horizontal cells should be driven to the more bright light state. This phenomenon is to be expected during a weak light, when the presynaptic membrane is slightly hyperpolarized and therefore its resistance is not too low as in darkness. This explanation was confirmed by the measurements in the model (Fig. 4b). Here the curve for the "brightest light" has no bend in its lower part because the resistance of the nonsynaptic membrane of horizontal cells was assumed linear in the model. But the approach of the curve for a "weak light" to the curve for a "bright light" with depolarization of horizontal cell is evident. This phenomenon is the result of feedback, because after its elimination (by short-circuiting R_g) the *V-I* curves for all lights became linear. It is interesting to note that this property of *V-I* curves of horizontal cells was not taken into account during construction of the model; it was reproduced automatically without any changes in the model.

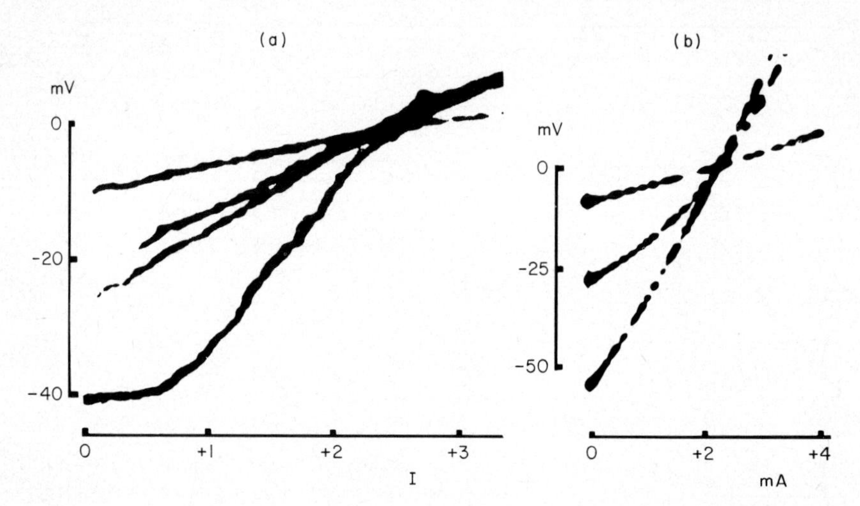

Fig. 4. The voltage-current relationships of the horizontal cell membrane in the pike retina (a) and in the automatic model (b). In both cases only depolarizing currents were passed. In (a) the upper curve was taken in darkness, two middle curves with light of low intensities, the lower curve with light of saturating intensity. Intensity of the current on horizontal scale in relative units. In (b) the upper curve is in "darkness" (low resistance of external segment membrane R_{es}), the middle curve in the "light" of moderate intensity, the lower curve in the "light" of maximal intensity (R_{es} is maximal). $R_s^{\min} < R_{hc} < R_s^{\max}$.

Thus the model reproduces qualitatively the main experimental observations concerning the manifestation of feedback between horizontal cells and receptors. Among the assumptions on which the model is based two are most important. The first is the leaking of some current generated by horizontal cells into the receptors, i.e. the existence of some electrical coupling between both cells. As shown previously, very small coupling is enough for the normal functioning of the model. More critical seems to be the second supposition about the nonlinearity of the presynaptic membrane of the cone. This supposition is not very original and is based on the analogy with other synapses.[4] Moreover we have now some direct experimental evidence showing the nonlinearity of cone synaptic membrane (these data will be published separately). It has to be noted that putting a Ca^{++} battery into the presynaptic membrane should only act to increase all the effects described. The function $R_p = f_1(V_p)$ (Fig. 1c) is also not very critical. Somewhat different exponents as well as linear functions were taken with qualitatively similar results. The only important point is the existence of a negative slope in the V-I relationship of the presynaptic membrane.

Let us consider now the role of the resistance of the cone cell membrane R_c in the mechanism of feedback. If R_c is relatively very small ($R_c \ll R_p$), the potential drops generated across it by a feedback mechanism should be also small and can be below experimental threshold. On the other hand, it is evident from the equivalent circuit of Fig. 1b that in this case the feedback, as manifested at the level of synapse, should be not less effective, because practically it doesn't depend on R_c. Probably, this situation is realized in fish retina, where the work of the feedback mechanism is manifested clearly at the level of horizontal cells,[8] but it has no reflection in cone responses of the type shown in the cones of the turtle.[2]

It is interesting also to note that in the model the potential difference in somatic (V_c) and presynaptic (V_p) membranes of the cone are different, and change differently during excitation. Figure 5 is taken in the model and illustrates the relationship between V_c and V_p with different conditions of illumination. Firstly, V_p is everywhere somewhat less than V_c (because of the potential drop across R_g). Secondly, with darkening of a single cone (by its depolarization from a separate source, and with a very low R_{hc} 0·2 kΩ—so as not to shift the membrane potential in horizontal cells from the light level), V_p decreases much more strongly than V_c. When the whole retina is made dark (in the model, with the same depolarization, but with higher R_{hc}—20 kΩ, so that $R_s^{min} < R_{hc} < R_s^{max}$), the difference between V_p and V_c is almost absent. This property of the model helps to understand why the "working range" of cone membrane potentials is so narrow (only 20–25 mV) and is shifted to the hyperpolarizing side (from about −40 mV in darkness to about 60 mV in bright light). On the presynaptic membrane, which is the output of the cone, this range is wider at the expense of stronger depolarization with local darkening. It can be seen from this that the system with feedback, as described, should discriminate well small dark objects against light backgrounds.

The important parameter of the model is the resistance of the synaptic gap R_g. It determines the fraction of current flowing through the cone. Thus even very small changes of R_g should significantly modify the effectiveness of feedback. Figure 6 illustrates the changes of input signal in horizontal cell (e.g. the change of R_s) with small alterations of R_g during some moderate constant "illumination" of the cone. The increase of R_g by 20% results in the decrease of R_s to less than one half of its original value; at the same time, the membrane potential of the cone (V_c) doesn't change significantly.

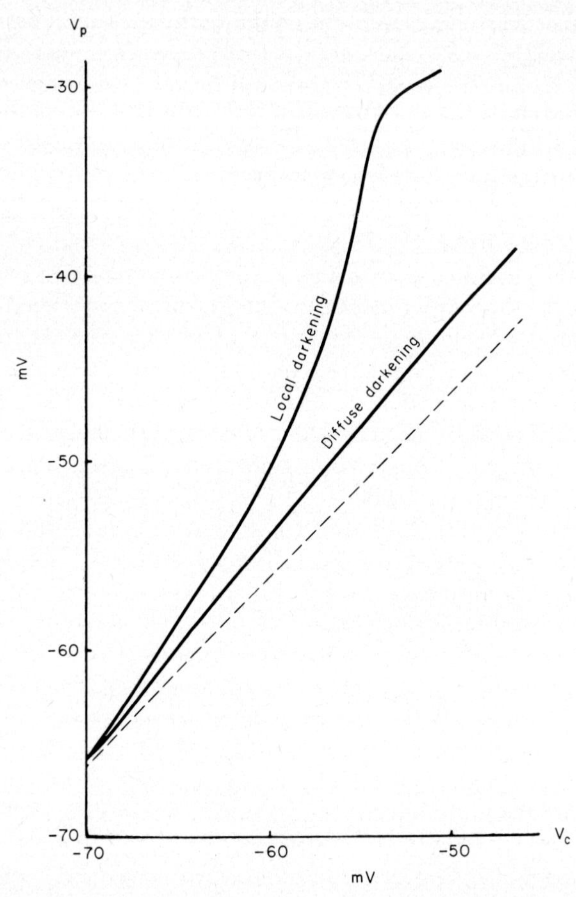

Fig. 5. The relationship obtained in the model between the potential difference in presynaptic (V_p) and somatic (V_c) membranes of the cone, when the latter is depolarized by the "darkening" (e.g. injecting of depolarizing current inside the cone). The upper curve corresponds to "local darkening" ($R_{hc} = 0{\cdot}2\ \mathrm{k}\Omega$, so that $R_{hc} \ll R_s^{\min}$), the lower curve to "diffuse darkening" ($R_{hc} = 20\ \mathrm{k}\Omega$, so that $R_s^{\min} < R_{hc} < R_s^{\max}$). The slope of the dashed line is 45°.

Thus, according to the hypothesis, the feedback between horizontal cells and receptors looks like a positive feedback with local stimulation and like negative feedback with stimulation of the surrounding receptors. The feedback is fulfilled by means of electric current, although without special electrical synapses. Their role is played by the same membranes (pre- and subsynaptic), which are responsible for the direct (chemical) transmission, and the partial isolation from outer media is realized by the resistance of synaptic gap. We can therefore say that the output signal of horizontal cells which

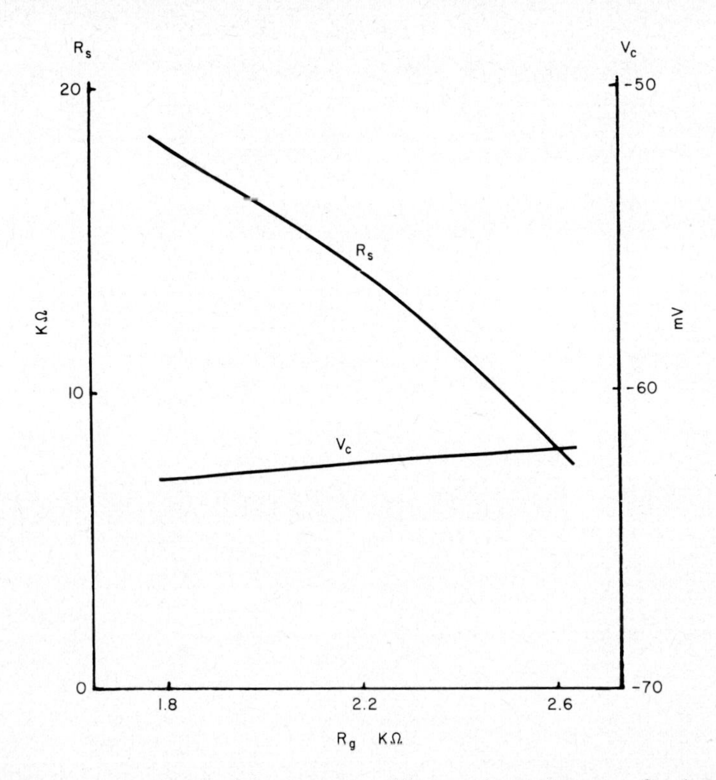

Fig. 6. The "input signal of horizontal cell" (i.e. the resistance of subsynaptic membrane R_s—the left ordinate) as a function of resistance of synaptic gap R_g (horizontal scale) with a constant "moderate illumination" ($R_p = 28$ kΩ; $R_s = 18$ kΩ). The increase of R_g from 1·8–2·6 kΩ results in the decrease of R_s less than one half; at the same time the cone membrane potential V_c (right ordinate) doesn't change significantly. R_{hc} is 0·2 kΩ.

controls the liberation of chemical transmitter by cones is not the potential but the currents. There is some evidence indicating that interaction between horizontal cells and bipolars is carried out by means of the same current.[8]

References

1. Yu. A. Trifonov 1968, *Biophysica* **13**, 809 (in Russian);
   Yu. A. Trifonov, A. L. Byzov and L. M. Chailahian 1974, *Vision Res.* **14**, 229;
   Yu. A. Trifonov, L. M. Chailahian and A. L. Byzov 1971, *Neurophysiology* **3**, 89 (in Russian).
2. D. A. Baylor, M. G. F. Fuortes and P. M. O'Bryan 1971, *J. Physiol. London* **214**, 265.

3. P. M. O'Bryan 1973, *J. Physiol. London* **235**, 207.
4. B. Katz and K. Miledi 1970, *J. Physiol. London* **207**, 789;
J. L. Blioch and E. A. Liberman 1968, *In* "Synaptic Processes", Kiev, "Naukova Dumka", p. 62 (in Russian).
5. A. L. Byzov 1977, *Neurophysiology* **9**, 68 (in Russian).
6. A. L. Byzov and L. Cervetto 1977, *J. Physiol. London* **265**, 85.
7. F. S. Werblin 1975, *J. Physiol. London* **244**, 53.
8. Yu. A. Trifonov and A. L. Byzov 1977, This volume, p. 252.

15

Synaptic Organization of Retinal Receptors

ARNALDO LASANSKY

Laboratory of Neurophysiology, National Institute of Neurological and Communicative Disorders and Stroke, National Institutes of Health, Bethesda, Maryland, U.S.A.

Rods and cones interact with second-order neurons at two kinds of specialized contacts that differ from one another most conspicuously, but not exclusively, by the presence or absence of an opaque ribbon in the adjacent receptor cell cytoplasm. When a ribbon is associated with the junction (Fig. 1), it usually bisects a wedge-shaped projection of the receptor cell ending, the synaptic ridge, its outer edge separated from the cell surface by another opaque body, the arciform density.[1] Depending on the fixation procedure, a variable number of vesicles is found near the ribbon. Thus, few of them are present at this location when the tissue is first fixed in aldehyde (Fig. 1), but they tend to form a continuous layer on each side of the ribbon when the initial fixative solution contains osmium tetroxide (Figs 9 and 12). Other features of the ribbon junctions are an increased width of the intercellular space, which may contain a small amount of opaque material, and an opaque cytoplasmic lining under the plasma membrane of the processes making contact with the sides of the synaptic ridge (Fig. 1). Such processes have been shown to belong to horizontal cells in all the instances thus far investigated.[2,3,4,5]

Those junctions not associated with a ribbon can be further subdivided into two varieties, basal and distal junctions, of which at least the former may be assumed to be a synapse because it is the only contact that some bipolar cells make with the receptors.[3,4] The basal junctions are found at the basal surface of the receptor cell endings,[6] and are marked by a widened and regular

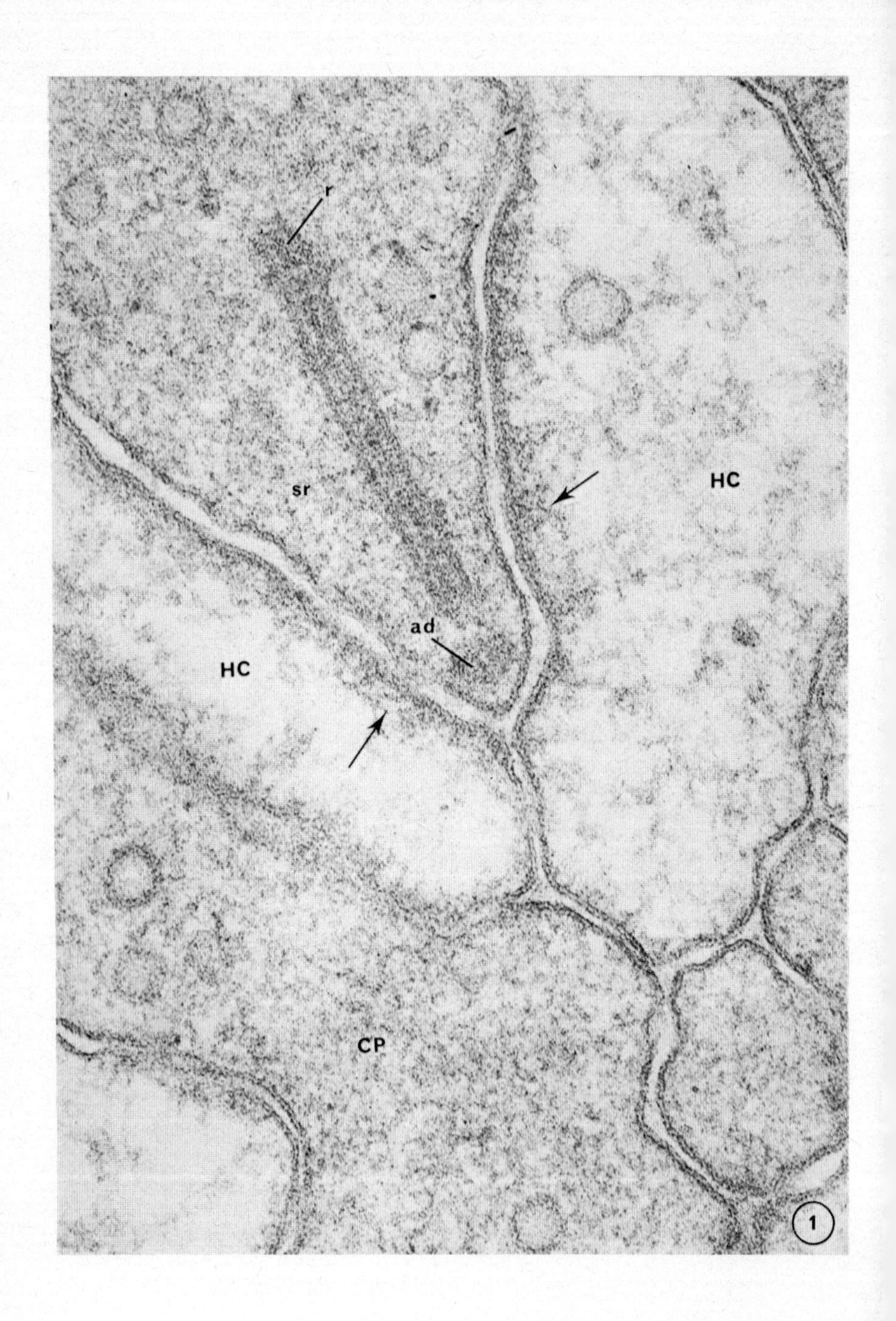
r
sr
HC
ad
HC
CP
1

intercellular gap occupied by an opaque material that in most electron micrographs appears as either homogeneous or faintly cross-striated (Fig. 2). In suitable planes of section, however, the intercellular material can be shown to be highly organized,[4,6] appearing in some cross-sections of the junctions as regularly spaced opaque bars which span the intercellular gap (Fig. 3) and probably represent a lateral view[4] of an orthogonal lattice revealed by tangential sections of the contact area (Fig. 4). The junctional membranes are often symmetrically lined by opaque cytoplasmic material (Fig. 2), but sometimes the opaque lining is seen only under the receptor cell membrane (Fig. 6). Conspicuously absent is any crowding of synaptic vesicles on either side of the junctions. The bipolar cell processes engaged in basal junctions are ordinarily devoid of vesicles (Fig. 6), although a few can be seen occasionally (Fig. 2), while the receptor cell cytoplasm adjacent to the basal junctions contains a lower concentration of synaptic vesicles, under any conditions of fixation, than the rest of the ending (Figs 2 and 6).

A good opportunity to compare the structure of ribbon and basal junctions is provided by the retina of the salamander, because some ribbon junctions engage only one horizontal cell process and therefore their general configuration is similar to that of basal junctions (Figs 5 and 6). The latter exhibit an opaque cytoplasmic lining only under the receptor cell membrane (Fig. 6), while at ribbon junctions it is seen only under the horizontal cell membrane (Fig. 5). At both contacts the intercellular gap is widened, but at ribbon junctions its opaque content is sparser and not obviously structured. Finally, there is no counterpart for the ribbon and arciform density at basal junctions (Fig. 7). The features of ribbon junctions appear to be equivalent to those of other chemical synapses, the receptor cell ending being the presynaptic element.[8] On the other hand, no such parallel can be drawn for the basal junctions, particularly because of their lack of close association with vesicles. Nevertheless, their polarity is generally accepted as being the same as for ribbon junctions, mostly because it would seem to be a reasonable assumption to make. Contacts which are structurally similar to basal junctions have been observed between basket cell axon terminals in the cerebellar cortex, but their function remains undetermined.[9]

Fig. 1. Electron micrograph of a cone ribbon junction in the retina of the turtle (*Pseudemys scripta elegans*). Each side of the synaptic ridge (*sr*) makes contact with a horizontal cell process (HC) that shows an opaque lining (arrow) under the plasma membrane. Fixed in 2% glutaraldehyde in 0·1 M phosphate buffer (pH 7·4) and post-fixed in 1% osmium tetroxide in the same buffer. Stained in block with uranyl acetate; sections stained with uranyl acetate and lead citrate. CP, cone pedicle cytoplasm; *r*, synaptic ribbon; *ad*, arciform density. From reference 4, by permission of the Royal Society of London. ×157,000.

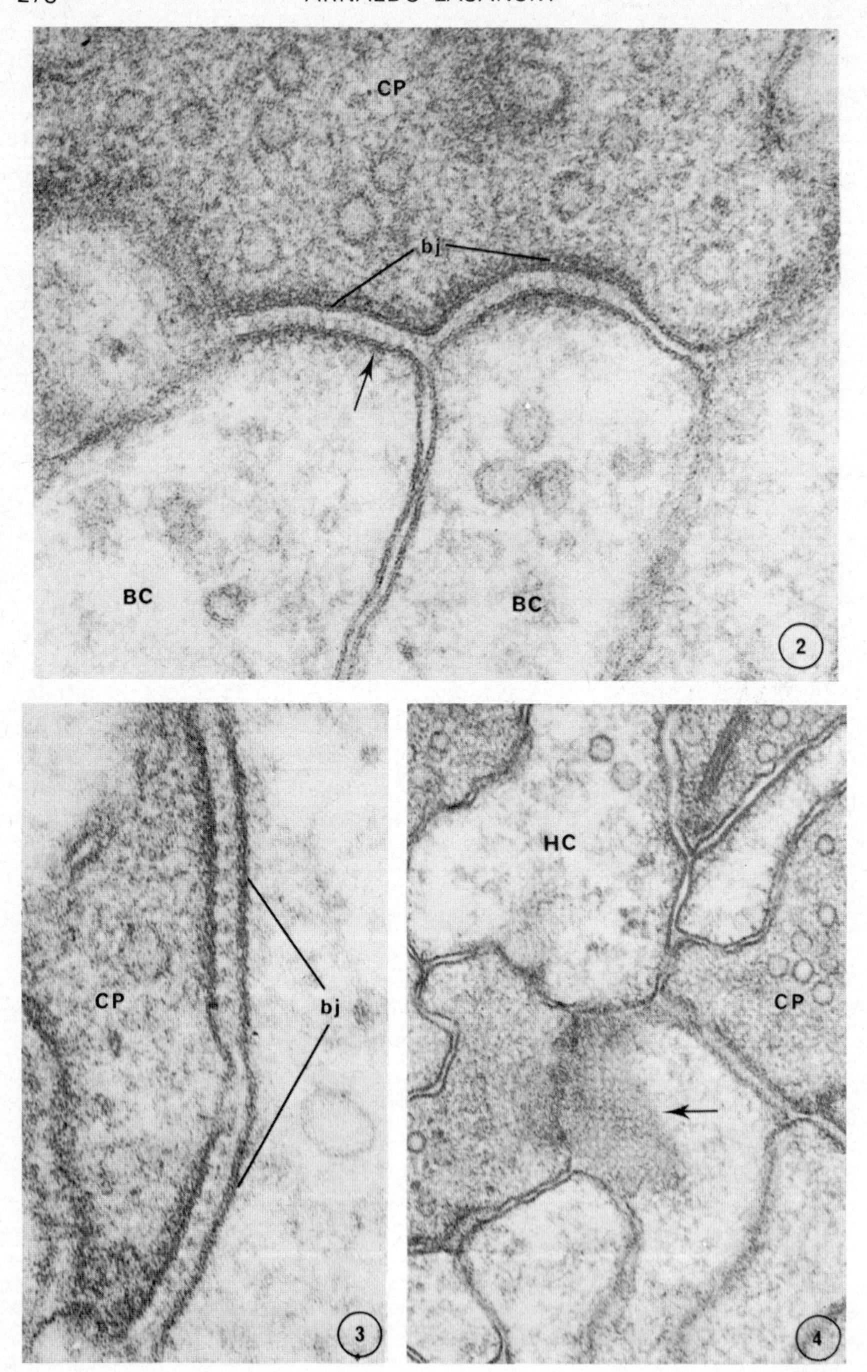
CP
bj
BC
BC
2
CP
bj
3
HC
CP
4

Variant forms of both types of junction are often observed. Basal junctions may exhibit a narrower and nonuniform gap width, but the specialized nature of the contact can still be discerned from the presence of opaque material in the gap and on the cytoplasmic surface of the receptor cell membrane (Fig. 8). These contacts have been referred to as of the narrow-gap type,[4, 5] to distinguish them from those described above, and constitute the whole extent of some of the basal junctions in salamander retina,[5] while in turtle, when present, they alternate with wide-gap segments within a single basal junction.[4] On the other hand, at some ribbon junctions there is no opaque lining under the membrane of the processes making contact with the receptor (Fig. 9). In salamander retina such processes may belong to horizontal or bipolar cells, the latter making only this kind of ribbon contact in all species studied thus far, whether their branchlets end on the sides of the synaptic ridge[5] or at the apex (Fig. 13).

The second variety of non-ribbon junction has been termed distal junction with reference to its position relative to the synaptic ribbon, since the processes that make distal junctions are also engaged at a ribbon junction by the same receptor ending.[4, 5] Distal junctions, therefore, cannot be assumed to be synapses in the way that basal junctions can, because they are not the only contact at which a given neuron or process may interact with a receptor. Nevertheless, their synaptic nature may be hypothesized in some instances because of their close structural resemblance to basal junctions. Such is the case for distal junctions between cones and horizontal cells in turtle (Fig. 10) and salamander (Fig. 11) retinae, while the same resemblance clearly does not exist at distal junctions between rods and horizontal cells in salamander retina

---

Fig. 2. Basal junctions (*bj*) of a turtle cone. A layer of opaque cytoplasm lines the inner surface of the cell membranes on both sides of the junctions, and the material within one of the junctional gaps shows some cross-striation (arrow). The adjacent cone cell cytoplasm (*CP*) contains vesicles, but they are not focused on the membrane. Vesicles are also seen within the bipolar cell processes (*BC*) involved in the junctions. Fixed and stained as for Fig. 1. From reference 4; by permission of the Royal Society of London. × 126,000.

Fig. 3. Transverse section of a basal junction (*bj*) in turtle retina. Opaque cross-bars are seen spanning the intercellular gap. *CP*, cone pedicle. Fixed and stained as for Fig. 1, except that the block staining was omitted. From reference 6; by permission of the Journal of Cell Biology. × 160,000.

Fig. 4. Face-on view of a basal junction in turtle retina. An orthogonal lattice pattern (arrow) is formed by two intersecting sets of equally spaced lines. Fixed and stained as for Fig. 1. *CP*, cone pedicle. *HC*, horizontal cell process making a ribbon junction with the cone pedicle. From reference 4, by permission of the Royal Society of London. × 65,000.

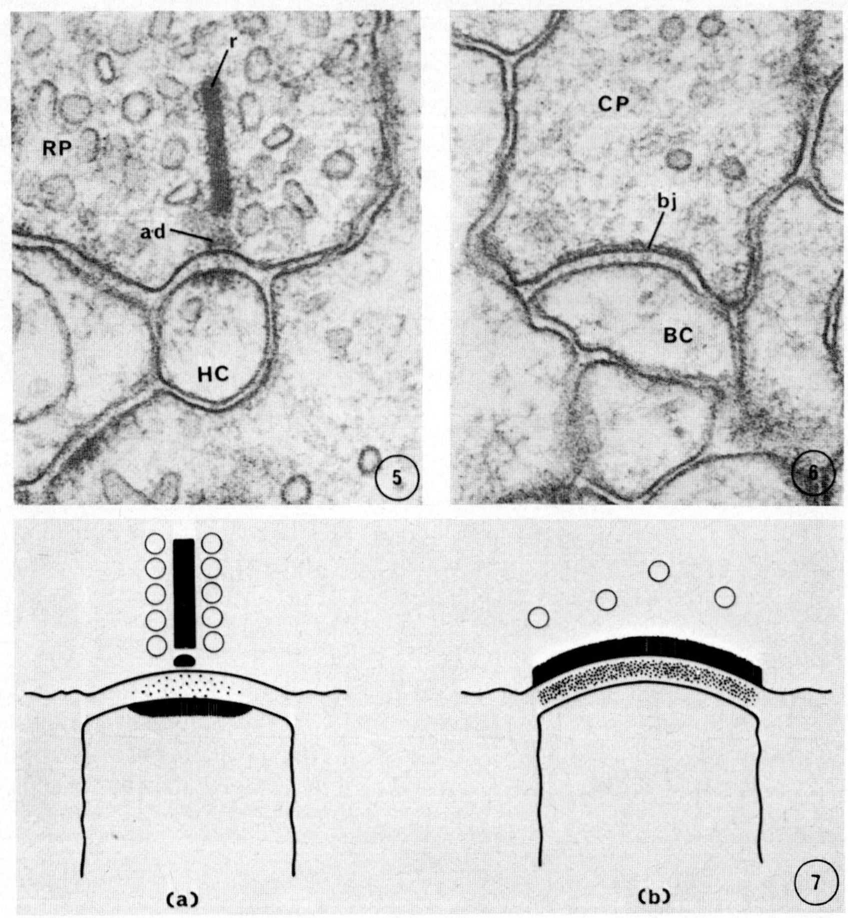

Fig. 5. Ribbon junction of a rod pedicle (*RP*) with a single horizontal cell process (*HC*) in the retina of the larval tiger salamander (*Ambystoma tigrinum tigrinum*). The junctional gap contains some material, and the plasma membrane of the horizontal cell process is lined by opaque cytoplasm. Fixed in a solution containing 2·5 % glutaraldehyde and 1 % osmium tetroxide in 0·1 M phosphate buffer (pH 7·4). Stained in block with uranyl acetate; sections stained with lead citrate. *r*, synaptic ribbon; *ad*, arciform density. From reference 5, by permission of the Royal Society of London. ×86,000.

Fig. 6. Basal junction (*bj*) of a cone pedicle (*CP*) in the retina of the larval tiger salamander. The junctional segment of the cone cell membrane is lined by a thick layer of opaque cytoplasm. Fixed and stained as for Fig. 5. *BC*, bipolar cell process. From reference 5, by permission of the Royal Society of London. ×86,000.

Fig. 7. Diagrammatic representation of ribbon junctions (a) and basal junctions (b) in salamander retina (see text). From reference 5, by permission of the Royal Society of London.

(Fig. 12). It seems possible then that cones, at least, make two kinds of synapses (ribbon and non-ribbon) with the same horizontal cell process.[4,5]

It has been suggested previously,[4] that this dual contact could represent the morphological counterpart of the synaptic feedback known to take place between horizontal cells and cones in turtle retina,[10] the horizontal cells being postsynaptic at ribbon junctions and presynaptic at distal junctions. Vesicular profiles were frequently found within horizontal cell processes involved in distal junctions, and the fact that they were not crowded near the junction did not seem to be a serious contradiction, as vesicle clusters are also absent on either side of basal junctions. Later studies on serial sections, however, revealed that some of the horizontal cell processes involved in distal junctions do not contain vesicles at all,[5] thus casting some doubt on the assumed polarity of the distal junctions, in the event that they are synapses. Finally, intracellular recordings from salamander receptors appear to support the earlier assumption, since only cones, but not rods, show signs of receiving horizontal cell feedback,[11] a finding that would seem to agree with the different appearance of cone and rod distal junctions with horizontal cells (Figs 11 and 12).

Distal junctions are also made by the processes of certain bipolar cells on their way to a ribbon junction. A good example is given in the monkey retina by the invaginating midget bipolar cells, as well as by the rod bipolar cells,[7] whose dendrites, before ending at the apex of a synaptic ridge,[3] make a distal junction identical in appearance to a basal junction of the narrow-gap type (Fig. 13). In turtle retina, processes ending at the apex of cone synaptic ridges, and probably belonging to bipolar cells,[4,7] also make distal junctions with the pedicles, in this instance entirely analogous to basal junctions of the wide-gap type (Fig. 14). The parallel between distal junctions of invaginating bipolar cells and basal junctions should perhaps not be emphasized in view that freeze-fractured preparations show differences in intramembrane structure between the two kinds of contacts.[12] Nevertheless, an analogy is also suggested by the observation in salamander retina that among the terminal branchlets of certain single processes—believed to belong to bipolar cells of the same kind as the invaginating type of other retinae—some end at ribbon junctions, while others end at basal junctions.[5]

Since some bipolar cells make contact with the receptors only at non-ribbon (basal) junctions, while others engage them at both non-ribbon (distal) and ribbon junctions, it follows that the non-ribbon junction is the prevalent type of contact between receptors and bipolar cells. To know why some bipolar cells make in addition ribbon junctions, it is necessary to understand the functional meaning of each type of junction. A good way to begin would be to identify the type of response that can be intracellularly recorded from each kind of bipolar cell.

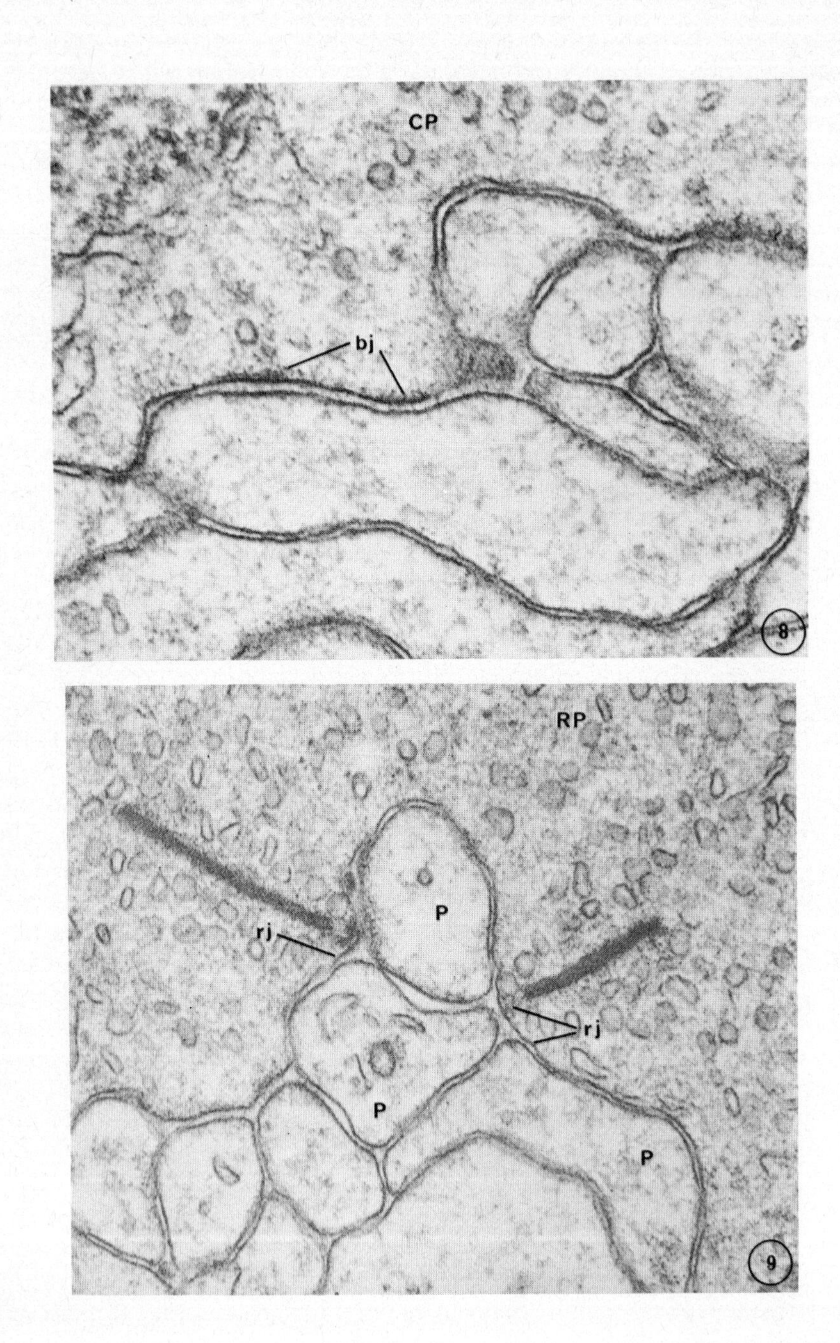
CP
bj
8
RP
P
rj
rj
P
P
9

Fig. 8. Cone basal junction (*bj*) of the narrow gap type in salamander retina. The inner surface of the cone cell membrane shows an opaque lining of uneven thickness. Fixed and stained as for Fig. 5. *CP*, cone pedicle. From reference 5, by permission of the Royal Society of London. ×86,000.

Fig. 9. Rod ribbon junctions (*rj*) in salamander retina. None of the processes (P) of the second order neurons shows an opaque membrane lining. Fixed and stained as for Fig. 5. *RP*, rod pedicle. From reference 5, by permission of the Royal Society of London. ×65,000.

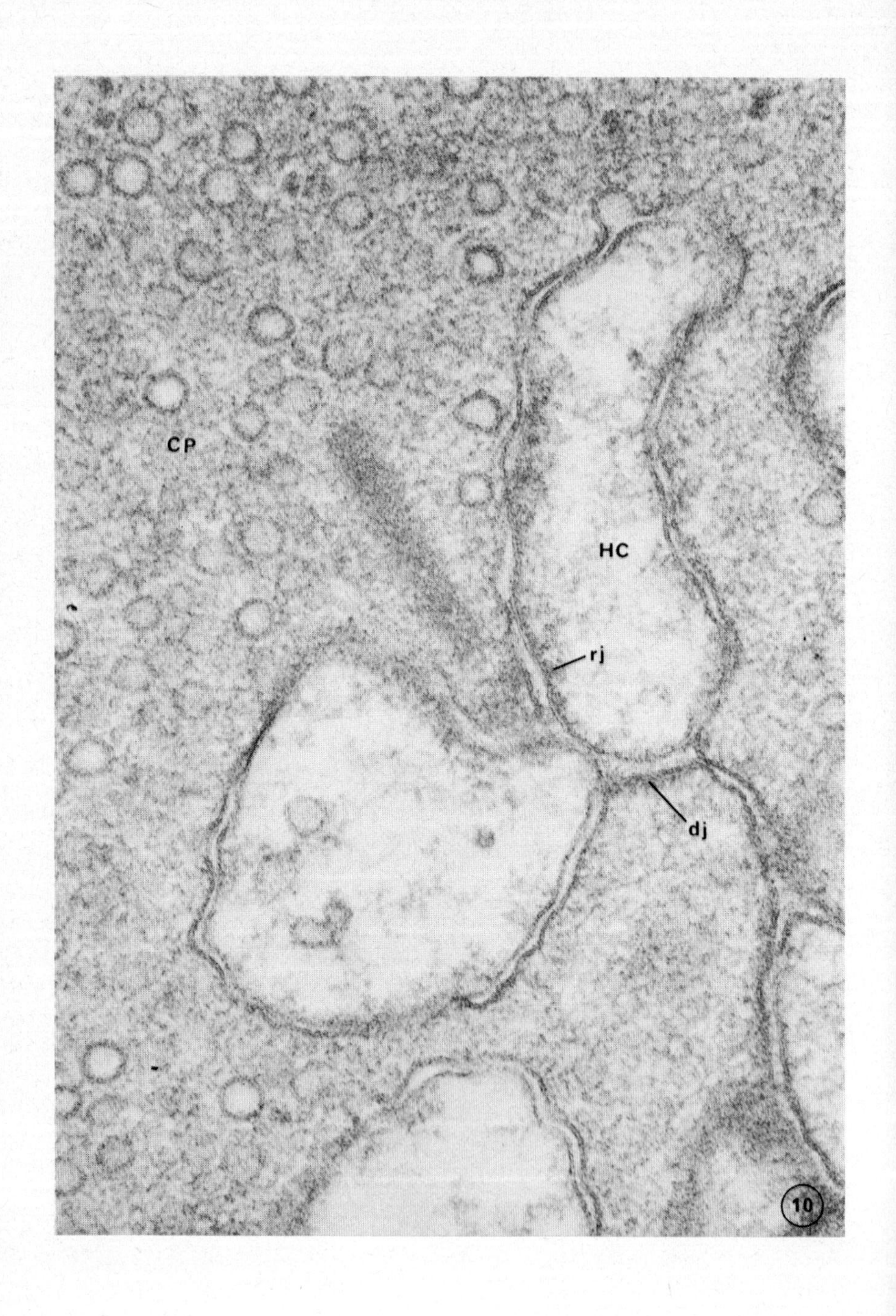
CP
HC
rj
dj
10

Fig. 10. Distal junction (*dj*) between a horizontal cell process (*HC*) and a turtle cone (*CP*). In addition to a uniformly wide gap filled by opaque material, the junction shows an opaque lining on the cytoplasmic surface of the cone cell membrane. The horizontal cell process also makes a ribbon junction (*rj*) with the cone pedicle. Fixed and stained as for Fig. 1. From reference 4, by permission of the Royal Society of London. × 118,000.

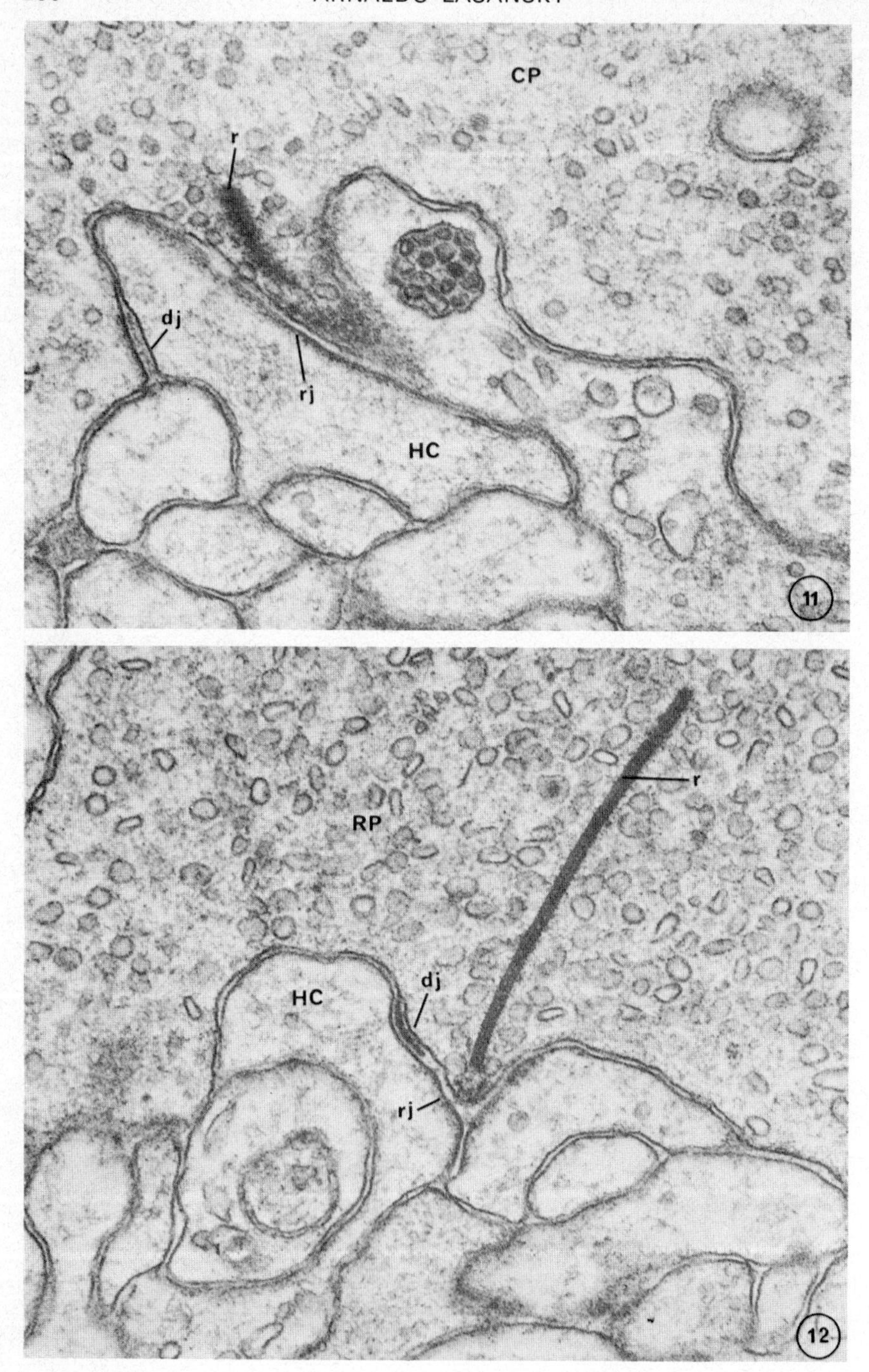
CP
r
dj
rj
HC
11
r
RP
dj
HC
rj
12

Fig. 11. Distal junction (*dj*) between a salamander cone pedicle (*CP*) and a horizontal cell process (*HC*). The junctional gap is wide and contains opaque material. The cytoplasmic surface of the cone junctional membrane is lined by an opaque layer. The pedicle and the horizontal cell process also make contact at a ribbon junction (*rj*). Fixed and stained as for Fig. 5. *r*, synaptic ribbon. From reference 5, by permission of the Royal Society of London. ×65,000.

Fig. 12. Distal junction (*dj*) between a salamander rod pedicle (*RP*) and a horizontal cell process (*HC*). The intercellular gap is widened at the junction and is bisected by a band of very opaque and homogeneous material; neither junctional membrane has an opaque lining. The pedicle also makes a ribbon junction (*rj*) with the horizontal cell process. Fixed and stained as for Fig. 5. *r*, synaptic ribbon. From reference 5, by permission of the Royal Society of London. ×65,000.

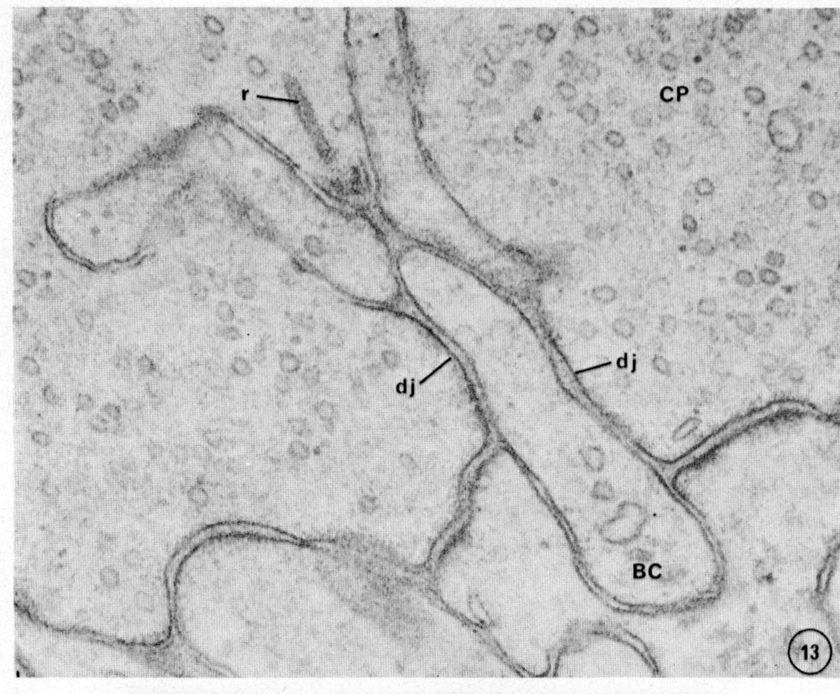
r
CP
dj
dj
BC
13

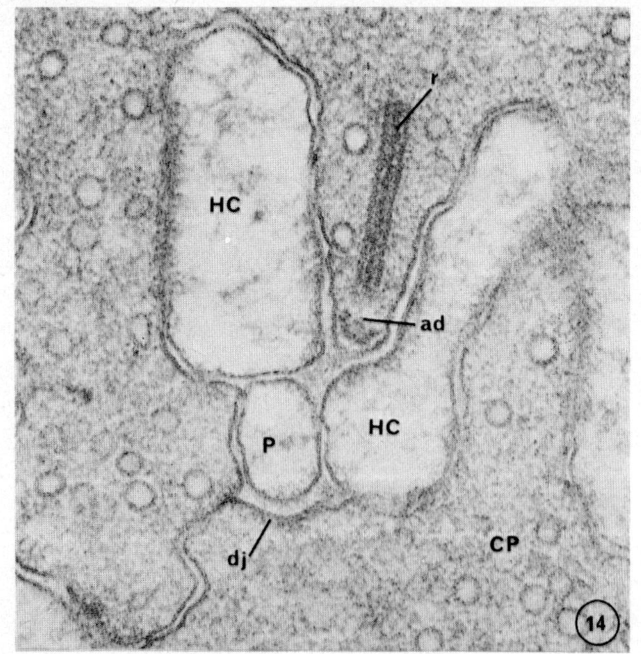
r
HC
ad
P
HC
dj
CP
14

Fig. 13. Ribbon junction at a cone pedicle (*CP*) in monkey retina. A process from an invaginating midget bipolar (*BC*) ends opposite to the apex of the synaptic ridge, while being encircled by a sleeve-like distal junction (*dj*) with the pedicle. Fixed and stained as for Fig. 1. *r*, synaptic ribbon. ×65,000.

Fig. 14. Ribbon junction of a turtle cone. A process thought to belong to a bipolar cell (*P*) ends opposite to the apex of the synaptic ridge, while making a distal junction (*dj*) with the same cone pedicle (*CP*). Fixed and stained as for Fig. 1. *HC*, horizontal cell process; *r*, synaptic ribbon; *ad*, arciform density. From reference 4, by permission of the Royal Society of London. ×84,000.

References

1. F. S. Sjöstrand 1953, Ultrastructure of the retinal rod synapses of the guinea pig eyê, *J. appl. Phys.* **24**, 1422;
   E. D. P. De Robertis and C. M. Franchi 1956, Electron microscope observations on synaptic vesicles in synapses of the retinal rods and cones, *J. biophys. biochem. Cytol.* **2**, 307;
   A. J. Ladman 1958, The fine structure of the rod- bipolar cell synapse in the retina of the albino rat, *J. biophys. biochem. Cytol.* **4**, 459.
2. W. K. Stell 1967, The structure and relationships of horizontal cells and photoreceptor-bipolar synaptic complexes in goldfish retina, *Am. J. Anat.* **121**, 401.
3. H. Kolb 1970, Organization of the outer plexiform layer of the primate retina: electron microscopy of Golgi-impregnated cells, *Phil. Trans. R. Soc. Lond. B.* **258**, 261.
4. A. Lasansky 1971, Synaptic organization of cone cells in the turtle retina, *Phil. Trans. R. Soc. Lond. B.* **262**, 365.
5. A. Lasansky 1973, Organization of the outer synaptic layer in the retina of the larval tiger salamander, *Phil. Trans. R. Soc. Lond. B.* **265**, 471.
6. A. Lasansky 1969, Basal junctions at synaptic endings of turtle visual cells, *J. Cell Biol.* **40**, 577.
7. A. Lasansky 1972, Cell junctions at the outer synaptic layer of the retina, *Inv. Ophthal.* **11**, 265.
8. E. G. Gray and H. L. Pease 1971, On understanding the organization of the retinal receptor synapses, *Brain Res.* **35**, 1.
9. S. Gobel 1971, Axo-axonic septate junctions in the basket formations of the cat cerebellar cortex, *J. Cell Biol.* **51**, 328;
   C. Sotelo and R. Llinas 1972, Specialized membrane junctions between neurons in the vertebrate cerebellar cortex, *J. Cell Biol.* **53**, 271.
10. D. Baylor, M. G. F. Fuortes and P. M. O'Bryan 1971, Receptive fields of cones in the retina of the turtle, *J. Physiol.* **214**, 265.
11. A. Lasansky and S. Vallerga 1975, Horizontal cell responses in the retina of the larval tiger salamander, *J. Physiol.* **251**, 145.
12. E. Raviola and N. B. Gilula 1975, Intramembrane organization of specialized contacts in the outer plexiform layer of the retina. A freeze-fracture study in monkeys and rabbits, *J. Cell Biol.* **65**, 192.

16

Electrical Noise in Turtle Cones

The late E. J. SIMON* and T. D. LAMB

Physiological Laboratory, University of Cambridge, Cambridge, U.K.

Intracellular recordings obtained from cones and rods of the retina of the turtle, *Pseudemys scripta elegans*, show that there are random fluctuations of the membrane potential in darkness and that these are suppressed by bright light.[1, 2] Such light-sensitive noise was observed consistently, but the size of its variance in different cells varied over about a fifty-fold range. Figure 1(a) illustrates a record obtained from an unusually noisy red-sensitive cone. The variance in darkness was about 0·40 mV^2 and was reduced to 0·025 mV^2 by a bright light which hyperpolarized the cell by 14 mV at the initial peak. A more typical cone is shown in Fig. 1(b) in which the noise amplitude of 0·018 mV^2 was reduced to 0·004 mV^2 in bright light.

At first it might seem surprising that light reduces the noisiness of a cone because it might be thought that the random nature of quantal absorption would increase the noise as it does in an invertebrate photoreceptor[3] or in a photomultiplier tube. However, the absorption of light by the visual pigment is believed to affect the cell's conductance by liberating an internal transmitter substance that diffuses to the cell membrane and there closes ionic channels (for review see reference 4). If a residual amount of such a transmitter were present in darkness, the cell might be noisy due to random closure of the light-sensitive ionic channels either spontaneously or as a result of fluctuations in transmitter concentration.[1]

* To the deep regret of his colleagues Elliott Simon died on 30th September 1976.

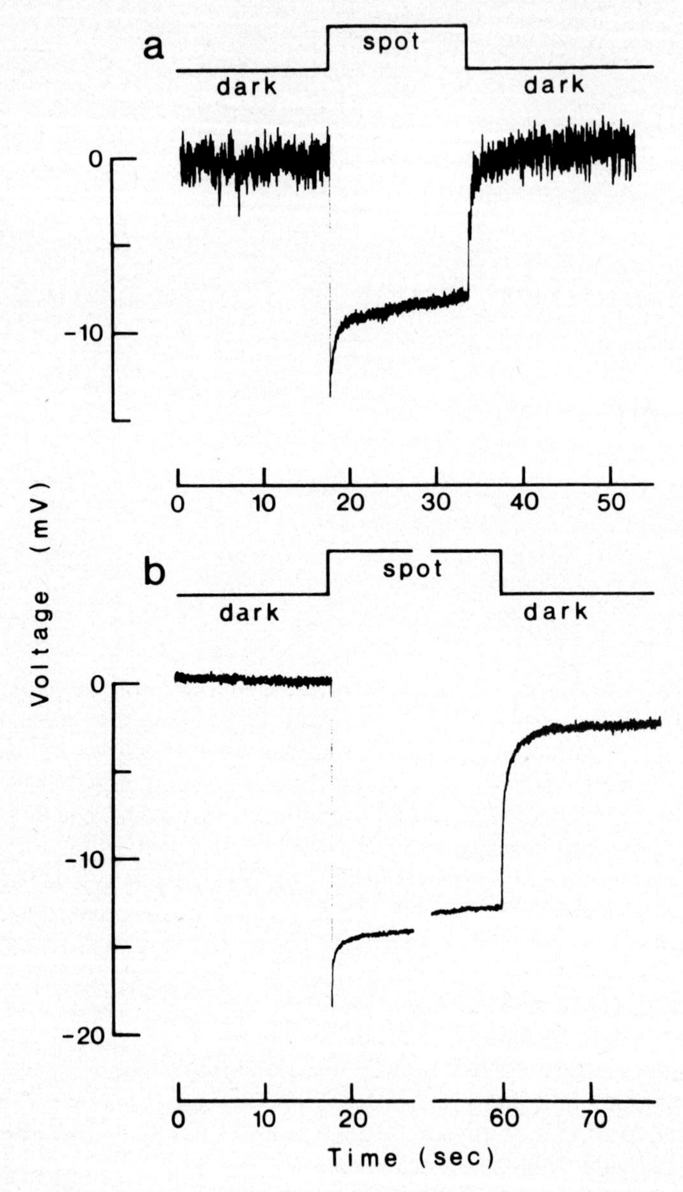

Fig. 1. Recordings from cones in darkness and during steady central illumination. (a) Response of an isolated red-sensitive cone to a 6 μm diameter spot delivering $8{\cdot}05 \times 10^5$ photons μm^{-2} sec^{-1} at 639 nm; (b) Response of a coupled green-sensitive cone to a 132 μm spot delivering $1{\cdot}67 \times 10^6$ photons μm^{-2} sec^{-1} at 559 nm. From Simon *et al.*[1] with permission.

Before examining this hypothesis it is necessary to explain why some cells are much noisier than others. As the quieter cells appear to have larger receptive fields,[1] it is possible that all cones possess a similar noise source and the coupling which is known to exist between cones[5] causes both a reduction in noise and an increase in receptive field size by averaging the responses of a number of cells.

In order to test this notion we measured both the variance and the tightness of coupling in a large sample of cones. Although it is straightforward to measure the variance, a method was required to obtain a quantitative measure of the degree of coupling. If the photoreceptors can be approximated as a passive linear two-dimensional network, then the decay of voltage with displacement from a long narrow slit of light will be exponential.[6] That is

$$V(x) = k \exp\left(-\left|\frac{x}{\lambda}\right|\right) \tag{1}$$

where V is voltage, x is displacement from the slit and the length constant λ is the desired quantitative measure of the tightness of coupling (the larger the λ, the tighter the coupling).

We illuminated the retina with brief flashes in the shape of a slit and moved the image across a cell's receptive field while recording the responses. The spatial profile of Fig. 2 is the result of such an experiment. When plotted semi-logarithmically (b) the descending limbs of the profile are indeed well fitted by straight lines with slopes corresponding to $\lambda = 23$ μm and 19 μm on the two sides, and corresponding exponentials are redrawn in the linear coordinates of (a). Near the receptive field centre the profile deviates from exponential shape probably as a result of the combined effects of finite slit width, finite cell dimensions and light scattering.

It can be seen from the double logarithmic plot of Fig. 3 that there is a strong correlation between intrinsic noise (variance in darkness minus variance in bright light) and λ. The circles are for 22 red-sensitive cones and the triangles are for four green-sensitive cones; values of λ obtained in opposite directions of slit movement for each cell are connected by a horizontal line. Ordinate is voltage variance (mV^2) and abscissa is length constant (μm). The two noisiest cells with variance of 0·43 mV^2 and 0·35 mV^2 had very steep spatial profiles indicating that they might have been completely isolated from other cones, and so are plotted at the left.

The experimental points are well fitted by the curves which are the expected relations between variance and λ for a discrete square grid array (solid curve) and for a distributed network (broken curve) of passively coupled identical noise sources. It is interesting to note that both models predict that variance is inversely proportional to λ^2 for tight coupling—the decrease in r.m.s. noise is directly proportional to λ. In order to compare the models with the

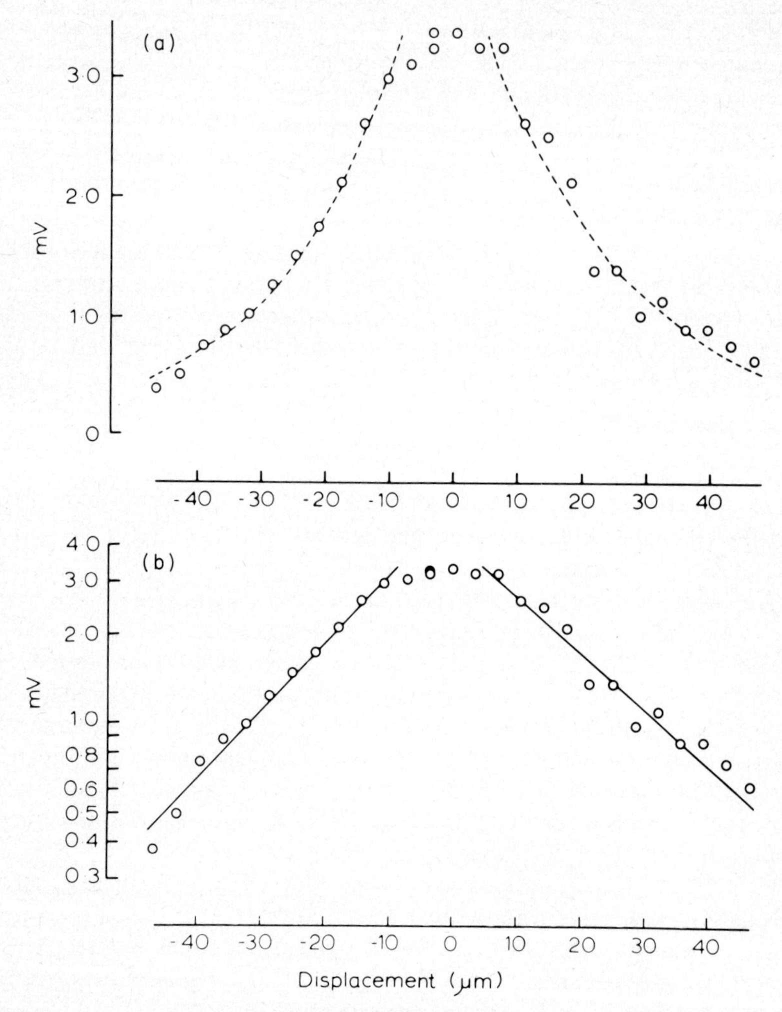

Fig. 2. Profile of spatial sensitivity in a coupled red-sensitive cone. Response to slit displacement is given in linear (a) and semi-logarithmic (b) coordinates. The straight lines in (b) were fitted to the points by eye and represent length constants of 23 μm and 19 μm; corresponding exponential curves are redrawn in (a). Flashes delivered 1·46 $\times 10^3$ photons μm^{-2} at 643 nm. From Lamb and Simon[2] with permission.

experiments it is necessary to scale λ by the average spacing D between interconnected cells and to scale the variance by the variance in an isolated cell, $\sigma^2(V)_{isol}$. The illustrated curves are for values of $D = 15$ μm and $\sigma^2(V)_{isol} = 0{\cdot}4\ mV^2$. These values appear quite reasonable as D is close to the mean cell spacing 17·4 μm of red-sensitive cones measured histologically[7] and as

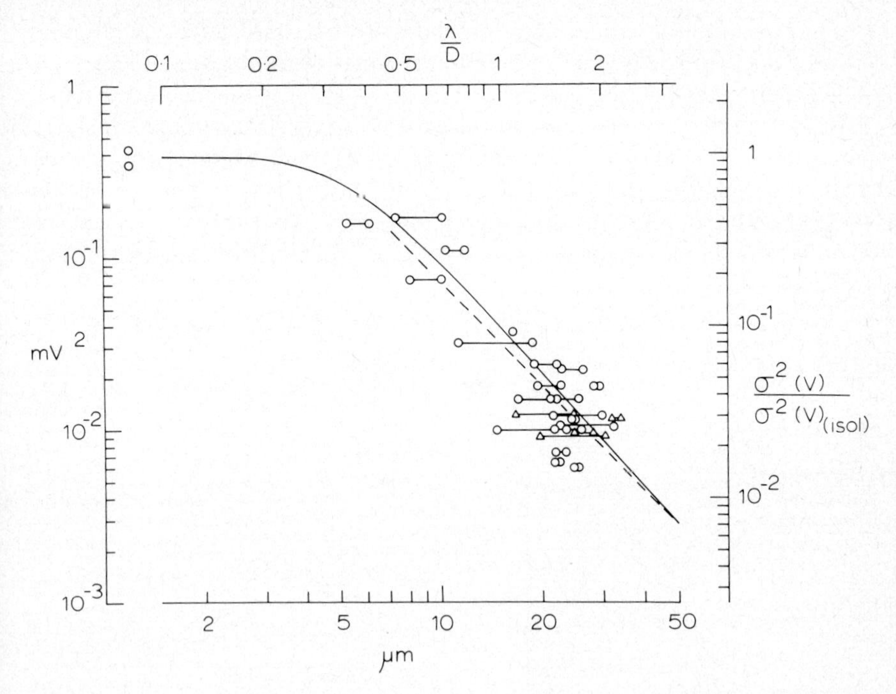

Fig. 3. Correlation of intrinsic voltage variance $\sigma^2(V)$ and length constant λ for cones. Values for red-sensitive (○) and green-sensitive cones (△) are plotted in double logarithmic coordinates. Horizontal lines connect λ measured in opposite directions of slit movement for each cell. The curves are the predictions of a square grid model (solid) and a distributed model (interrupted) of a coupled network with $\sigma^2(V)_{\text{isol}} = 0{\cdot}4\ \text{mV}^2$ and $D = 15\ \mu\text{m}$. The two noisiest cones are plotted at the top left. From Lamb and Simon[2] with permission.

$\sigma^2(V)_{\text{isol}}$ is about the same as the variance of the two noisiest cones which were thought to be isolated on the basis of their exceptionally steep spatial profiles.

We conclude that the variation in noise magnitude observed in different cones is adequately explained by differences in the tightness of intercellular coupling and that all turtle cones possess a noise source which would give rise to a variance of about $0{\cdot}4\ \text{mV}^2$ in the absence of coupling. Functionally, the interreceptor coupling will improve the signal-to-noise ratio when the stimulus is diffuse. For such stimuli the signal in a cell is independent of the coupling since no current flows between cells, and for tight coupling the r.m.s. noise is inversely proportional to λ, so that the signal-to-noise ratio is directly proportional to λ. This argument is not valid for small stimuli because the coupling then reduces the signal more than the noise and worsens the signal-to-noise ratio.

The next question to consider is whether the noise characteristic of all cones arises from synaptic transmission onto the cones or from an internal source. The experiments just presented rule out the possibility that the noise could arise in interreceptor junctions because the noisiest cells are those which are not coupled. It might be thought that the known synaptic action of horizontal cells onto cones[5] could cause the noise, but this mechanism is not consistent with a number of experimental observations. First, it is easily shown that the noise is fully suppressed by small spots of light[1] that fail to evoke responses in

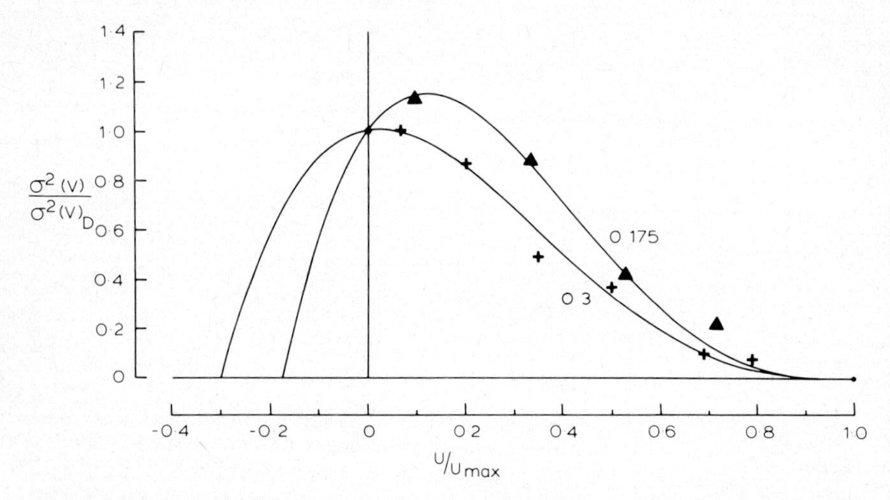

Fig. 4. The relation between intrinsic noise and light-evoked hyperpolarization U. Points are measurements from two cones for which $\sigma^2(V)$ is normalized to the dark noise $\sigma_D^2(V)$ and U is normalized to its limiting steady value in bright light U_{max}. Stimulus diameter of 105 μm was effectively full field for these coupled cells. The curves are calculated from a model in which the noise arises from fluctuations in the concentration of blocking molecules (see text). Numbers against the curves represent the residual concentration of z in darkness as a fraction of K (eqn (2)).

the horizontal cells, which have large receptive fields in the turtle.[8] Secondly, the degree of noise suppression depends only on the cone response and not on the stimulus pattern used to evoke that response.[9] And third, when the synaptic horizontal cell input to a green-sensitive cone is activated by a red annulus that gives no direct cone response,[10] the cone depolarizes but the noise is not changed.[9]

It is therefore possible to rule out synaptic input to the cones as the noise source and to conclude that the source is internal to the cone.

The observed variation of noise with the voltage arising from application of steady lights of different intensities is plotted in Fig. 4 for two cones. The residual noise in bright light has been subtracted from all variances, and the points have been normalized so that the variance in darkness and the voltage in bright light are unity. In one of the cells there is a noise increase in dim light while for the other there is a decrease at all intensities.

On the assumptions of Simon *et al.*[1] that the variance $\sigma^2(z)$ of the number of blocking molecules z is equal to z and that the relationship between z and hyperpolarization U is of the form

$$\frac{U}{U_{\max}} = \frac{z}{z + K}, \tag{2}$$

it may be shown[11] that

$$\sigma^2(U) = U_z U\left(1 - \frac{U}{U_{\max}}\right)^3 \tag{3}$$

where U_z is the hyperpolarization evoked by one transmitter molecule in the absence of others. By further assuming that in darkness there is a finite concentration of z,[1] the curves of Fig. 4 are obtained, where the voltage which would exist in the absence of transmitter molecules is the point where each curve leaves the abscissa. Results from 10 cells suggest that, on this model, cones in darkness are hyperpolarized by about 4 mV from the level which would apply if all the light-sensitive channels were open.

Power spectra of the noise were computed and Fig. 5 plots the difference spectrum (spectrum in darkness minus spectrum in bright light) for a red-sensitive cone in which the variance was reduced ten-fold by bright light. If the noise arose from elementary shot events having the shape of the small-signal light response, then the spectrum ought to correspond to the square of the Fourier transform of that response which is given by eqn (41) of Baylor *et al.*[12] However, the spectrum calculated in this way had a lower half-power frequency (of 2 Hz) and fell much more steeply than the measured points. The plotted curve which fits the points is, instead, the product of two Lorentzians given by the equation

$$S(f) = \frac{S_0}{[1 + (2\pi f\tau_1)^2][1 + (2\pi f\tau_2)^2]} \tag{4}$$

with a low-frequency asymptote $S_0 = 2{\cdot}2 \times 10^{-2}$ mV^2 Hz^{-1} and time constants τ_1 and τ_2 of 25 ms and 12 ms.[13] We associate the shorter τ with the cell's capacitive time constant and the longer τ with the temporal behaviour of the elementary conductance events. In other cones τ_1 ranged from 17–60 ms with a mean of 40 ms and τ_2 averaged 8 ms.

We have considered a stochastic scheme[9] based on the kinetic model of Baylor *et al.*[12] in order to explain the behaviour of the noise. In their model light produces, by a chain of reactions, a blocking molecule z which can combine with open ionic channels and close them:

$$\begin{array}{c} \text{light} \rightarrow \rightarrow \rightarrow \rightarrow z \rightarrow \rightarrow \rightarrow \\ + \\ r \\ \alpha \downarrow \uparrow \beta \\ b \end{array} \tag{5}$$

In the context of this scheme, two extreme possibilities are apparent. First, the binding might be rapid (α and β large), and the high frequency fluctuations in the number of channels blocked b would be obscured by the cell's capacitive filtering. In this case it can be shown that the elementary voltage event

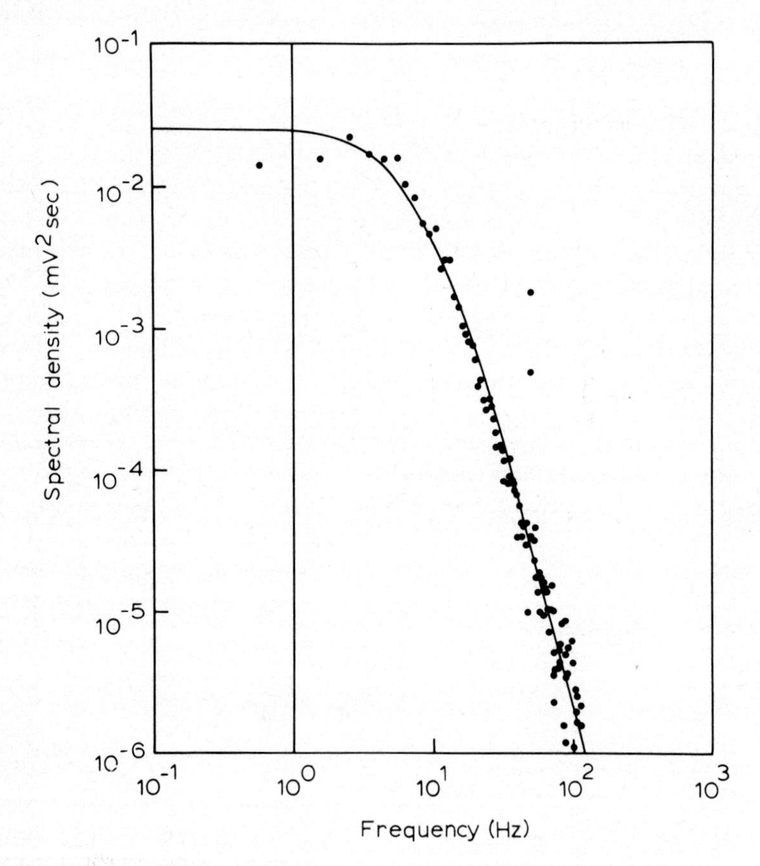

Fig. 5. Difference spectrum of voltage noise in a red-sensitive cone. Darkness, 113 s record length, variance 0·176 mV^2; bright light, 20 s record length, variance 0·017 mV^2. The two points close to 50 Hz are the result of mains interference. From Lamb and Simon[13] with permission.

corresponds approximately to the effect of one additional z molecule and that τ_1 corresponds to the mean time for which blocking molecules are in the active state.

On the other hand for comparatively slow binding (α and β small), the elementary voltage is the effect of closing a single ionic channel and τ_1 is related to the mean time for which a channel exists in the open and closed states.

There is little experimental reason to choose between these alternatives, but an examination of their implications is helpful. The first case with rapid binding and capacitive filtering is effectively the same as the model of Simon *et al.*[1] discussed on page 297. Ignoring the nonlinearity in eqn (3) the elementary voltage event U_z elicited by a single blocking molecule is of the order

$$U_z = \frac{0{\cdot}4 \text{ mV}^2}{4 \text{ mV}} = 100 \text{ µV}$$

and in darkness the average number of molecules is about

$$Z_D = \frac{4 \text{ mV}}{100 \text{ µV}} = 40.$$

When corrected for nonlinearity at quarter-maximal voltage, the elementary voltage event in darkness is about 130 µV and the dark level corresponds to roughly 22 simultaneous events. Taking the volume of a turtle cone outer segment to be 2×10^{-14} l, the concentration of free blocking molecules is then about 10^{-9} M. This model is only a reasonable approximation for weak binding, which requires the number of bound channels to be substantially less than 20. On the basis that about one-third of the total channels are blocked in darkness, the total number of channels must then be considerably less than 60. These estimates of channel number seem very low and cause us to suspect that "packets" of molecules may be involved in generating the individual noise events.

The second case of slow binding predicts that the elementary voltage reflects the lifetime of a blocked ionic channel. On this model it can be shown that the maximum noise occurs at a hyperpolarization of at least half-maximal suggesting that in darkness perhaps two-thirds of the channels are closed and that the cell may be hyperpolarized by some 10–20 mV from the voltage with all channels open. Rough calculations indicate that closure of a single channel in darkness would give a voltage of 50–100 µV and that the average number of channels blocked in darkness is perhaps 300. This would mean that the total number of channels is roughly 500 or somewhat greater, while the number of free blocking molecules depends on the equilibrium constant of binding (α/β) and could well be either much higher or much lower than 300. For weak binding (say $\alpha/\beta = 0{\cdot}1$) the free concentration of z might be in the region of 10^{-7} M.

An alternative explanation is that the noise arises from the spontaneous occurrence of events indistinguishable from photoisomerizations. However, the spectrum of the noise does not correspond to that expected if each such event gives rise to a response having the shape of the small-signal light response; instead the noise extends to higher frequencies. On the kinetic model of Baylor *et al.*[12] each isomerization leads to a chain of reactions (p. 298). If these reactions are of a stochastic nature then it may be shown that the power spectrum of the number of blocking molecules z is independent of the rate constants in the chain of reactions, and depends only on the rate constants of removal and of binding to channels. Thus the measured spectral time constant would again correspond to that of the removal or binding reactions. Such a source of events indistinguishable from photoisomerization could not arise from thermal isomerization of the pigment, because the extremely high predicted Q_{10} of about 15 should have been observable in preliminary temperature change experiments, but was not.

If the input resistance of an isolated cone is taken to be 200 MΩ,[2] the magnitude of the elementary current pulse is then about 0·5 pA. As the driving potential giving rise to this current is probably the same as the voltage required to reverse the light response, or about 40 mV,[14] the elementary conductance change would be of the order 10^{-11} S.

Finally, the time integral of one elementary voltage event, about 100 μV and 40 ms duration, is similar to the time integral of one photoisomerization, about 25 μV and 175 ms duration.[15] Therefore it seems that the absorption of one photon evokes on average approximately one of the elementary noise events.

References

1. E. J. Simon, T. D. Lamb and A. L. Hodgkin 1975, Spontaneous voltage fluctuations in retinal cones and bipolar cells, *Nature* **256**, 661.
2. T. D. Lamb and E. J. Simon 1976, The relation between intercellular coupling and electrical noise in turtle photoreceptors, *J. Physiol.* **263**, 257.
3. M. G. F. Fuortes and S. Yeandle 1964, Probability of occurrence of discrete potential waves in the eye of *Limulus*, *J. gen Physiol.* **47**, 443.
4. R. A. Cone 1973, The internal transmitter model for visual excitation: some quantitative implications, *In* "Biochemistry and Physiology of Visual Pigments" p. 275. Springer, Berlin.
5. D. A. Baylor, M. G. F. Fuortes and P. M. O'Bryan 1971, Receptive fields of cones in the retina of the turtle, *J. Physiol.* **214**, 265.
6. T. D. Lamb 1976, Spatial properties of horizontal cell responses in the turtle retina, *J. Physiol.* **263**, 239.
7. D. A. Baylor and R. Fettiplace 1975, Light path and photon capture in turtle photoreceptors, *J. Physiol.* **248**, 433.

8. E. J. Simon 1973, Two types of luminosity horizontal cells in the retina of the turtle, *J. Physiol.* **230**, 199.
9. T. D. Lamb and E. J. Simon 1977, Analysis of electrical noise in turtle cones, *J. Physiol.* **272**, 435.
10. M. G. F. Fuortes, E. A. Schwartz and E. J. Simon 1973, Colour dependence of cone responses in the turtle retina, *J. Physiol.* **234**, 199.
11. B. Katz and R. Miledi 1972, The statistical nature of the acetylcholine potential and its molecular components, *J. Physiol.* **224**, 665.
12. D. A. Baylor, A. L. Hodgkin and T. D. Lamb 1974, The electrical response of turtle cones to flashes and steps of light, *J. Physiol.* **242**, 685.
13. T. D. Lamb and E. J. Simon 1976, Power spectral measurements of noise in the turtle retina, *J. Physiol.* **263**, 103*P*.
14. D. A. Baylor and M. G. F. Fuortes 1970, Electrical responses of single cones in the retina of the turtle, *J. Physiol.* **207**, 77.
15. D. A. Baylor and A. L. Hodgkin 1973, Detection and resolution of visual stimuli by turtle photoreceptors, *J. Physiol.* **234**, 163.

Discussion

H. B. Barlow: You have analyzed and presented your results to show what receptor noise can tell one about the mechanism of membrane hyperpolarization, but receptor noise is also important as a factor that may limit the overall performance of the visual system. I think your measurements could be presented so that their implications in this regard emerge more clearly.

If there is a noise level of $\sigma(V)$ mV (rms), then it will not be possible to detect reliably brief light induced signals that are of magnitude less than a few times $\sigma(V)$. Your Fig. 4 shows how voltage noise $\sigma^2(V)$ varies with hyperpolarization, and it is these results one would like to see expressed as the light intensity, say ΔI, required to produce a signal equal to the noise, $\sigma(V)$, plotted against the light intensity I required to cause the steady hyperpolarization at which the noise was measured.

There is a very simple model of receptor noise that accounts for some, though not all, of the psychophysical facts around photopic and scotopic threshold, and my questions are best framed in relation to the model. It postulates that there is a source of events x in the receptor that cannot be distinguished from quantal absorptions n, and the total noise is simply $(n + x)^{1/2}$, the square root of the sum of these noise events and real quantal absorptions. In other words, the postulated noise is like the dark current of a photocathode. If I understand them correctly your results show that there is in addition a noise source of higher frequency. This is of interest with regard to mechanism, but the components of it that could be filtered out do not necessarily limit performance, so for these purposes one needs the noise measured over the frequency band of light-induced signals. In addition one must of course take into account the stimulus/response function relating light and hyperpolarization. When these two factors are taken into account, what do your results tell one about n and x in the psychophysical model?

First, if the results of Fig. 4 give evidence for a source increasing as $I^{1/2}$, this is likely to be quantal noise, and thus related to n in the model. It would be particularly interesting to know (a) what proportion of incident quanta appear to give rise to this noise, and (b) what is the highest quantum efficiency of detection you can achieve in its presence? Naturally we would very much like to know the comparable figures for rods.

Second, there is apparently noise in the absence of added light, and it would be of interest to relate this to x of the model. For instance, what would be the maximum allowable value of x in your cones? How does this compare with the corresponding value for rods? From psychophysical facts one would expect the greater sensitivity of rods to be mainly the result of their lower value of x, but is this supported by your results?

There are three other ways in which these results on receptor noise may relate to visual performance:

(1) If it is correct that x is much lower for rods than for cones, this may be a rather direct result of the shift of rod spectral sensitivity to shorter wavelengths. If an energy barrier E, which must be surmounted by an absorbed quantum, shifts with the absorption spectrum it will only increase about 10 % when the photopic peak sensitivity at 560 nm shifts to the scotopic peak at 507 nm. However because E is so large in relation to kT this will decrease the probability of E being surmounted thermally to about 1/4000 and thus enormously decrease the expected rate of thermal isomerization. If thermal isomerizations occur, they would contribute to x, and thereby reduce sensitivity. I know no other explanation for the selective advantage of the Purkinje shift to the blue in the high sensitivity rod system of receptors.[1]

(2) Another striking psychophysical difference between rods and cones is the much lower Weber fraction of the latter. A change of luminance of a fraction of 1 % can be detected by cones under favourable conditions, whereas for rods a change of 5 % or more is required. Now if x is counted over an interval comparable to the receptor's response time, then it will be necessary for a receptor to signal a change of order $x^{1/2}$, or else it will fail to operate at the sensitivity allowed by its intrinsic noise. This corresponds to a *relative* change $x^{-1/2}$, and it will be much lower when x is high. Thus an efficient cone may need to signal a small change in the *rate* of excitatory events, whereas for a rod to operate up to its noise limit it would probably be sufficient for it to signal every event, noise or quantal, as it occurs.

(3) We now know, however, that rods are not isolated, but interconnected. This may help to overcome the nonlinearity and noisiness of the rod-bipolar synapse,[2] or facilitate the detection of diffuse light stimuli, but the capacity of

cones and the cone-bipolar synapse to signal very small *relative* changes of light intensity sets one thinking along different lines. It is not only the sensitivity of vision that is impressive, but also its capacity to work over a very large dynamic range. Could the interconnection of rods be advantageous, not only for sensitivity, but also to improve the dynamic range of the rod system? Considering the range of luminances ($>10^4$) that must be signalled it would appear disadvantageous to use the rod-bipolar synapse in all-or-none fashion at low luminances. The interconnection of receptors may not only enable a weak stimulus to work through several rod-bipolar synapses,[2] but may also make it possible for transmitter release to be continuously graded up to high luminance levels.

References to discussion comment of H. B. Barlow

1. H. B. Barlow 1957, Purkinje shift and retinal noise, *Nature* **179**, 255–256.
2. G. Falk and P. Fatt 1972, Physical changes induced by light in the rod outer segment of vertebrates, pp. 200–244, *In* "Handbook of Sensory Physiology", vol. 7, part 1 (ed. H. J. A. Dartnall), Springer-Verlag, Heidelberg.

E. J. Simon and T. D. Lamb: In terms of the psychophysical model our results can be used to estimate the parameter x but not n.

The average response of an isolated turtle cone to a single photo-isomerization has a shape which may be described by eqn (43) of Baylor *et al.*[12]

$$\Delta v(t) = \Delta V \cdot n e^{-t/\tau}(1 - e^{-t/\tau})^{n-1}, \tag{6}$$

for $n \simeq 7$, $\tau \simeq 60$ ms and with a peak amplitude Δv_m of about 25 μV.[15] On the assumption that the cone dark noise is the cumulative effect of random events each having a size and shape identical with the average photon response, it is possible to calculate the rate of events to which this corresponds. Allowing for the "shape factor"[11] of eqn (6) it may be shown that

$$\text{rate} \simeq 0{\cdot}58 \frac{\sigma^2(V)}{\tau(\Delta v_m)^2} \tag{7}$$

if there is linear event summation. The variance of 0·4 mV^2 in an isolated cone extends to higher frequencies than the spectrum of the average photon response, and the variance of interest is about 0·2 mV^2. Putting $\tau = 60$ ms gives a rate of about 3000 sec^{-1} in the isolated cell. However nonlinear summation will have distorted both the measured variance and sensitivity, and correction for these at a polarization in darkness of quarter maximal will reduce the rate to about 2000–2500 sec^{-1}. Taking the collecting area of a cone to be 10 μm^2[15] this corresponds to an incident intensity of 200–250 photons μm^{-2} sec^{-1} in a "silent" isolated cone. In fact the value is unchanged in a coupled cone, and represents x in your formulation.

Unfortunately we have not measured both variance and sensitivity during steady illumination. Were this done then $(n + x)$ could be calculated from the variance divided by the square of sensitivity, if correction were also made for changes in τ during illumination. We have little data on rods, and are unable to make detailed comparisons between the two systems, but the existence of noise of similar magnitude together with a much greater sensitivity points to a much lower rate of equivalent isomerizations.

17

The Threshold Signal of Photoreceptors

GORDON L. FAIN

Jules Stein Eye Institute, University of California School of Medicine, Los Angeles, California, U.S.A.

One of the most challenging problems to the student of the visual system is to understand how the eye achieves its remarkable sensitivity. When fully dark-adapted, a human observer can detect a flash of light bleaching single pigment molecules in fewer than one out of every 500 to 1000 of our receptors.[1] Just as remarkable is our sensitivity to contrast. In the presence of bright background light, we can distinguish an increment in intensity which is about 1% of the mean ambient level. The way single cells in the visual system mediate the visual threshold is still poorly understood. The absorption of light by the photopigment produces a change in the receptor membrane potential. At absolute threshold, the signals of the rods are summed by retinal interneurons so that, when 5–10 rods each absorb a quantum of light, the signals these cells produce are large enough to cause a significant change in the firing of ganglion cells and hence of cells in the central nervous system.[2] In the presence of bright background light, the system presumably functions in a similar fashion. An increment of light superimposed upon the background produces a change in receptor membrane potential large enough to be transmitted reliably to higher-order neurons. When the sum of the receptor signals exceeds some threshold value, there is a change in the firing of ganglion cells and of visual neurons in the brain. Somehow these changes produce our conscious sensation of the stimulus.

In this paper I shall consider one step in the train of events leading to visual detection. I shall attempt to estimate the amplitude of the receptor signal at the behavioral threshold. By comparing the sensitivity of single receptors to behavioral responses, I hope to demonstrate that the receptor response at threshold is only a few microvolts in amplitude. The small value of the threshold response suggests that the sensitivity of synaptic transmission between receptors and second-order cells in the retina makes an important contribution to the sensitivity of vision.

The Single Quantum Responses of Receptors

One approach for estimating the amplitude of receptor responses at threshold would be to measure their single quantum responses. Hecht, Schlaer and Pirenne[2] first showed that visual detection at absolute threshold is mediated by a small group of rods each of which need absorb only a single quantum of light. Single quantum responses are thus large enough to be transmitted to second order cells and to be integrated by the central nervous system to produce a threshold sensation.

In an attempt to measure the amplitude of single quantum responses, I recorded intracellularly from the large "red" rods of the toad, *Bufo marinus*. In these experiments,[3] the intensity of the stimulating light was reduced so that only a fraction of rods absorbed any quanta. At dim intensities the proportion (P) of flashes bleaching pigment molecules in any given photoreceptor follows the Poisson distribution,

$$P(n) = \frac{\lambda^n}{n!} e^{-\lambda}, \tag{1}$$

where n is the number of molecules bleached in a flash, and λ is the mean of n.

Since the number of pigment molecules bleached in the outer segment of each receptor varies from one flash to the next, the responses generated by receptors should also vary. It thus came as a surprise when I first recorded the responses of toad rods to dim light intensities to discover little detectable variation in response amplitude in successive flashes. The rod of Fig. 1, for example, gave nearly identical responses to flashes which bleached, on average, only a fraction of a pigment molecule per receptor. For the responses in the furthest column to the right in Fig. 1, the Poisson distribution predicts that only 20% of the flashes will bleach any pigment molecules in the outer segment of the receptor. Nevertheless the receptor responded to every flash at this intensity, as if each flash contained the same energy.

We now know the reason for this unexpected result. Rods in the toad are connected to one another by an extensive network of junctions.[4] Just distal to the outer limiting membrane, the rods project fine processes called fins, which extend from the inner segments of the rods and interdigitate between glial cell

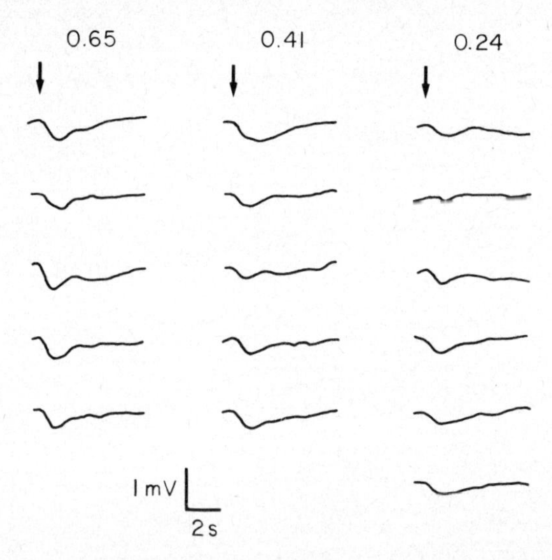

Fig. 1. Responses of toad rod to flashes bleaching fewer than one pigment molecule per receptor. In each column responses are to consecutive, 9 ms flashes of diffuse 502 nm light. Mean intensities of the flashes are given above each column in units of rhodopsin molecules bleached per receptor. Arrows indicate onset of flashes. (Reprinted with permission from reference 3. Copyright 1975 by the American Association for the Advancement of Science.)

processes in the region of the fiber basket. Fins from adjacent rods make large, close-apposition contacts with one another that closely resemble gap junctions seen elsewhere in the nervous system (Fig. 2). There are four to six junctions and about 1 μm^2 of gap junctional membrane between any two neighboring rods. Gap junctions observed in other parts of the nervous system are thought to function as electrical synapses, providing for the passive electrotonic flow of current between neurons.[(5)] In a similar fashion, the gap junctions between red rods appear to couple the signals of the receptors. This probably explains how responses can be recorded from a toad rod when no pigment molecules have been bleached in its outer segment.

Imagine now what happens at threshold. When a pigment molecule is bleached in a rod, there is a change in the permeability of its membrane which produces a photocurrent. Some of this photocurrent will fall across the resistance of the receptor's own plasma membrane, but a substantial proportion of it will pass through the gap junctions to other receptors. As a result, voltage responses will be produced in many receptors, with the amplitude of the voltage response presumably becoming smaller the further away the receptor is from the receptor which absorbed the quantum. It is possible to show that, as a result of the coupling between receptors, the

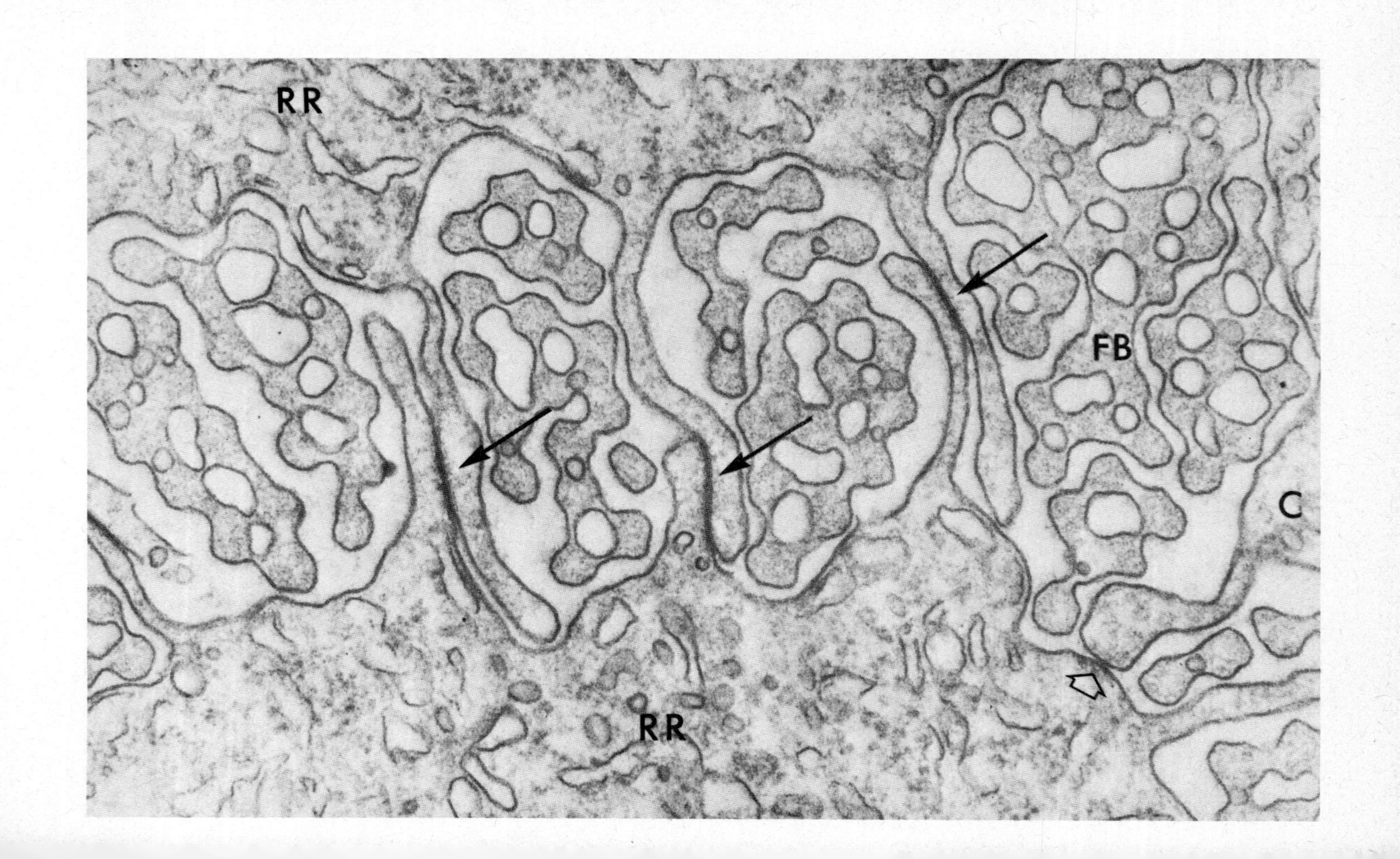
RR
FB
C
RR

maximum voltage response produced by the absorption of one quantum, even in the receptor which actually caught the quantum, is only 50–100 μV.[3] The responses in the cells not catching quanta will presumably be even smaller.

It should now be clear that recordings of receptor responses at dim light intensities have told us much less about visual detection than we might have hoped. If in the toad, as in the human, 5–10 rods must absorb a single quantum to produce a threshold detection, the photocurrents produced by these rods will spread across the retina producing voltage responses in many other receptors. The amplitudes of these responses will vary widely from receptor to receptor, being large in those rods near to the ones which have caught the quanta and small in those far away. It is possible that only a few of the receptors produce responses large enough to be transmitted to second-order neurons—perhaps only those which actually absorb a quantum. On the other hand, even the smallest signals may be reliably transmitted. If we are to understand which receptors contribute to threshold, we must find out the minimum signal receptors are able to communicate through their synapses.

In the following, I shall describe a method for estimating the minimum signal receptors can communicate to higher-order neurons. This method relies upon a comparison of receptor sensitivity to behavioral responses at the visual threshold. Suppose it were possible to record intracellularly from an animal's photoreceptors and, under the same stimulus conditions, register the animal's behavior. At a certain intensity, the animal will tell us that he perceives the stimulus. If under the same conditions we can measure the sensitivity of his photoreceptors, then we can estimate the amplitude of the receptor response at threshold. It is, of course, not possible to do this experiment at the absolute threshold, since the amplitudes of receptor responses are non-uniform when only a few receptors are catching quanta. Hence there is no unique value for the receptor sensitivity. However at the increment threshold in the presence of bright background light, the variation in the amplitude of increment responses from one receptor to the next will be minimal. There are two reasons for this. First, the quantum noise becomes much less important: the ratio of the standard deviation of the number of quanta caught per receptor to the mean number caught decreases as the flash intensity increases. Second, the connections between the photoreceptors will tend to minimize the variation in their response amplitudes.

Fig. 2. Electron micrograph showing large close-apposition junctions (arrows) between fins of red rods (*RR*)·*FB*, fiber basket formed by processes of glial cells. A much smaller focal junction between a cone fin (*C*) and a rod is indicated by the open arrow. Junctions between rods and cones do not appear to make a significant contribution to the responses of rods.[8] Magnification, approximately ×45,000. (Reprinted with permission from reference 4. Copyright 1976 by the Cold Spring Harbor Laboratory of Quantitative Biology.)

In the following I shall first describe the sensitivity of photoreceptors in the presence of background light. I shall then describe behavioral measurements of the increment threshold for the red-eared turtle, *Pseudemys scripta elegans*. Finally I shall compare these two and estimate the response in the receptors at the behavioral threshold.

Light Adaptation in Photoreceptors

The effect of background light on the sensitivity of receptors has been investigated by intracellular recording for the rods of the nocturnal gecko,[6] the cones of the turtle,[7] and the rods of the marine toad.[8] The results from these receptor types are qualitatively similar. The onset of background light produces a large initial hyperpolarization of the receptor membrane potential, which decays to a steady-state plateau voltage (Fig. 3). Test flashes superimposed upon the background produce a further hyperpolarization which departs from and returns to the maintained plateau voltage, with *off*

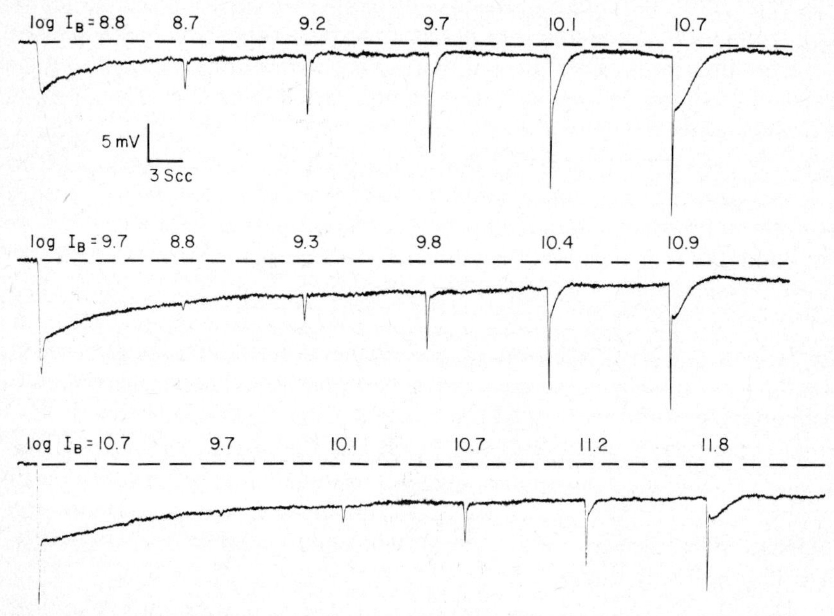

Fig. 3. Light adaptation of toad rods. Records show responses of rod to diffuse background light at three intensities, given in the figure in units of log equivalent quanta cm^{-2} s^{-1} at 502 nm. Incremental stimuli consisting of 9 msec flashes of diffuse light were superimposed upon the background. Intensities of incremental flashes, in units of log equivalent quanta cm^{-2} flash^{-1} at 502 nm, are shown above their respective responses. Dashed lines show dark resting membrane potential. (Reprinted with permission from reference 8. Copyright 1976 by the Physiological Society.)

transients appearing in both rods and cones at bright backgrounds. The amplitudes of the responses to test flashes depend upon the intensity of the test flash, as well as the intensity of the background light. As Fig. 3 illustrates, the background decreases the sensitivity of the receptor. As the background light is made brighter, the intensities of the test flashes must also be made brighter to produce responses of equivalent amplitude.

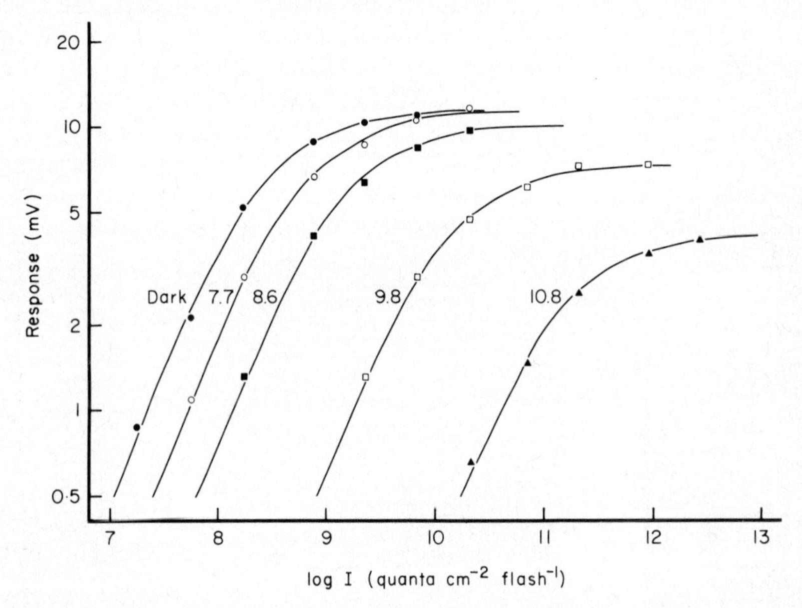

Fig. 4. Incremental intensity-response curves of toad rod. Peak response amplitude has been plotted on a log scale versus log incident flash intensity at 502 nm. Data points give dark-adapted responses (●) or increment responses (○, ■, □, and ▲), measured from the steady-state plateau level. Intensities of the 505 nm diffuse background lights are given to the left of each series of data in units of log quanta cm^{-2} s^{-1}. Data have been fitted with eqn (2). (Reprinted with permission from reference 8. Copyright 1976 by the Physiological Society.)

Figure 4 shows quantitatively the effect of background light on receptor sensitivity. The closed circles show the intensity-response relationship for a dark-adapted rod, and other symbols show similar data for the same rod in the presence of background lights of increasing intensity. To facilitate the analysis of these data, they have been fitted with the equation

$$\Delta V = \frac{\Delta V_{\max} \Delta I}{\Delta I + \sigma}, \tag{2}$$

where ΔV is the increment response amplitude measured from the steady plateau potential to the peak of the increment response, ΔV_{max} is the maximum amplitude of ΔV, ΔI is the increment flash intensity, and σ is a constant. Since for small amplitude responses ΔI is much smaller than σ, eqn (2) can be simplified to

$$\Delta V \approx \frac{\Delta V_{max}}{\sigma} \Delta I. \tag{3}$$

Thus near the increment threshold, response amplitude is proportional to test flash intensity. The constant of proportionality $(\Delta V_{max}/\sigma)$ is the increment sensitivity which, following Baylor and Hodgkin,[7] will be denoted S_F. Since eqn (2) is formally identical to the Michaelis–Menton equation, S_F can be determined from intensity-response data using a Lineweaver–Burke plot.[8]

I will now show that the dependence of S_F on the intensity of background light follows a modification of the Weber–Fechner relation over a considerable range of background intensities. The Weber–Fechner relation has the form

$$\Delta I_T = k(I_B + I_0), \tag{4}$$

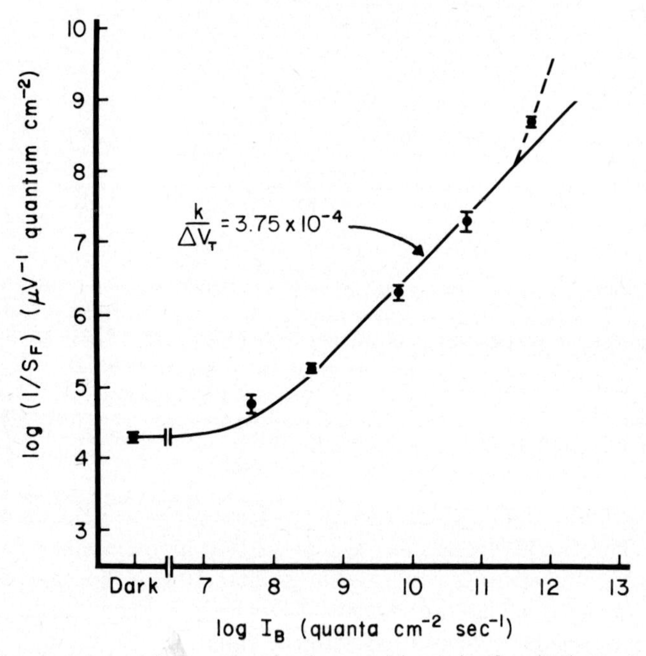

Fig. 5. Increment sensitivity versus background intensity for single toad rods. Data points give the means and bars, the S.E. of the log inverse increment sensitivity at λ_{max} (502 nm), measured in 505 nm continuous full-field background light. Means have been fitted with eqn (6), using $k/\Delta V_T = 3{\cdot}75 \times 10^{-4}$ s flash^{-1} μV^{-1} and $I_0 = 1{\cdot}47 \times 10^7$ incident quanta cm^{-2} s^{-1}.

where ΔI_T is the intensity of the test stimulus at the behavioral threshold, k is the Weber constant, I_B is the background intensity, and I_0 is a constant, usually referred to as the "dark light". From eqn (3)

$$\Delta I_T = \frac{\sigma \Delta V_T}{\Delta V_{max}} = \frac{\Delta V_T}{S_F}, \tag{5}$$

where ΔV_T is the voltage signal in the receptor at threshold. Provided ΔV_T is invariant with background intensity, eqns (4) and (5) can be combined to give

$$\frac{1}{S_F} = \frac{kI_0}{\Delta V_T} + \frac{k}{\Delta V_T} I_B \tag{6}$$

where $\Delta V_T/kI_0 = S_F{}^D$ is the sensitivity of the receptor in the dark.[7]

The mean sensitivities in Fig. 5 have been fitted with eqn (6) using a value for $k/\Delta V_T$ of $3{\cdot}75 \times 10^{-4}$ s flash^{-1} μV^{-1}. The increment sensitivity of toad rods is fairly well fitted with this equation for background intensities up to about 10^{11} quanta cm^{-2} s^{-1}. At brighter backgrounds the sensitivity decreases more quickly than predicted by eqn (6), and eventually the rods become saturated (for a more complete description of increment saturation in toad rods, see reference 8).

The increment sensitivities of Gecko rods and turtle cones are also satisfactorily described by eqn (6). It is of some interest, however, that $k/\Delta V_T$ is only $2{\cdot}0 \times 10^{-5}$ s flash^{-1} μV^{-1} for red-sensitive turtle cones.[7] In Fig. 6, the increment-sensitivity curves for toad rods and red-sensitive turtle cones have been plotted together on the same axes. The curves for the single receptors bear a striking resemblance to the rod and cone branches of the human increment-threshold curve.[9, 10] In fact, the ratios of $k/\Delta V_T$ and of I_0 for toad rods and turtle cones are similar to the ratio of the Weber constants and dark lights for the rod and red-sensitive cone systems in the human retina.[9, 11]

The increment sensitivity data for single photoreceptors provides a measure of $k/\Delta V_T$, the ratio of the Weber constant to the voltage response in the photoreceptor at the behavioral threshold. If we could determine the Weber constant independently, we could estimate ΔV_T. The measurement of k for the red-sensitive cone system of the turtle will be described in the following section.

The Increment-Threshold Curve of the Turtle

Of the cold-blooded vertebrates which have been used for single-cell recordings, the turtle is perhaps the best suited for quantitative measurements of visual behavior. Turtles are easy to train, and their responses are remarkably reliable. The turtle *Pseudemys scripta elegans* offers the additional advantage that its behavior is dominated by the red-sensitive cone system.

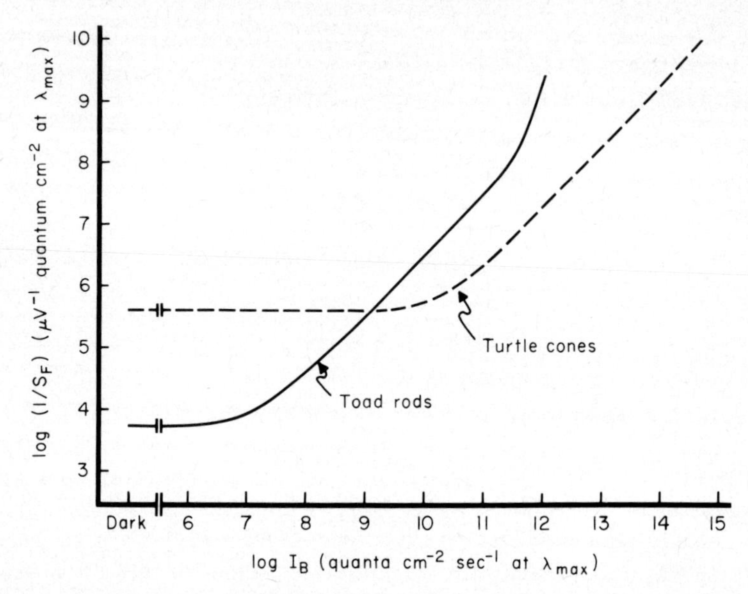

Fig. 6. Comparison of increment sensitivity curves for toad rods and turtle cones. Curves were constructed from eqn (6) using, for toad rods, values of $S_F^D = 1{\cdot}89 \times 10^{-4}$ μV quantum^{-1} cm^2 at λ_{max}, $I_0 = 1{\cdot}47 \times 10^7$ quanta cm^{-2} s^{-1} at λ_{max}, and $k/\Delta V_T = 3{\cdot}75 \times 10^{-4}$ s flash^{-1} μV^{-1} [8]; and for turtle cones, $S_F^D = 2{\cdot}5 \times 10^{-6}$ μV quantum^{-1} cm^2 at λ_{max}, $I_0 = 2{\cdot}0 \times 10^{10}$ quanta cm^{-2} s^{-1} at λ_{max}, and $k/\Delta V_T = 2{\cdot}0 \times 10^{-5}$ sec flash^{-1} μV^{-1}.[7, 13] (Reprinted with permission from reference 8. Copyright 1976 by the Physiological Society).

Figure 7 compares the wavelength dependence of the behavioral sensitivity of light-adapted *Pseudemys* (open circles) to the spectral sensitivity curve of single red-sensitive cones (filled circles). The behavioral measurements were taken from Granda, Maxwell and Zwick,[12] and the mean sensitivities of *Pseudemys* cones were calculated from the data of Baylor and Hodgkin.[13] The large standard deviations for the cone sensitivities at short wavelengths (bars in Fig. 7) are probably caused by variable absorption by the oil droplets.[13] The close correspondence of the two curves in Fig. 7 indicates that the luminosity system in *Pseudemys* is dominated by a single cone type. Intracellular measurements of spectral sensitivities from luminosity-type horizontal[14] and bipolar cells[15] also provide evidence that turtle sensitivity is dominated by red-sensitive cone input. The predominance of this single spectral class is not surprising, given its greater relative abundance[16] and larger cross-sectional area[17] than for other spectral cone types in *Pseudemys*. One consequence of this predominance is that direct comparisons can be made between the *input* to the visual system, that is the signals of red-sensitive cones, and their *output*, the turtle's visual behavior.

To estimate the value of k, behavioral increment-threshold measurements were made for the turtle *Pseudemys scripta elegans*—the same species Baylor and Hodgkin used in their study of light adaptation in red-sensitive cones. In these experiments,[18] turtles were placed in an experimental chamber similar to that used by Granda, Maxwell and Zwick.[12] Background lights of various intensities were projected from behind the animal onto a styrofoam integrating hemisphere. The test stimulus was superimposed upon the background

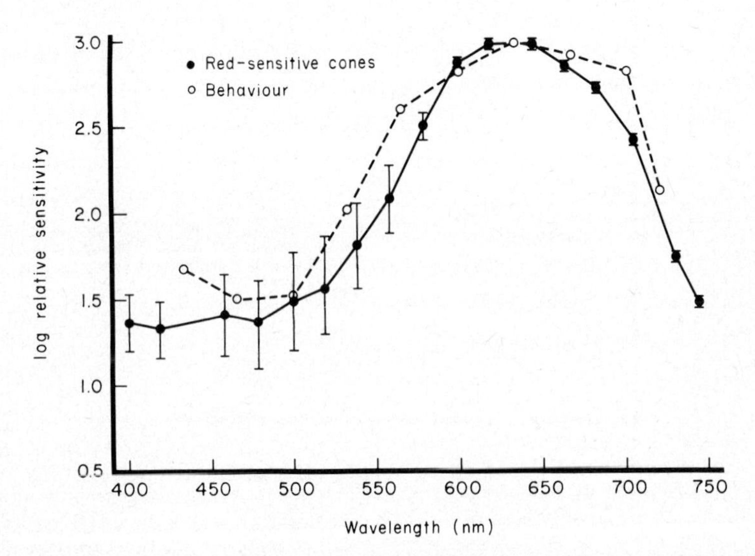

Fig. 7. Comparison of the wavelength dependence of the light-adapted behavioral sensitivity (broken curve) and of the spectral sensitivity of single red-sensitive cones (continuous curve) from *Pseudemys scripta elegans*. Behavioral measurements (open circles) were taken from the 90 s data of Fig. 9 of Granda, Maxwell and Zwick.[12] Filled circles give the means and bars, the standard deviations of red-sensitive cone sensitivities, calculated from the data of Baylor and Hodgkin.[13]

and consisted of a 4·9 mm diameter piece of milk glass, which was illuminated from outside the chamber. The target subtended 2°21′ and produced a spot in focus on the retina 300 μm in diameter, according to the schematic eye developed for *Pseudemys* by D. P. M. Northmore (personal communication). Both test and background stimuli were white lights provided by tungsten lamps, whose unattenuated intensities were measured with a calibrated thermopile.

The animal's head was placed in the same position before the presentation of each test stimulus, so that the test flash always fell on the same part of his retina (the linear area centralis). The animal was trained to withdraw his head in order to avoid an electric shock (for a detailed description of the training procedure, see reference 12). Stimuli 10–50 ms in duration were then presented at one minute intervals against a dark background and light backgrounds of various intensities. Threshold measurements were made by the "up-down" procedure[19] and were quite reliable: the mean 95 % confidence interval was ±0·08 log unit.

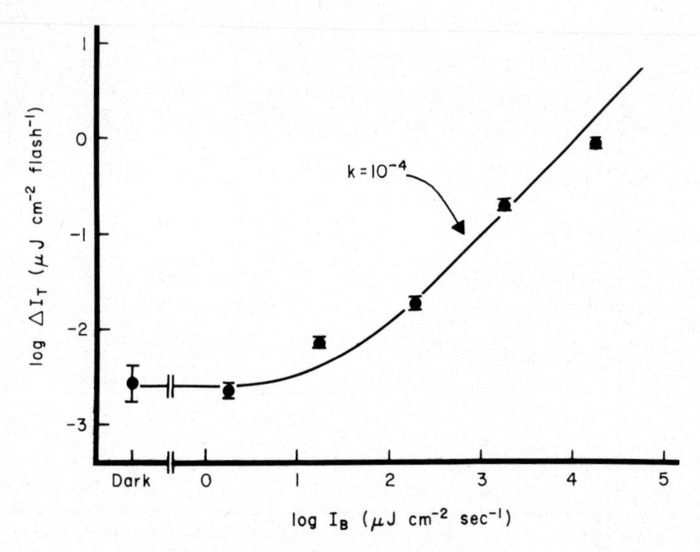

Fig. 8. Increment-threshold curve for the behavioral response of the turtle *Pseudemys scripta elegans*. Data points give means and bars, standard deviations for thresholds measured against backgrounds of varying intensity. Thresholds measured with flashes between 10 and 50 ms have been corrected for the difference in flash length and averaged together, since the behavioral threshold in turtle shows complete temporal summation for flashes up to at least this level. (A. M. Granda and J. H. Maxwell, unpublished data). Points have been fitted with eqn (4) using a value for k of 10^{-4} s flash^{-1}. (Reprinted with permission from reference 18. Copyright 1977 by Macmillan Journals Ltd.)

The results of the threshold measurements are shown in Fig. 8. The data have been fitted with eqn (4), using a value for k of 10^{-4} s flash^{-1}. From this value and from the value for $k/\Delta V_T$ given by Baylor and Hodgkin,[7] ΔV_T can be calculated to be 5 μV. There is some uncertainty in this value, since the fit of eqn (4) to the threshold data is not completely satisfactory. This disparity may be the result of experimental error, or alternatively ΔV_T may vary with background intensity. Even taking these possible deviations into account, the threshold signal of the red-sensitive turtle cones is almost certainly less than 10 μV. There are at least two reasons for believing this value to be an overestimate. First, the test target used in these experiments may not have been optimum for eliciting the threshold response. That is, a larger or differently-shaped target might have given responses at even lower increment intensities. Second, the intracellular recordings of turtle cone sensitivity of Baylor and Hodgkin were most probably made from the nuclear region of the receptor cell body. The voltages in the pedicle, where synaptic transmission occurs, are probably somewhat smaller.

In summary, the measurement of the Weber constant for the turtle *Pseudemys scripta elegans*, together with the intracellular measurements of receptor sensitivity of Baylor and Hodgkin in the same species,(7) have provided an estimate of the voltage signal in the photoreceptors at the visual threshold. This signal is 5–10 μV in amplitude. The estimate of this value depends upon two assumptions: that the amplitudes of receptor responses are linearly related to light intensity for small amplitude responses,(13) and that the behavioral responses of light-adapted *Pseudemys* are mediated by the red cones. The latter of these assumptions is probably not very important, since $k/\Delta V_T$ appears to have approximately the same value for both green- and red-sensitive turtle cones.(7) The blue-sensitive receptors almost certainly had a negligible effect on the measurements in Fig. 8, given their small size, their relative scarcity, and their insensitivity to the white light of the tungsten lamps used in the threshold measurements.

Visual Sensitivity and Transmission at the Receptor Synapse

If at the increment threshold the signals in the receptors are 5–10 μV in amplitude, then signals at least this small must be reliably transmitted to second-order cells. Photoreceptor synapses are remarkably sensitive: the threshold signals of turtle cones are more than three orders of magnitude smaller than the minimum signal transmitted across the giant synapse in the squid stellate ganglion.(21,22) There is recent evidence that the gain of transmission at the receptor synapse may be much larger than in other systems.(23) The unusual properties of the receptor synapse may have much to do with the sensitivity of vision.

If we are to understand how such small signals are transmitted across the receptor synapse, we need to know much more about the physiology of this synapse than we presently do. Even if we make the reasonable assumption that transmission from receptors is mediated by a mechanism similar to that for synapses in the rest of the nervous system, this tells us nothing about the voltage dependence of calcium entry into the receptor presynaptic terminal, the efficiency of vesicle release, or the dependence of the post-synaptic membrane voltage on the amount of released transmitter. It is not improbable that photoreceptor synapses have characteristics different from synapses in other areas of the nervous system, given their unusual anatomy.[24] Photoreceptors contain a highly specialized presynaptic system including the synaptic ribbon with its regular arrangement of vesicles and the arciform density, which may serve to direct the release of synaptic transmitter in a way that is more efficient than for other synapses. It is of some interest that electroreceptors and receptors in the acoustico-lateralis system have a similar presynaptic network. Transmission at the synapses of these receptors appears to be just as sensitive as at the photoreceptor synapse. In some species of fish the threshold signal in the electroreceptors is known to be less than 10 μV,[25] and the threshold signals of auditory and lateral line receptors are probably even smaller.

Although we know very little about receptor synapses, one recent finding suggests a partial explanation for their sensitivity. There is now good evidence that photoreceptors release transmitter continuously in the dark.[26] Presumably the low resting membrane potential of photoreceptors caused by the steady leakage of Na^+ into their outer segments produces a tonic activation of the calcium channels at the presynaptic terminal, and so a continuous entry of calcium ions and release of transmitter. The effect of light is to modulate this release, so that small changes in presynaptic voltage can produce fairly large changes in the flow of transmitter.

To illustrate this effect, let us suppose that the input-output curve for the photoreceptor synapse is similar in shape to that for the giant synapse of the squid. This curve, from the data of Katz and Miledi,[21] is given in Fig. 9. In this figure, the pre- and postsynaptic voltages of the squid neurons are plotted relative to their resting membrane potentials. Figure 9 shows that, in the giant synapse, the presynaptic terminal must be depolarized by 25–30 mV before any postsynaptic response is recorded. This initial polarization is presumably that required to activate the calcium channels in the presynaptic membrane. Once the calcium channels are open, the postsynaptic response is very sensitive to changes in presynaptic voltage. We may postulate that one of the functions of the dark current of photoreceptors is to bring the receptor membrane potential to the steep part of the curve in Fig. 9, thus vastly increasing the sensitivity of synaptic transmission. A similar suggestion has been made for tonic electroreceptors,[25,27] although the evidence for the continuous release of transmitter from these cells is less direct.

In this paper I have examined only one step in the chain of events leading to the visual threshold. The small size of the threshold signal in the receptors suggests that the synapses they make with second-order cells may be unusually sensitive, but we cannot as yet explain in much detail how this sensitivity is produced. Our understanding of the other events in visual detection is even more rudimentary. For example, how do second-order cells detect the small signals of receptors above the background noise of continuous quantal

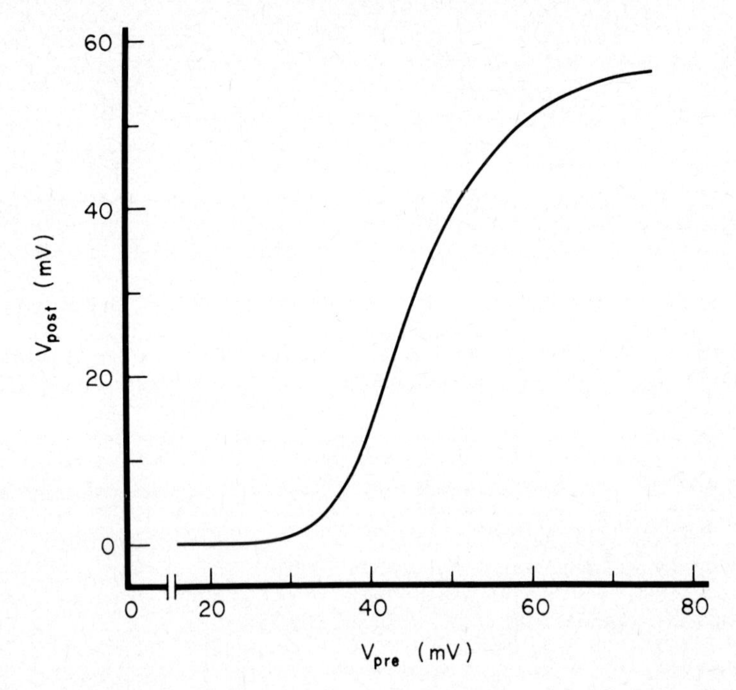

Fig. 9. Input–output curve for the giant synapse in the squid stellate ganglion. Data replotted from Fig. 9 of Katz and Miledi.[21] Voltages of pre- and postsynaptic terminals are given with respect to their resting membrane potentials.

release? What is the role of the bipolar cell, whose presynaptic terminals contain anatomical specializations similar to those seen in receptors? How are the threshold responses of ganglion cells integrated in the central nervous system? These questions are only now beginning to be asked, and we are still very far from providing adequate answers to them.

Acknowledgments

I am indebted to J. E. Dowling, G. H. Gold, A. M. Granda, and J. H. Maxwell for their contributions to this research. Many of the ideas presented in the discussion owe a large and hopefully obvious debt to M. V. L. Bennett. This work was supported in part by the Norton Simon Foundation, by the Grass Foundation and by NIH Grant EY 00331.

References

1. M. H. Pirenne 1962, Absolute thresholds and quantum effects, *In* "The Eye", vol. 2 (ed. H. Davson), pp. 123–140, Academic Press, New York.

2. S. Hecht, S. Shlaer and M. H. Pirenne 1942, Energy, quanta, and vision, *J. Gen. Physiol.* **25**, 819–840.
3. G. L. Fain 1975, Quantum sensitivity of rods in the toad retina, *Science, N.Y.* **187**, 838–841.
4. G. L. Fain, G. H. Gold and J. E. Dowling 1976, Receptor coupling in the toad retina, *Cold Spring Harb. Symp. Quant. Biol.* **40**, 547–561.
5. G. D. Pappas and S. G. Waxman 1972, Synaptic fine structure—morphological correlates of chemical and electrotonic transmission, *In* "Structure and Function of Synapses" (eds. G. D. Pappas and D. P. Purpura), pp. 1–43, Raven, New York; M. V. L. Bennett 1972, A comparison of electrically and chemically mediated transmission, *In* "Structure and Function of Synapses" (eds. G. D. Pappas and D. P. Purpura), pp. 221–256, Raven, New York.
6. J. Kleinschmidt 1973, Adaptation properties of intracellularly recorded *Gekko* photoreceptor potentials, *In* "Biochemistry and Physiology of Visual Pigments" (ed. H. Langer), pp. 219–224, Springer-Verlag, New York;
   J. Kleinschmidt and J. E. Dowling 1975, Intracellular recordings from gecko photoreceptors during light and dark adaptation, *J. Gen. Physiol.* **66**, 617–648.
7. D. A. Baylor and A. L. Hodgkin 1974, Changes in time scale and sensitivity in turtle photoreceptors, *J. Physiol.* **242**, 729–758.
8. G. L. Fain 1976, Sensitivity of toad rods: dependence on wavelength and background illumination, *J. Physiol.*, **261**, 71–101.
9. W. S. Stiles 1959, Colour vision: the approach through increment-threshold sensitivity, *Proc. Nat. Acad. Sci. USA* **45**, 100–114.
10. M. G. F. Fuortes, R. D. Gunkel and W. A. H. Rushton 1961, Increment thresholds in a subject deficient in cone vision, *J. Physiol.* **156**, 177–192.
11. H. B. Barlow 1958, Intrinsic noise of cones, *In* "Visual Problems of Colour", National Physical Laboratory Symposium 8, London: H.M. Stationery Office, pp. 615–630.
12. A. M. Granda, J. H. Maxwell and H. Zwick 1972, The temporal course of dark adaptation in the turtle, *Pseudemys*, using a behavioral avoidance program, *Vision Res.* **12**, 653–672.
13. D. A. Baylor and A. L. Hodgkin 1973, Detection and resolution of visual stimuli by turtle photoreceptors, *J. Physiol.* **234**, 163–198.
14. M. G. F. Fuortes and E. J. Simon 1974, Interactions leading to horizontal cell responses in the turtle retina, *J. Physiol.* **240**, 177–198.
15. S. Yazulla 1976, Cone input to horizontal cells in the turtle retina, *Vision Res.* **16**, 727–735.
16. K. T. Brown 1969, A linear area centralis extending across the turtle retina and stabilized to the horizon by non-visual cues, *Vision Res.* **9**, 1053–1062.
17. D. A. Baylor and R. Fettiplace 1975, Light path and photon capture in turtle photoreceptors, *J. Physiol.* **248**, 433–464.
18. G. L. Fain, A. M. Granda and J. H. Maxwell 1977, The voltage signal of photoreceptors at the visual threshold, *Nature*, **265**, 181–183.
19. W. J. Dixon and F. J. Massey 1957, "Introduction to Statistical Analysis," pp. 318–325, McGraw-Hill, New York.
20. E. A. Schwartz 1977, This volume.
21. B. Katz and R. Miledi 1967, A study of synaptic transmission in the absence of nerve impulses, *J. Physiol.* **192**, 407–436.
22. K. Kusano 1968, Further study of the relationship between pre- and postsynaptic potentials in the squid giant synapse, *J. Gen. Physiol.* **52**, 326–345.

23. J. F. Ashmore and G. Falk 1976, Absolute sensitivity of rod bipolar cells in a dark-adapted retina, *Nature* **263**, 248–249.
24. A. Lasansky 1977, Synaptic organization of retinal receptors, This volume, pp. 275–290.
25. M. V. L. Bennett 1970, Comparative physiology: Electric organs, *Ann. Rev. Physiol.* **32**, 471–528.
26. J. E. Dowling and H. Ripps 1973, Effect of magnesium on horizontal cell activity in the skate retina, *Nature* **242**, 101–103;
L. Cervetto and M. Piccolino 1974, Synaptic transmission between photoreceptors and horizontal cells in the turtle retina, *Science* **183**, 417–419;
R. F. Dacheux and R. F. Miller 1976, Photoreceptor-bipolar cell transmission in the perfused retina eyecup of the mudpuppy, *Science* **191**, 963–964;
A. Kaneko and H. Shimazaki 1976, Synaptic transmission from photoreceptors to bipolar and horizontal cells in the carp retina, *Cold Spr. Harb. Symp. Quant. Biol.* **40**, 537–546;
H. Ripps, M. Shakib and E. D. MacDonald 1976, Peroxidase uptake by photoreceptor terminals of the skate retina, *J. Cell Biol.* **70**, 86–96;
S. Schacher, E. Holtzman and D. C. Hood 1976, Synaptic activity of frog retinal photoreceptors, A peroxidase uptake study, *J. Cell Biol.* **70**, 178–192;
Yu. A. Trifonov and A. L. Byzov 1977, This volume, pp. 251–263.
J. Toyoda, M. Fujimoto and T. Saito 1977, This volume, pp. 231–250;
F. S. Werblin 1977, This volume, pp. 205–230.
27. A. B. Steinbach 1974, Transmission from receptor cells to afferent nerve fibers, *In* "Synaptic Transmission and Neuronal Interaction", pp. 105–140, Raven, New York.

Discussion

P. B. Detwiler: Behavioural measurements of the Weber fraction used stimuli projected on the area centralis. Baylor and Hodgkin's measurements of sensitivity as a function of background intensity involved cones in the peripheral retina. Is it not conceivable that the Weber fraction, and/or the effect of background on sensitivity, would be different in the central and peripheral portions of the retina?

G. L. Fain: It is certainly conceivable that the effect of background on receptor sensitivity is different in different parts of the turtle's retina, but it is doubtful that such an effect would alter the calculation of ΔV_T, the voltage response in the receptor at the increment threshold. We know, for example, that receptors in different parts of the turtle retina have different cross-sectional areas, different pigment densities, different outer segment orientations, and so on, so that, in the presence of a diffuse background, different receptors (even spatially separated receptors of the same pigment class) will catch different numbers of quanta and will therefore be adapted to varying degrees. However this would not affect the value of $k/\Delta V_T$, since the sensitivity of the receptor to both the background and increment test intensities will be altered by exactly the same amount. Equation (6) of my paper shows that, for bright backgrounds ($I_B \gg I_0$), $k/\Delta V_T \sim (I_B S_F)^{-1}$. Clearly a change in the quantum catch will not alter the product of I_B and S_F. For $k/\Delta V_T$ to be different in different red-sensitive turtle cones, there would have to be some fundamental change in the mechanism of light adaptation for receptors in different parts of the retina. This is, of course, possible, but in my opinion, unlikely.

R. Meech: Dr Fain has suggested a model to account for the high sensitivity of transmission of the receptor synapse. A similar situation exists in electroreceptors (Clusin, Spray and Bennett 1974, *Nature* **256**, 425–427). If continuous release of transmitter from the receptor synapse in the dark is associated with an appreciable calcium conductance of the presynaptic membrane there must be in addition an outward current for the membrane to remain at rest. The characteristics of this outward current, presumably potassium, are critical. If the outward current is large the membrane will hyperpolarize away from the range of high calcium conductance. If it is small "all-or-nothing" calcium spikes will be generated, as in TEA treated axons (Katz and Miledi 1969, *J. Physiol.* **203**, 459–487). This coupling between g_{Ca} and g_K could be achieved if g_K were voltage dependent in a rather specific fashion but g_K can also be regulated by changes in $[Ca^{++}]_i$ resulting from calcium influx (Meech and Standen 1975, *J. Physiol.* **249**, 211–239). Regulation of membrane conductance by intracellular calcium has been reported in electroreceptors (Clusin *et al.*, 1974).

Coupling between g_K and g_{Ca} is not necessarily "tight" and fluctuations in conductance may be expected. This may account for the voltage sensitive noise reported at this meeting by Schwartz in the rods of the turtle retina and also for some of the noise reported by Simon and Lamb in the cones.

G. L. Fain: We now have evidence that vertebrate photoreceptors contain voltage-sensitive Ca^{++} and K^+ conductances, which probably play an important role in synaptic transmitter release. TEA-treated photoreceptors can generate calcium spikes like those recorded from squid presynaptic membrane (G. L. Fain, F. N. Quandt, and H. M. Gerschenfeld 1977, *Nature*, in press). Steady-state current-voltage curves from rods in normal Ringer show a region of outward-going rectification just positive to the dark membrane potential, which is markedly reduced by TEA (F. N. Quandt and G. L. Fain, in preparation). This rectification is probably produced by a voltage-dependent potassium conductance. It appears not to be affected by external Co^{++}, suggesting that the g_K of photoreceptors is not regulated by $[Ca^{++}]_i$. It is likely that a significant proportion of both the calcium and potassium channels are open at the dark membrane potential, with the result that both g_K and g_{Ca} would contribute to receptor membrane noise in the dark,

regardless of the "tightness" of the coupling between them. Since both g_K and g_{Ca} decrease as the receptor hyperpolarizes, they could be responsible for a significant proportion of the reduction in receptor noise which has been observed during the light response.
(*Reply added in proof.*)

18

Comparison of the Voltage Noise and the Response to One Photon in the Rods of the Turtle Retina

E. A. SCHWARTZ

Department of Neurobiology, Harvard Medical School, Boston, Massachusetts, U.S.A.

The immediate concern of this work was to determine the reliability with which rods in the turtle retina signal the absorption of one photon. To reach the conclusion, I will proceed in three steps. First, I will describe two types of experiments which allow the average response to one photon to be estimated. Then, I will describe the intrinsic noise of rods. Finally, I will compare the amplitude of the response produced by a single photon with the amplitude of the noise and describe how synaptic filtering may effect the reliability with which the signals produced by dim lights are detected. The results are published in more detail elsewhere[1,2,4,6] and the present account outlines the major findings.

The Response to One Photon

The signal produced by the absorption of a single photon has been estimated in two ways.

The first method involves recording intracellularly from one rod the average responses produced by a small diameter spot at several different light intensities (these experiments are reported in detail in references 2 and 4). The shape of the average response to the flash of a dim, small spot is a slow hyperpolarizing wave which reaches its maximum after approximately a half second and then slowly recovers (Fig. 1). The shape of the response can be described by an empirical equation which yields the solid line. This mathematical description will be useful later. If responses are determined at several different dim light intensities the shape of each response remains the same but the amplitude changes in proportion to the light intensity. With

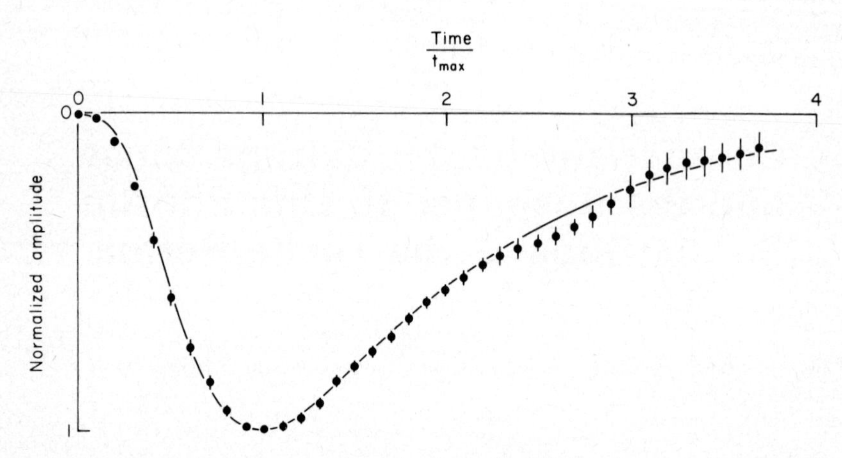

Fig. 1. Mean response of seven rods to the flash of a 100 μm diameter, dim spot. The vertical lines through the points are ±1 s.e. of mean. The continuous curve was calculated from $2{\cdot}87[\exp(-0{\cdot}163t/t_{max}) - \exp(-3{\cdot}237t/t_{max})]^5$. The amplitudes of the responses were between 1 and 2 mV. The mean peak voltage in each experiment was scaled to unity and then the average time course determined. The mean peak voltage was 24 ± 6 μV $(Rh^*)^{-1}$ and t_{max} was 512 ± 93 ms (from reference 1).

several simple assumptions the number of porphyropsin molecules bleached can be estimated from the light intensity.[4] In Fig. 2, the peak amplitudes of several average responses have been plotted along the ordinate against the number of porphyropsin molecules bleached in the impaled rod as abscissa. The amplitude of the responses increases at first in proportion to the number of porphyropsin molecules bleached; at intensities brighter than sufficient to bleach approximately 100 porphyropsin molecules in the impaled rod, the shape is altered and the amplitude increases more slowly. The linear relationship at dim intensities allows the data to be extrapolated to the response produced by the photoisomerization of one porphyropsin molecule (abbreviated Rh^*). The slope of the line in Fig. 2 is 24 μV $(Rh^*)^{-1}$. The same experiment repeated with twenty-six cells yielded a range of 16–31 μV $(Rh^*)^{-1}$. Recently, Copenhagen and Owen[3] have performed a similar experiment and found a response of 28 μV $(Rh^*)^{-1}$.

The second method for estimating the size of the response produced by a single photoisomerization involves a quantal analysis similar to that which is used to estimate the size of the voltage produced by one quantum of transmitter at a chemical synapse (these experiments are reported in more detail in references 1 and 4). The responses produced by a rod when the flash of a dim, *small* spot is repeated many times are not all identical. They vary, in part, due to fluctuation in the number of photons that are absorbed. If a series

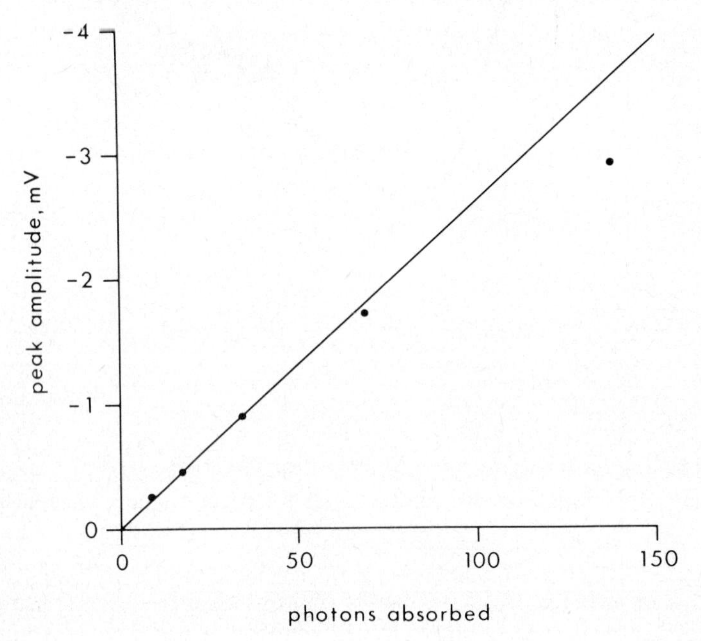

Fig. 2. Responses to the flash of a dim, small diameter spot are proportional to light intensity. The mean peak amplitude is plotted as ordinate against photons effectively bleached in the impaled rod as abscissa. The stimulus was 25 μm in diameter. The number of molecules bleached was estimated from the irradiance of the light, the duration of the flash, and an effective collecting area of 10 μm^2 (see reference 1). The nominal intensity of the 25 μm spot was increased four times to compensate for light attenuation at the centre of the small diameter image (see reference 2). The slope of the straight line is 24 $\mu V\ (Rh^*)^{-1}$.

of reponses are recorded, the mean and the variance of the change in the voltage may be measured at each moment of time following the flash. If the variance is determined by a single Poisson process, such as the number of photons absorbed by the impaled cell, then its square root will have the same time course as the mean. This is the case (Fig. 3). For a Poisson process the ratio of the variance to the mean at the peak of the response gives the size of the quantized event. For the experiment of Fig. 3 the ratio is approximately $(0{\cdot}2\ mV)^2/1{\cdot}36\ mV = 29\ \mu V$. Similar values were obtained in four experiments. Note that this method of estimating the response to a photon does not

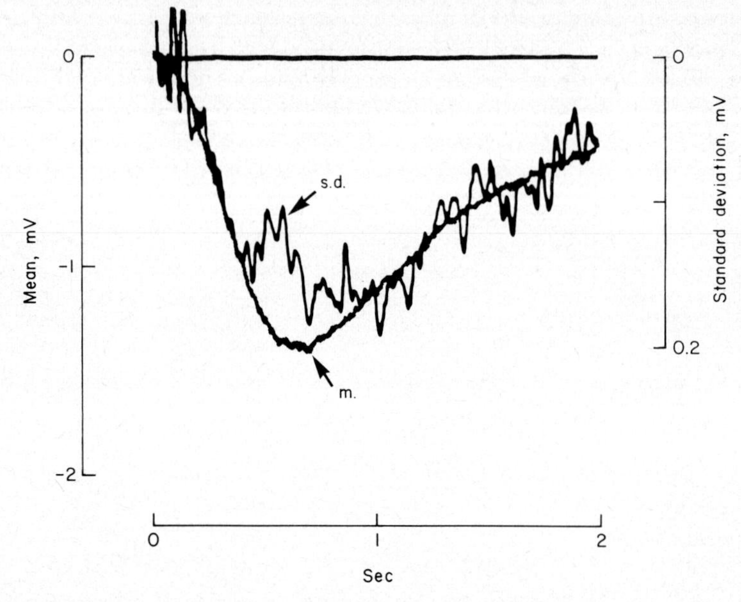

Fig. 3. Comparison of the time course of the mean hyperpolarization and its standard deviation determined from 100 repetitions of a 100 μm stimulus which delivered a mean 4·0 photons μm^{-2} $flash^{-1}$ (from reference 1).

require a knowledge of the number of photons absorbed. For the same cell, dividing the peak mean voltage by the number of porphyropsin molecules bleached gives 1·36 mV/40 $Rh^* = 34\ \mu V\ (Rh^*)^{-1}$. The estimates of the single photon response provided by both methods are in good agreement and indicate an amplitude of approximately 30 μV.

During the previous experiments it was important to study the responses of a rod with a *small* spot of light for it is now known that rods interact over a large retinal area.[1-5] The interaction appears to be mediated by electrical connexions between the rods themselves.[4] Consequently the hyperpolarization produced by an absorbed photon will decrement to neighbors. The response to a photon is therefore produced by a group of adjacent rods, the largest voltage occurring in the rod that absorbs the photon and smaller voltages being produced in neighboring rods (for a quantitative description see reference 1).

Intrinsic Noise

The effectiveness of the signal generated by one photon will depend upon the intrinsic noise from which it must be discriminated. It has recently been possible to observe the intrinsic noise of rods (these experiments are reported in detail in reference 6). A noise which is probably similar has also been observed in cones.[7] An example of the noise in rods is illustrated in Fig. 4. For these records recording was capacitatively coupled to pass the frequencies above 0·2 Hz and only the fluctuation about the mean level is indicated. The

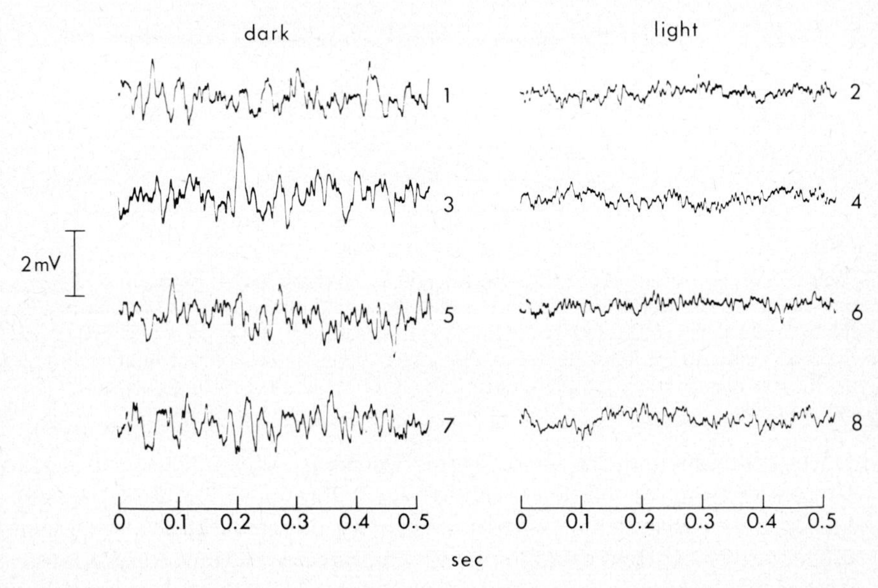

Fig. 4. Voltage fluctuations observed during darkness or continuous light. The intracellular voltage of a rod was recorded during alternate one minute periods of dark or illumination by 13 photons $\mu m^{-2}\ s^{-1}$. Illustrated are 0·5 s segments chosen arbitrarily from the steady state within each period. The order of the experiment is indicated by the number to the right of each trace. Data was filtered to pass frequencies between 0·2 and 180 Hz (from reference 6).

cell was exposed to alternate periods of darkness and dim light. During each period of darkness the voltage appeared noisy. During each period of illumination the cell hyperpolarized (not shown) and the noise markedly decreased.

The noise could be decreased in the absence of illumination by hyperpolarizing a cell with current injected through the impaling micropipette. In the example of Fig. 5, the intracellular voltage (upper record) was first recorded during darkness. After a short period in which the noise was observed, a hyperpolarizing current was passed through the impaling pipette. During current injection the noise decreased. After cessation of current injection and a return of the noise to its previous high amplitude, the retina was illuminated.

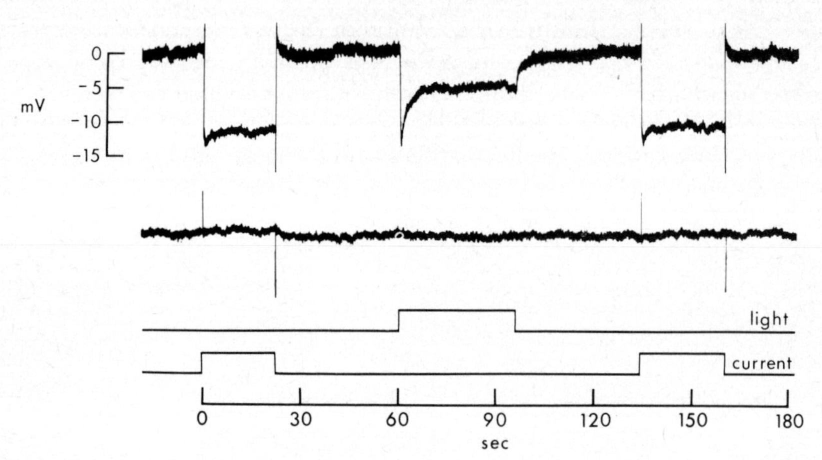

Fig. 5. Voltage noise was reduced by both light and hyperpolarizing current. The light, indicated by the upper timing trace, delivered 102 photons μm^{-2} s^{-1}. The hyperpolarizing current, indicated by the lower timing trace, was $1{\cdot}4 \times 10^{-10}$ A. The upper record is the intracellular response. The lower record is the extracellular response. Voltage is measured as the change from the level maintained during darkness (from reference 6).

During illumination the noise again decreased. When the pipette was withdrawn from the rod (lower record), the voltage noise recorded extracellularly was less than that observed intracellularly during darkness and was not influenced by a current equal to that which decreased the intracellular noise. The last observation indicates that a decrease in noise with hyperpolarization was a property of the rod membrane rather than an artifact of the recording equipment.

In order to compare the noise to the response produced by light a quantitative description of the noise is necessary. For this purpose power density spectra of the noise were determined during dark and light and during current injection. An example for one rod is shown in Figs 6 and 7. In these figures the log of the power is plotted along the ordinates and the log of the frequency along the abscissae. The log-log coordinates allow a large range of data to be conveniently displayed. In 6(a) the spectrum determined during dark is indicated by the closed symbols and the spectrum determined during light is indicated by the open symbols. Both spectra include the noise of the recording system, the micropipette, and a component of cell noise which was

---

Fig. 6. Power density spectra of the noise eliminated by light. In (a) are spectra of the noise recorded during dark (closed circles) and during illumination by $4{\cdot}8 \times 10^{2}$ photons μm^{-2} s^{-1} (open circles) plotted along log-log coordinates. In (b) is the difference between these two spectra. The difference is a noise generated by the cell and eliminated by light (from reference 6).

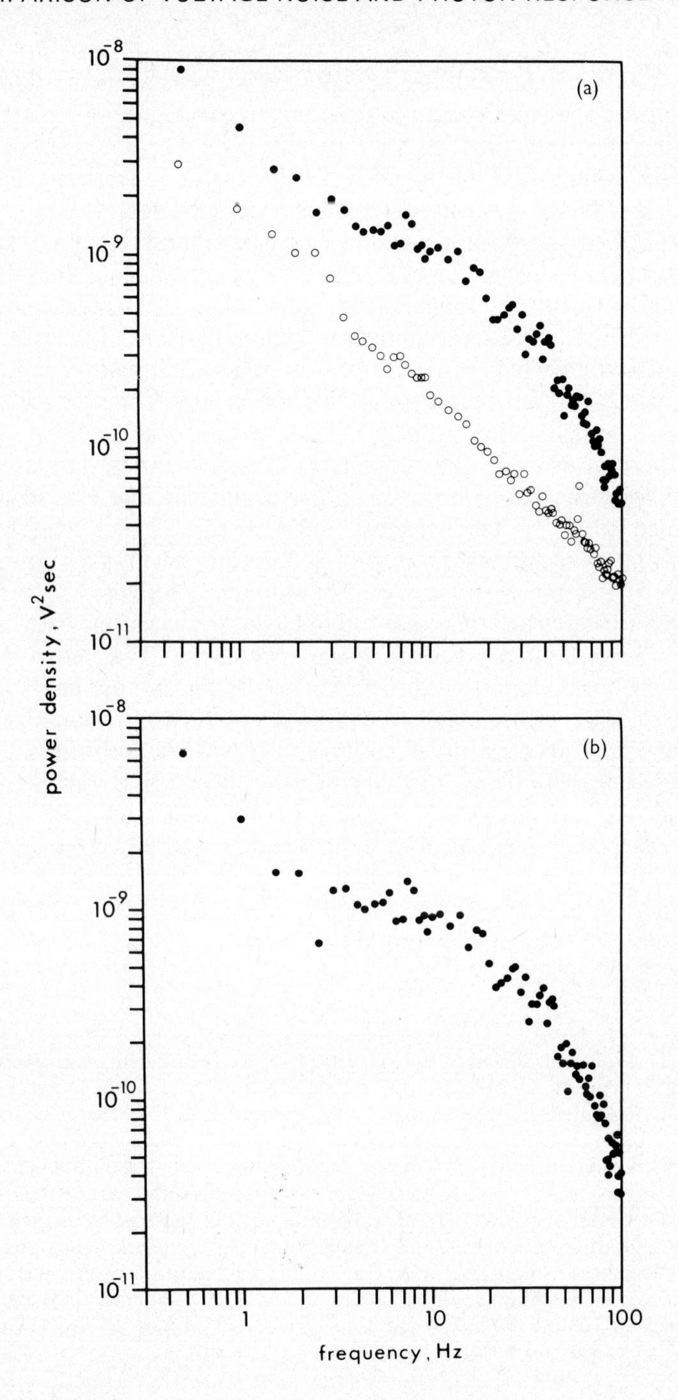

10^{-8}
10^{-9}
10^{-10}
10^{-11}
(a)
(b)
power density, V^2sec
1
10
100
frequency, Hz

unmodified by illumination. Assuming these forms of noise are the same during dark and light, the difference between the two spectra of Fig. 6(a) yields the spectrum of a noise which was generated by the cell and eliminated by light. The difference spectrum is shown in Fig. 6(b). The spectrum reveals the frequency composition of the noise. In this case it is apparent that a large amount of the noise was in low frequency components.

The noise which was eliminated by light can be separated into two parts by an experiment which combines the use of both light and current injection. In 7(a), the spectrum determined during dark is again indicated by the closed circles. When during dark a continuous hyperpolarizing current was injected, the spectrum indicated by the triangles was obtained. Injecting a larger current did not alter this spectrum. However, when current was injected and the retina was also illuminated, the spectrum indicated by the open circles was obtained. The difference between the two spectra obtained during darkness without current injection (closed circles) and with current injection (triangles) is attributed to a "voltage-sensitive" noise that was decreased by hyperpolarization. The difference spectrum for this component of the noise is shown in Fig. 7(b) as the closed circles. The difference between the two spectra obtained during current injection without light (triangles) and with light (open circles) is indicated in Fig. 7(b) by the open circles. The component of the noise, which was eliminated by light but not by current, will be termed "light sensitive" noise. Hence, there appears to be two forms of noise with different spectral compositions: voltage-sensitive noise with the spectrum indicated by the closed circles in Fig. 7(b) and light-sensitive noise with the spectrum indicated the open circles in 7(b). The normal effect of light (as in Fig. 6b) was to decrease the light-sensitive noise directly and to produce a hyperpolarization which in turn decreased the voltage-sensitive noise.

For both components of the noise, the power density, G, at a frequency, f, can be described by an equation of the form

$$G(f) = \frac{\alpha}{1 + (2\pi\tau f)^2} \tag{1}$$

Fig. 7. The noise eliminated by light is the sum of two components. Same cell as Fig. 6. In (a) the spectrum observed during the dark is replotted as the closed circles; the spectrum indicated by the triangles was measured during the injection of $1{\cdot}8 \times 10^{-10}$ A hyperpolarizing current; the spectrum indicated by the open circles was measured during the continued injection of the same current and illumination by $4{\cdot}8 \times 10^2$ photons $\mu m^{-2}\ s^{-1}$. In (b), the closed circles are the difference between the closed circles and triangles of part (a) and represent a noise eliminated by current; the open circles are the difference between the triangles and open circles of part (a) and represent a noise eliminated by light but not by current. In (b), integrating each spectrum yields a variance associated with the closed circles of $3{\cdot}38 \times 10^{-8}\ V^2$ and associated with the open circles of $0{\cdot}88 \times 10^{-8}\ V^2$. The solid line is the solution of eqn 1 with $\alpha = 1{\cdot}1 \times 10^{-9}\ V^2$ s and $\tau_m = 8$ ms. See text (from reference 6).

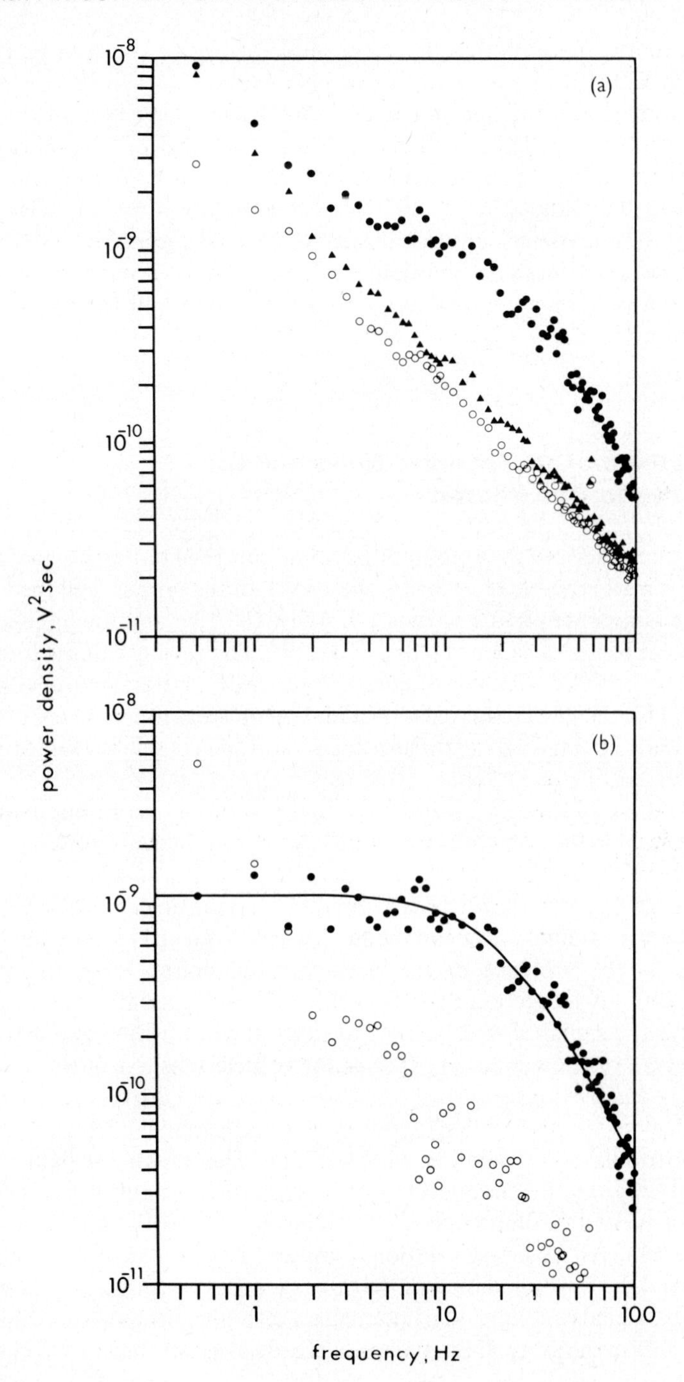
(a)
(b)
10^{-8}
10^{-9}
10^{-10}
10^{-11}
power density, V^2 sec
1
10
100
frequency, Hz

where α and τ are constants. In the example of Fig. 7, the voltage-sensitive noise can be fitted to eqn (1) with $\tau = 8$ ms and $\alpha = 1{\cdot}01 \times 10^{-9}$ V^2 s (solid line). By repeating the analysis under conditions which minimize voltage-sensitive noise, the light-sensitive noise can be studied in relative isolation. Under these conditions light-sensitive noise can be fit by the same equation with $\tau = 230$ ms and $\alpha = 7 \times 10^{-9}$ V^2 s (details are given in reference 6). There are several molecular mechanisms which yield spectra described by eqn (1). I will not consider these molecular models here. For our purpose, the formal description permits a comparison between the noise and the response to a single photon.

Comparison of the Intrinsic Noise and the Response to One Photon

Having characterized the amplitude and time course of both the signal and the noise it is now possible to compare these two. In the noisiest cell the voltage-sensitive component had a variance of $6{\cdot}5 \times 10^{-8}$ V^2 and the light-sensitive component had a variance of $0{\cdot}87 \times 10^{-8}$ V^2. The variance of the total noise was $7{\cdot}4 \times 10^{-8}$ V^2. Taking a square root yields a standard deviation of 271 μV. This may be compared with a maximum response of 30 μV produced by one photon. That is, the intrinsic noise was nearly 9 × the single photon response.

Now I do not know the ultimate sensitivity of turtle vision. But in man it is well-known that the absorption of only a few photons reaches consciousness.[8] I will assume that the evolutionary pressure of natural selection has been almost as generous to the turtle as to man and that the turtle can also perceive a small quantity of light. How can the small event produced by very dim light be detected in the presence of the large intrinsic noise? Detection may be improved if advantage were taken of the different spectral properties of the noise and the signal. The difference in spectral composition is illustrated in Fig. 8 where the power density spectra for voltage-sensitive noise (solid line) and light-sensitive noise (dashed line) as given by eqn 1 are compared with the power density spectrum calculated from the mathematical description of the response to a flash shown in Fig. 1 (dotted line). The response to light contains power at only very low frequencies. At high frequencies only noise is present. Detection would be improved if the influence of noise at high frequencies could be removed. A simple solution is suggested by the recent work of Baylor and Fettiplace[9] which indicates that the synapses of rods transmit the low frequency signals of light and attenuate the higher frequencies which are present only in the noise. If the synapse were to act as an efficient low pass filter

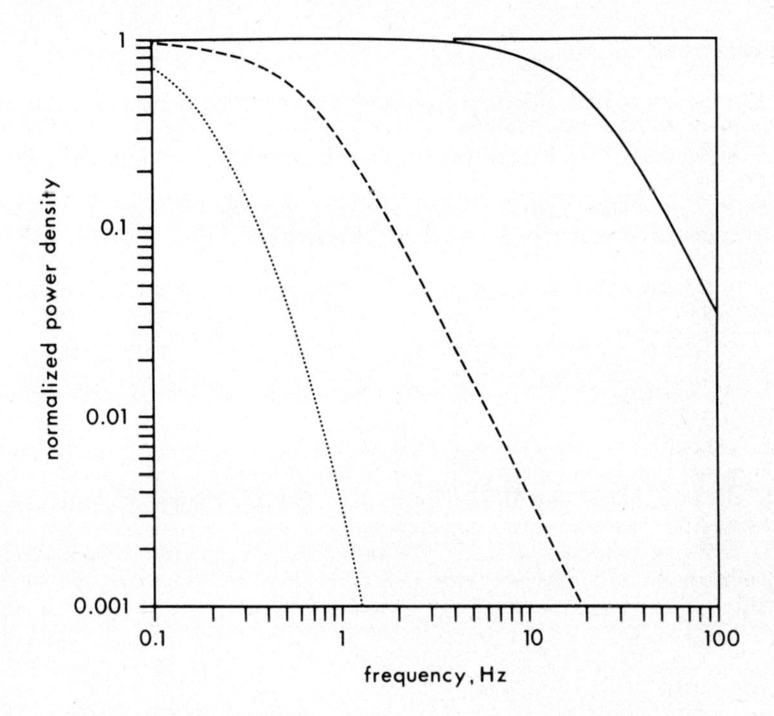

Fig. 8. Comparison of the power density spectra for voltage-sensitive noise (solid line), light-sensitive noise (dashed line) and response to a dim flash of light (dotted line).

the variance of each component of the noise transmitted would be approximately given by the equation

$$\text{var } U_j = \frac{\alpha_j}{4\tau_s} \tag{2}$$

where τ_s is the time constant of the synaptic filter and α_j is the zero frequency asymptote of either the voltage or light sensitive component. For example, if the synaptic time constant were 600 ms, then the standard deviation of the noise transmitted would be 65 μV rather than 271 μV and the noise would be reduced by a factor of four. At the same time the average response to a dim flash would be little affected.

In summary, rods possess two forms of noise. The predominent component is voltage-sensitive and extends to high frequencies. The total noise is much larger than the responses to one photon. This noise can be decreased by synaptic filtering without affecting the responses to light.

References

1. E. A. Schwartz 1976, Electrical properties of the rod syncytium in the retina of the turtle, *J. Physiol.* **257**, 379–406.
2. E. A. Schwartz 1973, Responses to single rods in the retina of the turtle, *J. Physiol.* **232**, 503–514.
3. D. R. Copenhagen and W. H. Owen 1976, Functional characteristics of lateral interactions between rods in the retina of the snapping turtle, *J. Physiol.* **259**, 251–282.
4. E. A. Schwartz 1975, Rod-rod interaction in the retina of the turtle, *J. Physiol.* **246**, 617–638.
5. G. L. Fain 1975, Quantum sensitivity of rods in the toad retina, *Science, N.Y.* **187**, 838–841.
6. E. A. Schwartz 1977, Voltage noise observed in rods of the turtle retina, *J. Physiol.* **272**, 217–246.
7. E. J. Simon, T. D. Lamb and A. L. Hodgkin 1975, Spontaneous voltage fluctuations in retinal cones and bipolar cells, *Nature* **256**, 661–662.
8. S. Hecht, S. Shlaer and M. H. Pirenne 1942, Energy, quanta and vision, *J. gen. Physiol.* **25**, 819–840.
9. D. A. Baylor and R. Fettiplace 1975, Transmission of signals from photoreceptor to ganglion cells in the eye of the turtle, *Cold Spring Harbor Symp. Quant. Biol.* **40**, 529–536.

Discussion

P. B. Detwiler: In view of the problems associated with passing current through high resistance microelectrodes, how certain are you of the membrane potential change evoked by applied current?

E. A. Schwartz: The micro-pipettes used behaved linearly with a resistance of approximately 200 MΩ for currents less than approximately $0{\cdot}5 \times 10^{-10}$ A (reference 1). For currents between 0·5 and $2{\cdot}5 \times 10^{-10}$ A the resistance deviated from linearity by a few percent. The intensity of current injected to reduce voltage-sensitive noise was usually $1\text{–}2 \times 10^{-10}$ A. Hence the membrane voltage of the impaled cell during current injection was uncertain by perhaps ± 4 mV. In addition, because rods are electrically connected, injected current spreads to neighboring rods. The portion of the noise generated within each of these neighbors was therefore influenced by a smaller voltage than that produced in the impaled rod. Hence during current injection the observed noise was produced by many sources which were each at slightly different potentials. For these reasons, it has not been possible to investigate the relationship between membrane potential and the amplitude of the voltage-sensitive noise.

19

Retinal and Central Factors in Human Vision Limited by Noise

H. B. BARLOW

The Physiological Laboratory, Cambridge, U.K.

When I started research in vision, Rose and de Vries' papers[1] had just appeared, and it was an enormously attractive thought to me that the limiting performance of the eye could, under many conditions, be accounted for simply by the quantum fluctuation of the absorbed light. Now the simplest prediction from such a hypothesis is that increment threshold ΔI will increase as the square root of background luminance I. Though ΔI usually rises more rapidly, conditions can be found where this Rose-de Vries Law, $I \propto I^{1/2}$, holds well over a wide range of background luminance.[2] This occurs when the test stimulus is of small area and short duration, and the suggested explanation is, of course, that the quantum fluctuations of the background are proportional to $I^{1/2}$ and set a lower limit to the number of quanta that can be detected from the test flash. For this to be a valid explanation the light must be summated over a fixed area and time, but unfortunately this is probably not true; over the range where the square root law holds, the product of summation time and area is approximately proportioned to $I^{-1/4}$.[3] It is therefore hard to accept that the Rose-de Vries hypothesis, in its elementary form, is the explanation of the Rose-de Vries law.

Another way of stating the Rose-de Vries hypothesis would be to say that the quantum efficiency is constant. Rose[4] has very forcefully stated the case for this, and it is obviously a useful guide to instrument designers and others to be told that quantum efficiency stays between 10 % and 1 % over a wide range of mean luminance of the visual scene. However to those interested in how the eye works, the concept of quantum efficiency is chiefly of interest in leading one to factors *other than* quantum fluctuations that limit performance. To use the concept this way, one finds where quantum efficiency declines, for that is where the other factors enter. It is thus not surprising that my emphasis is on its lack of constancy, whereas Rose and others emphasize its constancy.

It is also, for the physiologist, particularly important to stick to a strict definition of the overall quantum efficiency of visual performance. It is the ratio of numbers of quanta required by an ideal device to the number required by an actual subject performing *exactly the same task*. To simplify the calculation for the ideal performance, unjustified assumptions are sometimes made about the way the eye performs the task.[4] The eye, considered as an instrument, has many tricky design features, and it is by no means safe to assume that it can be characterized by a few simple parameters, or that one knows what they are. One can detect a highly redundant stimulus, such as a flickering light or a grating, when the elements of which it is composed (a single flash or a single bar of the grating) are far below the limit of visibility. Also, partial summation continues well beyond a summation parameter defining a reciprocity limit. There are also likely to be parallel pathways with different parameters subserving different detection tasks. For such reasons the eye may seem to have a higher quantum efficiency than is in fact found to be the case when measurements are critically conducted.[5]

In spite of these difficulties it is a useful tool to explore how much we actually know about the factors limiting vision, and in this paper I propose to review estimates of the fraction of quanta absorbed (absorbtive or photometric quantum efficiency) and compare them with the best measurements of the quantum efficiency of visual performance of the whole organism (psychophysical or behavioural quantum efficiency), or of a neurophysiological preparation.

Psychophysical and Photometric Efficiencies

The situation at and near absolute threshold will be considered since this is where there is the smallest gap between photometric and psychophysical quantum efficiencies, but the gap is not as small as it appeared to be when the original work was done. Hecht, Shlaer and Pirenne[6] estimated that no more than 10 % of 507 nm light incident at the cornea was absorbed in the rods; their measured thresholds averaged 112 quanta at the cornea, so they concluded that a flash visible on 55 % of occasions caused, on average, about 10 quantal absorptions. They then showed that the slope of frequency-of-seeing curves was such that at least 5 quanta, and sometimes 8, must be utilized at threshold. They were well pleased with how close this figure approached the direct estimate of 10, and reached the physiologically satisfying conclusion that "biological variability" and "psychological factors" were not disturbing their measurements; quantum fluctuations were the dominant factor.

Now this argument rests heavily on that figure of $<10\%$ for the fraction of corneal quanta absorbed. It was reached by a beautiful and judicious argument that brought in many facts no one else had considered, but I think the conclusion is undoubtedly wrong, perhaps by as much as a factor of 3. Table 1 shows the most judicious estimates I can now make. The main factor

Table 1
Photometric, psychophysical and neurophysiological estimates of the proportion of quanta at the cornea that excite rods. For fuller discussion of the factors see references 18, 21

	Human	Cat
Photometric		
Trans. of media	0·5 –0·75	0·5 –0·85
Prop. entering rods	0·7 –0·8	0·7 –0·9
Abs. by rhodopsin	0·5 –0·55	0·55–0·7
Fraction exciting	0·6 –1·0	0·6 –1·0
Product:	0·11–0·33	0·12–0·54
Psychophysical		
Human absolute threshold		
assuming noise $X_c =$ 0	0·054	
10	0·059	
120	0·11	
580	0·33	
Neurophysiological		
Cat retinal ganglion cell		
Measured $X_c = 10$		0·15

that has changed is the figure for absorption in the receptors in the third line, which Hecht *et al.* thought was under 20%. The figure of 50% or more given here is that derived by Alpern[7] and colleagues by several methods, and I think it is strongly supported by the results of Marks[8] and Liebman[9] on photometry of single photoreceptors. (See also MacNicholl's discussion of Rose's paper: this volume, p. 12.)

A factor which has changed in the other direction is the fraction exciting, in the fourth line. It is usually assumed that only quanta causing isomerization can excite, and I have therefore included the factor of 0·6.[10] But we do not know this for certain; even without isomerization the energy from the quantum is locally available, and the amount is comparable to the mechanical energy of a threshold acoustic stimulus.[11] Even if it does not isomerize, it is all available in a single subcellular particle, whereas the threshold acoustic stimulus is presumably spread out among many hair cells underlying the basilar membrane, presenting a much harder detection task.

I think one must admit the possibility that as many as 33 % of quanta entering the cornea are available to excite rods, but you don't have to believe the higher figure to be worried by the discrepancy between these estimates of photometric efficiencies, and the behavioural quantum efficiencies. In the bottom part I have summarized these. The figure for the human is obtained by taking the average threshold from the results of Hecht *et al.*,[6] namely 112 quanta, and dividing this into the average of the number of quanta utilized, obtained from the frequency-of-seeing curves, namely 6.

Now some estimates of human quantum efficiency above 5 % have been published, but as mentioned earlier there has usually been a flaw capable of leading to spuriously high results. For example, invalid assumptions were made about the spatial and temporal integration of the retina; or the number of quanta involved was calculated for a single eye, whereas the psychophysical data were obtained with two; or the estimate of the signal/noise ratio of the human subject needed refinement. Hallett[20] developed a method to correct for drifts of sensitivity and obtained quantum efficiencies up to 10% by using this correction. However Hecht's figure of 5–6%, subsequently confirmed by Baumgardt[12] and myself[5], is the highest one that is fully convincing.

Some phychophysical estimates below 5 % have also appeared. Sakitt[13] suggested that subjects looking at very weak flashes of light can give responses graded according to the actual number of quanta exciting rods, this number being thought to vary from zero to about 6 under her conditions. This explained many aspects of her data, and it required a dark light X_c at the cornea (see below) of 10 to 40, and quantum efficiencies of about 3 %. The latter figure is in acceptable agreement with other psychophysical estimates considering that she used 7° eccentricity, but X_c is much lower than that estimated below. The weakness of her argument is that she cannot say "At *most* one, or two, or three . . . rod excitations were involved for the corresponding ratings", whereas quantum efficiency arguments lead to firm limits. It therefore seems preferable to take 3 % at 7° eccentricity as a minimum rather than as a definitive estimate.

Van Meeteren and Bougard[14] used an interesting method in which the efficiency is obtained by comparing a subject's performance when limited by quantal fluctuations with his performance at a similar task when it is limited by the fluctuations of a random dot pattern within which the test pattern appears. It is a more quantitative version of Rose's[1] original method. Their figure for rod vision is 1 %, though it appears that the exact nature of the task performed affects the result, and 1 % may not be the figure for an optimal task.

Also included in Table 1 is the result of estimates done on retinal ganglion cells in the cat. The figure given is the average for the 5 most sensitive ganglion cells out of 11 thoroughly studied and, in contrast with the human, this figure *is* compatible with the photometric absorption, though it is at the lower limit. Now there are two obvious differences between the cat and human estimates: first, the cat figure includes an allowance for intrinsic retinal noise; second, the cat method is done at a level preceding the central processing and decision making, and hence avoids any losses of efficiency incurred there.

The quantitative effects of visual noise and central inefficiency are pursued in the following sections of this paper.

Intrinsic Noise

Hecht *et al.*[6] matched their psychophysical data to a model in which a "seen" response was given if c' or more quanta were absorbed, "not seen" if less than c'. When the frequency of seeing curves predicted by this model are plotted with log (stimulus intensity) as abscissa, the slope at $P = 0{\cdot}5$ is approximately given by[15]

$$\mathrm{d}P(c' \text{ or more})/\mathrm{d}(\log I) = c'^{1/2}/(2\pi)^{1/2} \log e. \tag{1}$$

Their experimental frequencies of seeing suggested values of c' ranging from 5 to 8, averaging close to 6. It is also clear that, on their model, the 50% threshold $I' = c'/F$ (approx.), where F is the proportion of quanta at the cornea that is absorbed in rods and contributes to visibility. For their experiments I' averaged 112 quanta at the cornea, and hence the best estimates of F from their results is $6/112 = 0{\cdot}054$. By optical methods the proportion of quanta at the cornea that excite rods is now thought to be at least twice as high (Table 1).

It is now abundantly clear that there is noise in the retina, and this might account for the difference. The simplest way to explore its effect on visual performance is to introduce a new parameter x, which represents a number of random independent events liable to be confused with the effective absorption of a quantum of light.[15] It is true that this way of expressing noise rather suggests it is very peripheral in origin, for instance thermal isomerization, but wherever it occurs it should be possible to express it in this form to a first approximation, and it is the simplest way to express it if one is interested in its effect on visual performance. Figure 1 shows how a set of predicted frequency of seeing curves is generated by supposing a fixed criterion c and varying x. These become flatter and shift to the left as x is increased, in rather the same way that the set without noise (dotted curves) shifts and flattens as c' is decreased. For instance, the curve for $c = 64$, $x = 40$ has a slope nearly equal to that for $c' = 9$ without noise, whereas that for $c = 64$, $x = 50$ would be nearly equal to $c' = 3$ without noise. The experimental curves for Hecht *et al.*, characterized by a slope for $c' = 6$, would require a noise of $x = 44$ if the

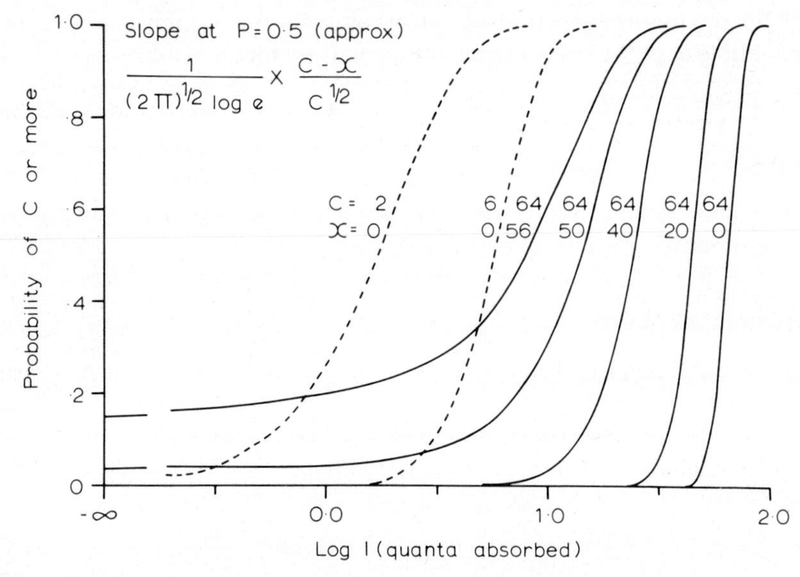

Fig. 1. The effect of intrinsic noise on frequency of seeing curves. A number x of noise events is postulated. These are confusable with quantal absorptions l, and a response is given when $l + x \geq c$. Log l is plotted as abscissa and the ordinate is the probability of c being exceeded. The five continuous curves are for a fixed, very high, criterion $c = 64$, and noise x varying from 56 to 0. They form a series rather like those of the noiseless Hecht, Shlaer and Pirenne model with the criterion c varied; two of these (dotted) are shown for comparison.

criterion was $c = 64$. Notice that the number absorbed at threshold would then be about $64 - 44 = 20$, three times what the noiseless case requires.

Approximate relations can be derived[15] for these quantities. Thus the slope at $P = 0{\cdot}5$ is given approximately by:

$$\mathrm{d}P(c \text{ or more})/\mathrm{d}(\log I) = (c - x)/c^{1/2}(2\pi)^{1/2} \log e. \tag{2}$$

There is also a relation between I', the 50 % threshold, F the fraction of quanta at the cornea absorbed, c and x. The absorbed quanta make up the difference between c and x, hence

$$c = x + FI'. \tag{3}$$

There are differences in shape between the simple cumulative Poisson curves and these theoretical curves in the presence of noise, and this second series is actually preferable on empirical grounds for the curves flatten out at low light intensities and intersect the ordinate at finite probabilities for zero flash intensities: in other words, false positives occur on the noise model, as they do in reality, whereas they do not on the other model. But in practice it is very difficult to use anything from frequency of seeing data except estimates of mean and slope, and as you see these depend on both c and x. Clearly, to make any progress one needs an independent method of estimating x.

One can obtain this as follows. Suppose the absolute threshold experiment is modified slightly. Stimuli are delivered containing either L, or $L + K$, quanta (on average), but instead of asking the subject if he sees the flash or not, he is told that "dim" (containing L quanta) and "bright" (containing $L + K$) will occur with equal probability, and it is his task to say which an unknown one is. He can see samples of the two classes whenever he likes, but whenever he receives an unknown he must of course classify it as dim or bright. When L and $L + K$ are small, of the order of 100–200 quanta, this is very much like a threshold task, for the subject rapidly finds that he rarely sees the dim ones, so he classifies all the seen ones as bright. For higher values a subject can find a dividing criterion among the visible stimuli, and he is aided in doing this not only by samples of "dim" and "bright" obtained when required, but also by being told whether a judgement was right or wrong, thus receiving feedback.

We define K^* as the value of K which allows the two classes to be discriminated with a d' (or internal signal/noise ratio) of 1. As L is increased, initially K^* is little effected or may even decrease, but eventually K^* increases steeply. If L corneal quanta cause l absorptions, these will be added to x, the noise events, and one can get information about the noise by finding the value of L at which it starts to impair detection of additional quanta.

Figure 2 shows the result for two subjects. It is hard to see any deleterious effect of L on K^* until L is greater than about 200 quanta in one subject, 300 in another. Define X_c as the number of quanta at the cornea that would cause x quantal absorptions. Then this result is incompatible with a value of X_c below about 200. However, one cannot press the accuracy of this result for the following reason. If the situation was simply as I've described it, K^* should increase as $\sqrt{L}$ at high values, but clearly it goes up more steeply. Also the small but definite sensitization effect is unexplained. Nevertheless, if X_c was small, say about 10, then values of L in that range should affect K^*, and they do not.

This type of experiment clearly suggests that X_c is of the order 200 or more, and this is also fully compatible with the value of I_0, the "dark light" or "eigengrau", derived from increment threshold experiments. I_0 was found to be about 1000 quanta s^{-1} deg^{-2} in a review of many such experiments,[2] and if summation time and area are taken as 0·2 s and 1 deg^2, this agrees well with the above result.

To find what this value of X_c implies about F one reasons as follows. The slope on the noise model (eqn 2) is equal to that determined experimentally by Hecht *et al.* (eqn 1) so we know that:

$$(c - x)/c^{1/2} = (c')^{1/2}. \qquad (4)$$

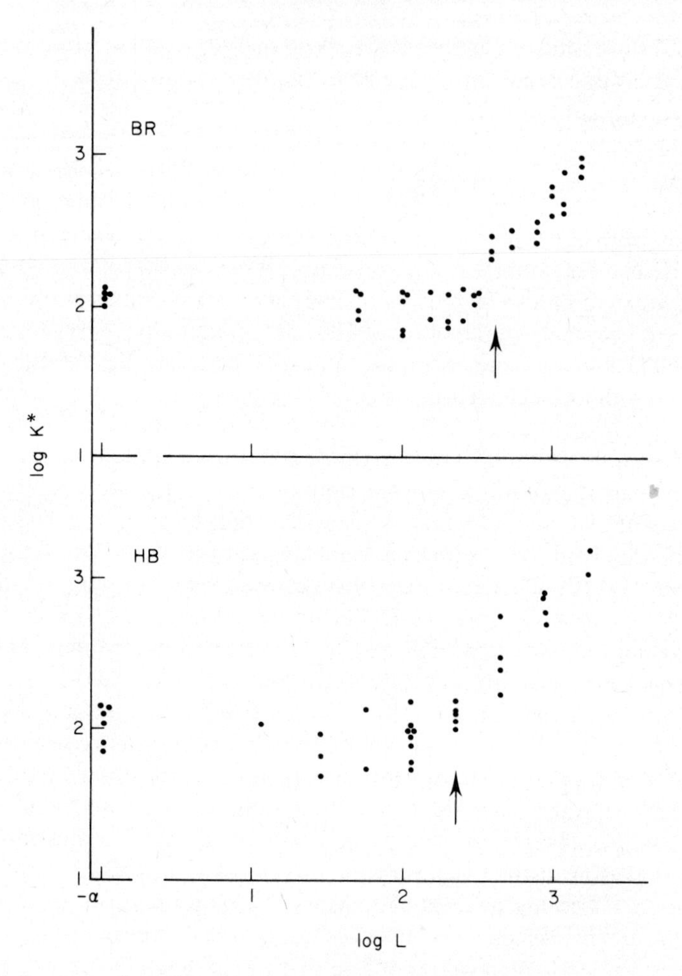

Fig. 2. Estimation of X_c. The subject's task was to discriminate between light flashes containing (on average) L and $L + K$ quanta. K^* is the value of K (obtained by interpolation) for which the discriminability between L and $L + K$ has a $d' = 1$. It is very little influenced by L until this reaches a value of 200–300, which suggests that X_c has a value as high as this.

The number of events x confusable with isomerization is simply:

$$x = FX_c. \tag{5}$$

Combining eqns 3, 4 and 5 and rearranging terms gives:

$$F = x/X_c = c'(I' + X_c)/(I')^2. \tag{6}$$

Substituting Hecht, Shlaer and Pirenne's values of $c' = 6$ and $I' = 112$, and the above value of $X_c = 200$, one finds that $F = 0{\cdot}15$. Values of F of 0·11 and 0·33, which are the lower and upper estimates of photometric efficiency in Table 1, require values of X_c of 120 and 580, respectively.

My conclusion at this point is that intrinsic noise can account with reasonable quantitative accuracy for the difference between photometric and psychophysical quantum efficiencies. The notion of intrinsic noise was advanced with the idea that it originated somewhere in the retina, that is in the synapses or receptors or even by thermal isomerization of photopigment molecules. Some of the relations between psychophysical performance and visual noise as measured by x, X_c, or I_0 are discussed elsewhere in this volume following the reported measurements of the electrical noise of cones (reference 16; see also reference 17). However it would be wrong to assume that the high values of X_c reported above arise solely in the receptors, as shown in the following section.

Sources of Intrinsic Noise

In a system with a noisy input it is pointless to reduce the noise of later stages far below that of the input for, owing to the fact that variances add, the largest source of noise dominates and smaller sources contribute a negligeable amount to the total noise. Thus one may not get a clear answer to the question "What causes visual noise?" anymore than one does to the question "What limits acuity?". There is, however, mounting evidence that the high value of X_c of 200–300 arrived at above is not all retinal in origin. For the cat retinal ganglion cell X_c is only of the order 10, sometimes less. This was derived as follows.[18] The maintained discharge, and the variance of the maintained discharge, are both thought to be due to x, the events indistinguishable from quantum absorptions. If a stimulus flash is given, extra impulses are generated, and variance is increased. By extrapolating backwards one obtains estimates of X_c at the intersection with the abscissa. The results from mean and variance are slightly different, and there is also a good deal of variability from cell to cell. But the average of the best 5 cells was 4·6, of all 11 cells 11·8. These are clearly very much lower than the human psychophysical estimate, namely over 200.

From those measurements one can also obtain an estimate of the quantum efficiency, which is given in Table 1. From that type of measurement one obtains estimates from 5% to 18%, but for the best 5 it averaged over 12%, which was the lower limit for photometrically estimated absorption in the cat (see Table 1).

The human periphery is not necessarily like the cat, but suppose it was: How then could one account for our estimate of $X_c = 200$? The answer is "Only too easily", for so far no allowance has been made for imperfections of the central mechanisms. Suppose, for instance, that central threshold, equivalent to the criterion number of events c, fluctuates; the psychophysical methods do not readily distinguish between fluctuations of c, and fluctuations of signal size resulting from the random fluctuations of x, which were originally conceived as being purely *retinal* in origin. This means that the formal explanation in terms of intrinsic noise is rather hollow; without more information we don't know what the noise is intrinsic to.

In Table 2 some of the possible sources of noise are listed, together with visual phenomenon that may be related. Obviously the methods of choice for getting information about the first four are now physiological, but recent work does bring out the very great importance of the fifth one. High level phenomena may have effects exactly like retinal noise, and what we thought was the latter may in fact be largely the former.

Table 2
Possible Sources and Effects of Intrinsic Noise

Thermal isomerisation	Quantum bumps in dark
	Purkinje shift
Intermediate stages	Receptor noise
Synaptic noise	Bipolar cell noise
	Receptor interconnections
Impulse quantisation error	Ganglion cell output
	Weber law
Threshold fluctuation and central inefficiency	Troxler type effects
	Random dot efficiencies

Threshold Variation and Central Efficiency

When one does measurements close to absolute threshold one is aware that funny things are happening. For instance, if one is to decide whether a stimulus belongs to a "dim" or "bright" class one would like to see many members of each class before deciding about an unknown. But this turns out to be a virtually impossible procedure, for as one repeatedly delivers the "bright" ones, they become invisible! At first one blames oneself for poor fixation, but paying attention to this shows that the stimulus reappears if you deliberately fixate $\frac{1}{2}^\circ$ away from the correct fixation position. These effects are not small, they are very large, as has been shown by Frome, MacLeod, Buck and Williams.[19] They found a $5X$ rise of threshold as a stimulus was repeated every $2\frac{1}{2}$ s for about 50 min. The total number of stimuli delivered was not much larger than the number required for frequency of seeing curves. The rate of delivery was quite fast, but one cannot slow it down more than four or five times when doing frequency of seeing curves without the risk of finding an unresponsive subject asleep on his bite bar.

You do not see such marked habituation effects in ganglion cells, and MacLeod has proved convincingly that they are central; cone stimuli habituate rod-detected flashes, and vice versa, and the effects are size-specific. I expect the habituation is related to the Troxler effect, the disappearance of flicker and also to the fading that occurs in stabilized images. But whatever it is due to, clearly one must take it into account in interpreting data on absolute threshold. Hallett's calculations[20] also suggest that the instability of the criterion over long periods is important.

The current line of thinking makes one doubt the assumption that the cortex deals efficiently with the evidence provided to it by the sense organs—an assumption that has been implicit in much of the work on the threshold. There is a method whereby this efficiency can be measured and this will now be described.

Estimating Central Efficiency

In these experiments the subject looks at an oscilloscope screen on which dots appear at random with controllable density and frequency of occurrence. They are, however, bright enough and few enough in number for it to be reasonable to assume that *retinal* factors do not limit their perception. The dots are not placed entirely at random, but some pattern is superimposed, such as an increased density in a particular region. One hopes, with such a stimulus, to break through the retinal barrier and study the ability of central mechanisms to detect these patterns. With suitable arrangements for delivering dots at random or in a controlled pattern, and for counting the numbers that appear, one can apply an absolute statistical measure to the ability of the higher centres to detect the stimulus. This is exactly analogous to Rose's measure of quantum efficiency, but in this case we aim to be limited, not by the quantum-catching powers of optics, receptors and pigments, but by the pattern-catching powers of the central mechanisms.

Figure 3 shows examples of such random dot patterns. The top left has only the uniform random dot background, while the bottom right has an obviously suprathreshold number of extra dots in the centre. The other two are intermediate. Note that we know the *external* signal/noise ratio of the stimulus for any given average number of dots added to the centre. If we find the *internal* signal/noise ratio, or d', the ratio of the two gives a measure of how efficiently one can make discriminations on the basis of the internal representation. Actually, to make it analogous to quantum efficiency, one uses the square of this ratio, and this can be regarded as an estimate of the

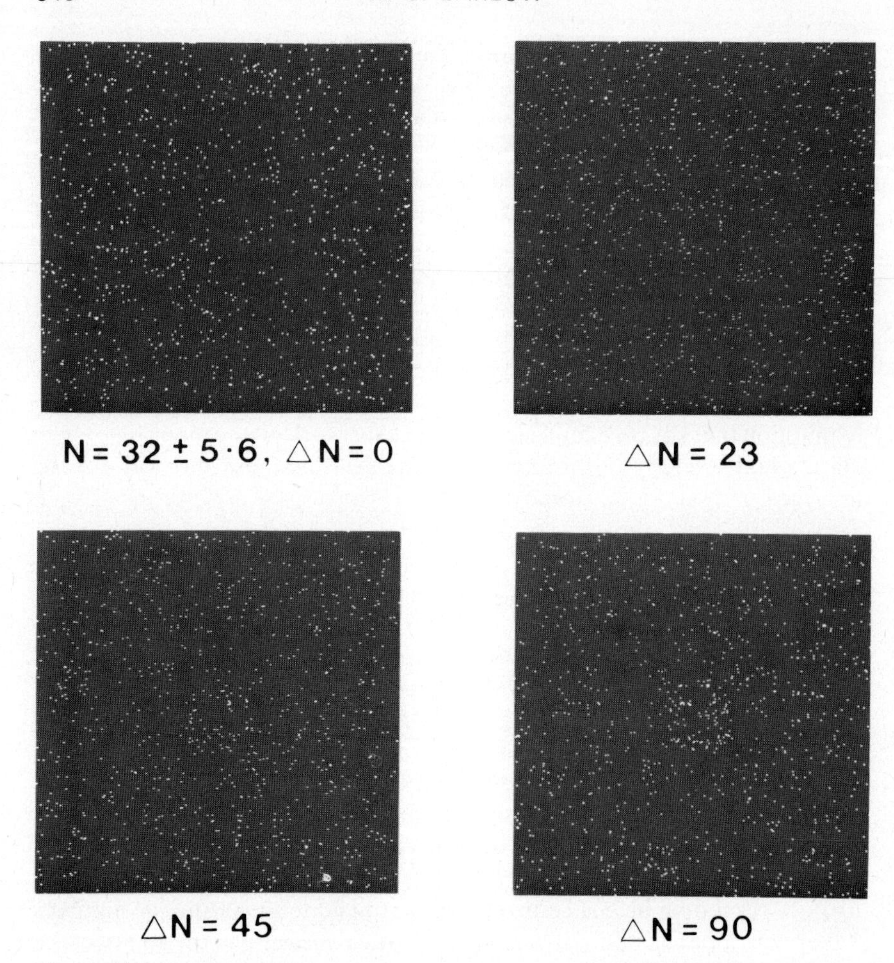

Fig. 3. Eight hundred dots are randomly spread over the whole square, and viewed under conditions where they can readily be seen. A small central square of 1/25 the total area has 32 dots expected, with standard deviation $\sigma = 5{\cdot}6$. ΔN dots are added to this central square, the four values corresponding to about 0, 4, 8 and 16 times σ. From the results plotted in Fig. 4 it will be seen that an excess of 23 dots could be detected with high confidence on about 70 % of trials, with some uncertainty in about 95 %. Without experience much larger values of ΔN are required for confident detection.

proportion of the sample of dots that is effectively utilized in deciding about the pattern.

Figure 4 shows one method of measuring *internal* signal/noise ratio, and it is clearly similar to Hecht, Shlaer and Pirenne's method. There were 800 dots, 32 expected in the central square. One varies the extra dots in the central square and does frequency of seeing curves to which regression lines are fitted. From

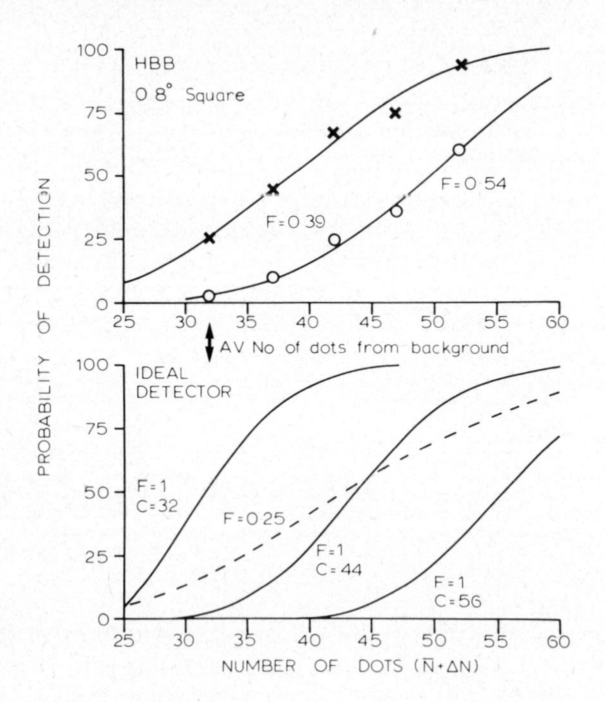

F = N(THRESH) / σ^2 WHERE:
N(THRESH) = TOTAL No OF DOTS AT THRESHOLD
σ = RECIPROCAL OF SLOPE OF PROBIT REGRESSION LINE

Fig. 4. Measuring perceptual efficiency. The average number of dots falling in the small central square (see Fig. 3, lower right) was varied, and the subject's task was to report if he definitely saw the excess (circles) or probably detected them (crosses). The lower curves show what could ideally be achieved in this experiment if every dot in the central square had been counted (i.e. $F = 1$), and criteria of 32, 44 and 56 had been taken for responding "seen". If only 1/4 the dots had been counted ($F = 0{\cdot}25$), the curves would have been flattened to the slope of the dashed one. By probit analysis, parameters of curves are found to fit the experimental points, and these curves are drawn through them at top. By the calculation indicated the selected parameters give the perceptual efficiencies of 0·39 and 0·54 for the two criteria used by the subject.

this the fraction of statistical information utilized is determined. In this case the subject judged "probables" as well as "seen" patterns, and the efficiencies were 39% and 54%.

The figure obtained varies of course with the stimulus parameters, but even after seeking out the optimum it is not easy to better 50%, and at present this seems the most likely way of reconciling the 5·5% quantum efficiency obtained from psychophysics with the 11% lower limit of photometric quantum efficiency.

Conclusion

Psychophysical estimates of "noise" are of the right magnitude to reconcile photometric and behavioural data. But although some noise comes from the retina, and probably the receptors, there is also an important central component. Perhaps after 30 years investigating these central mechanisms we shall understand why they perform as well and as poorly as they do, just as the studies on the receptors that have been presented at this conference are beginning to show what underlies the capacity of a rod to respond to a single quantum, a fact deduced from psychophysical measurement more than 30 years ago.

Acknowledgments

My thanks are due to Dr A. van Meeteren and Mr B. Reeves for help in the experiments shown in Figs 4 and 2.

References

1. A. Rose 1942, The relative sensitivities of television pick-up tubes, photographic film and the human eye, *Proc. Inst. Rad. Eng.* **30**, 293–300;
   A. Rose 1948, The sensitivity performance of the human eye on an absolute scale, *J. opt. Soc. Am.* **38**, 196–208;
   H. de Vries 1943, The quantum character of light and its bearing upon the threshold of vision, the differential sensitivity and acuity of the eye, *Physica* **10**, 553–564.
2. H. B. Barlow 1957, Increment thresholds at low intensities considered as signal/noise discriminations, *J. Physiol.* **136**, 469–488.
   M. A. Bouman 1961, History and present status of quantum theory in vision, *In* "Sensory Communication", pp. 377–401 (ed. W. A. Rosenblith), M.I.T. Press and J. Wiley, New York and London.
3. H. B. Barlow 1958, Temporal and spatial summation in human vision at different background intensities, *J. Physiol.* **141**, 337–350.
4. R. Clark Jones 1959, Quantum efficiency of human vision, *J. opt. Soc. Am.* **49**, 645–653;
   A. Rose 1977, Vision: Human versus electronic, This volume, pp. 1.
5. H. B. Barlow 1962a, A method of determining the overall quantum efficiency of visual discriminations, *J. Physiol.* **160**, 155–168;
   H. B. Barlow 1962b, Measurements of the quantum efficiency of discrimination in human scotopic vision, *J. Physiol.* **160**, 169–188.
6. S. Hecht, S. Shlaer and M. Pirenne 1942, Energy, quanta and vision, *J. gen. Physiol.* **25**, 819–840.
7. M. Alpern and E. N. Pugh, Jr. 1974, Density and photosensitivity of human rhodopsin in the living retina, *J. Physiol.* **237**, 341–270.
   F. Zwas and M. Alpern 1976, The destiny of human rhodopsin in the rods, *Vision Res.* **16**, 121–128.
8. W. B. Marks 1965, Visual pigments of single goldfish cones, *J. Physiol.* **178**, 14–32.

9. P. A. Liebman and G. Entine 1968, Visual pigments of frog and tadpole (*Rana pipiens*), *Vision Res.* **8**, 761–775.
10. H. J. A. Dartnall 1968, The photosensitivities of visual pigments in the presence of hydroxylamine, *Vision Res.* **8**, 339–358.
11. L. J. Sivian and S. D. White 1933, Minimum audible sound fields, *J. acoust. Soc. Am.* **4**, 288–321;
H. de Vries 1948, Brownian movement and hearing, *Physica* 14, 48–64.
12. E. Baumgardt 1960, Mésure pyrometrique du seuil visuel absolu, *Optica Acta* **7**, 305–316.
13. B. Sakitt 1972, Counting every quantum, *J. Physiol.* **223**, 131–150.
14. A. van Meeteren and J. Boogaard 1973, Visual contrast sensitivity with ideal image intensifiers, *Optik* **37**, 179–191;
A. van Meeteren 1973, Visual aspects of image intensification. Doctoral thesis, Riksuniversiteit Utrect. Report of the Institute for Perception TNO, Soesterberg, The Netherlands.
15. H. B. Barlow 1956, Retinal noise and absolute threshold, *J. opt. Soc. Am.* **46**, 634–639.
16. E. J. Simon and T. D. Lamb 1977, Electrical noise in turtle cones, This volume, pp. 291–304. Discussion following above paper.
17. E. A. Schwartz 1977, Comparison of the voltage noise and the response to one photon observed in rods of the turtle retina. This volume, pp.323–334.
G. L. Fain 1977, The threshold signal of photoreceptors, This volume, pp. 305–322.
18. H. B. Barlow, W. R. Levick and M. Yoon 1971, Responses to single quanta of light in retinal ganglion cells of the cat, *Vision Res.* **11**, Supplement 3, 87–102.
19. F. Frome, D. I. A. MacLeod, S. L. Buck and D. R. Williams 1976, Visual habituation to flashed peripheral targets, ARVO 1976.
20. P. E. Hallett 1969, Quantum efficiency and false positive rate, *J. Physiol.* **202**, 421–436.
21. A. B. Bonds and D. I. A. MacLeod 1974, The bleaching and regeneration of rhodopsin in the cat, *J. Physiol.* **242**, 237–253.

Discussion

Gordon L. Fain: Recordings from single rods and cones during light adaptation suggest that the difference in the dark lights of the scotopic and photopic systems may arise, at least in cold-blooded vertebrates, from differences in the photoreceptors themselves. That is, the ratio of dark lights for single rods and cones is nearly as great as that for the scotopic and photopic systems. This finding is of some interest, since it suggests that the difference in the dark lights of the rod and cone systems may be in large part the result of differences in the process of transduction within the two photoreceptor types.

If the difference in the dark lights of the rod and cone systems were due to a difference in the noisiness of the two systems then, at least in cold-blooded vertebrates, one would expect to find the major source of this noise in the photoreceptors themselves. If the noise occurred somewhere within the process of phototransduction, as for example rhodopsin molecules spontaneously bleaching, then one would expect it to contribute to the voltage noise of the receptors. In fact, the measured noise of receptors gives some credence to this idea, since the magnitude of the noise in rods and cones is nearly the same (see Simon and Lamb, this volume, p. 303) but the sensitivities of rods and cones are different. Thus the number of presumed spontaneous events would have to be much greater in cones than in rods to produce the same amplitude of noise, and this would correspond to the larger dark light in cones.

It is important to realize, however, that the difference in the dark lights of single rods and cones can be explained without any reference to their voltage noises. It is possible to show, for example, that the dark light of a photoreceptor, I_0, is given by

$$I_0 = \frac{S_F^D \Delta V_T}{k},$$

where S_F^D is the sensitivity of the dark-adapted receptor (measured in voltage per incident quantum), ΔV_T is the voltage response in the receptor at the behavioural increment threshold, and k is the Weber constant of the behavioural increment-threshold curve (see G. L. Fain, this volume, p. 305). Thus the differences in dark lights between single rods and cones can be explained simply by their different absolute sensitivities and by a difference in the mechanism of light adaptation in the two receptor types. Although rods and cones are both inherently noisy, this noise must be carefully distinguished from the dark light of light adaptation, since these two may be completely different phenomena.

H. B. Barlow: We agree with each other that the dark lights of single rods and cones differ approximately as predicted from psychophysics, and I would concede that the relationships can be *described* as Dr Fain outlines in his last paragraph. But this is not really an *explanation*, or is very incomplete as such, for it says nothing about why k, S_F^D and ΔV_T in his expression for the dark light I_0 have the values they do for rods and cones.

As background illumination is decreased the toad rods and turtle cones both increase in sensitivity, but when the background illumination drops to a value around their respective dark lights the sensitivities reach their plateaus and do not improve any further. One can regard these plateau values of S_F^D as the ultimate factors responsible for the difference between rods and cones, as I think Dr Fain does. Alternatively you can take the dark light as more than a mere abstraction and suppose that sensitivity fails to go on improving as real background is decreased because it is the sum (dark light + real light), that controls sensitivity, and this cannot be reduced below the value of the dark light even when real light drops to zero.

My proposal is that the cones in darkness have a much higher on-going signal than rods, possibly because of greatly increased isomerization due to the shift of their peak sensitivity to the red (see Discussion following Simon and Lamb's paper, this volume, p. 301). This is the intrinsic noise, which I would identify with the dark light I_0, and threshold signals must be detected as increments to it. On this view both the poor absolute sensitivity of cones, and their high incremental sensitivity (low Weber fraction) relative to rods are explained: cones do not have as high an *absolute* sensitivity as rods because their intrinsic noise stops them improving their sensitivity as much (if they did have as high sensitivity they would have an absurdly high electrical noise level); and they have better *incremental* sensitivity than rods because at their absolute threshold they must respond to small changes in the rate of noise events plus photoisomerizations if they are to signal the weakest lights that are statistically significant. This consideration does not apply to rods because the smallest significant number of photoisomerizations is not small compared to the number of noise events.

With regard to light adaptation I agree that the dark light, or equivalent background light, whose slow decline accompanies the slow process of dark adaptation, may best be kept separate from the irreducible dark lights that limit absolute sensitivities. The view that the changes in dark adaptation are paced or controlled by changes in a peripheral source of "equivalent background light" does not now look tenable. However, it may still be the case here again that sensitivity is controlled by a process occurring at a more peripheral point in the chain of transduction than the one at which it is measured.

Finally it is worth saying that human rods may not behave like toad rods, for their electrical response characteristic may not change much until saturation values of the background are reached. Here changes of sensitivity over most of the scotopic range presumably occur more centrally.

P. B. Fellgett: I would like to raise the following three points about Dr Barlow's paper.

(1) The term "quantum efficiency" is often used in a loose and poorly defined sense. In relation to the concepts described by Professor Rose (Chapter 1 of this volume), we have to begin with *noise-equivalent photon number*. This is defined as the minimum number of photons that could in principle enable a particular task to be performed. The ratio of this number to the actual mean number of photons available to a detector by use of which the task can just be performed, is defined as the *noise-equivalent quantum efficiency* of the detector in that task.

(2) One should consider more precisely the nature of the task in relation to which noise-equivalent quantum efficiency is evaluated. The only safe procedure is to evaluate the *information gain* in performing the chosen task, since the noise-equivalent quantum number is then precisely fixed. In simple cases, it may indeed be possible to work in terms of signal-to-noise ratio, frequency of seeing etc., but information-theoretic verification is always advisable.

Just because informational estimates are safe they always give a lower limit, and if any part of possible information gain is neglected the noise-equivalent quantum efficiency will be underestimated. This will happen, for example, if we neglect colour information or the ability of the observer to say in which part of the visual field a test flash occurred.

(3) My understanding of the figures that have been presented is that safe estimates of the noise-equivalent quantum efficiency of the eye are about half the supposed quantum efficiency of the primary response to light. Bearing in mind the difficulty of constructing an n-stage image-processing system in which each stage causes a loss of only $2^{1/n}$ on average, it seems to me that so far from there being any "discrepancy" to explain, the human visual apparatus has been extraordinarily well designed and constructed to give so small a ratio between primary and noise-equivalent performance.

H. B. Barlow: Professor Fellgett raised three points which I shall comment on sequentially.

(1) Yes, quantum efficiency is used in the following four different ways, at least; (i) the average proportion of quanta entering the eye that is absorbed in receptors, or a particular type of receptor; (ii) as above, but for isomerization or bleaching rather than absorption; (iii) the average proportion of quanta absorbed in a photopigment that cause isomerization or bleaching; (iv) the minimum proportion of quanta entering the eye that would enable the task to be performed. The last is, as Fellgett points out, the "noise equivalent quantum efficiency", and it involves a different principle, for in the first three one is only concerned with the energy of photons, whereas in the fourth one is also concerned with their uncertainty properties. The context usually makes it clear which sense is used and I hope no one was in doubt about my meaning.

(2) I do not think Professor Fellgett is right in saying "information gain" is the only safe quantity to evaluate. The important point when calculating *noise equivalent photon number* is to make sure that one calculates the minimum number of photons for exactly the same task the subject is performing. If the subject is neglecting colour information, or may receive a stimulus in a number of positions but does not have to specify where it occurs, then he is likely to be ignoring some of the information available to him. But this does not invalidate the estimate of noise equivalent quantum efficiency provided that we calculate

noise equivalent photon number for the same task as he performs. The value obtained may of course vary with the task, and if it does this is of special interest, for the reasons given below.

I have tried using entropy-informational measures, but the situation rapidly becomes too difficult to calculate (for me at least).

(3) Yes, the system as a whole is quite efficient at certain tasks, but one cannot know in advance what tasks a biological system is adapated to perform well. It is therefore specially interesting to see what tasks are performed efficiently, what tasks inefficiently. But there is another point. One observes that selective pressure can lead to a reflecting tapetum that causes at most a 50% or so increase in quantal absorption, and must carry the penalty of much greater intraocular scattered light. Consequently one also wants to know about the causes of minor losses of efficiency, because apparent inefficiency may point to important limiting factors that we are not yet aware of.

S. Yeandle: Remarks on the noise of invertebrate photoreceptors. At this meeting we have heard that in dim light the vertebrate photoreceptor develops hyperpolarizing potentials of the order of tens of microvolts per absorbed photon. Furthermore, its intrinsic noise is not small compared to signals generated by single photo absorptions. This seems a strange device to detect dim light. I think it should be pointed out that the arthropod receptor has what appears to be a more reasonable way to detect dim light. Although the essentials of how it does this have been known for some time, I think, for two reasons, it would be worthwhile to review the present knowledge of what occurs in a dark adapted arthropod photoreceptor exposed to dim light. First, it works differently from the vertebrate photoreceptor and, second, it has many features analogous to man-made devices as, for example, photomultipliers.

Briefly, in the dark-adapted arthropod photoreceptor very weak pulses of light evoke a response consisting of a variable number of discrete waves of depolarizing potential, originally called quantum bumps.[1,2,3] Available evidence suggests that each photon captured by the visual pigment has a certain probability, p, of triggering a bump. I will confine my remarks to Limulus, the horseshoe crab, because of my familiarity with this creature and because, of all arthropods, most is known about how its photoreceptors react to dim light.

In Limulus one can record intracellularly, from ommatidia in the lateral eye, bumps whose amplitudes can be in the millivolt range. In each ommatidium there are many primary receptors, called retinula cells, which capture photons and one or occasionally two secondary neurones, called eccentric cells, which generate the nerve impulses that travel up the optic nerve.[4] The retinula cells are electrically coupled to each other and to the eccentric cell.[5,6] When sufficient bumps occur so that the eccentric cell depolarizes below a threshold level, a nerve impulse is generated.[1]

The number of bumps following a short pulse of light is a random variable with a Poisson distribution whose parameter is proportional to the average energy of the pulse. This finding is consistent with the idea that each photon captured by the visual pigment triggers a bump with probability p.[7]

One can also record bumps from receptors on the ventral side of the animal called ventral photoreceptors that do not generate nerve impulses and have no known function.[8] Despite this, the responses to light of the ventral photoreceptor have been extensively studied because its responses are very similar to those of the retinular cell and because its large size and the absence of screening pigments, such as occur in the lateral eye, make many difficult experiments technically feasible. Estimates of p, defined above, for ventral receptor bumps range from 0·02 to 0·5.[9]

Both the amplitude of bumps and the latency between photon absorption and bump occurence vary randomly.[10, 11] The latency ranges from 60 millisecond to over 2 seconds at room temperature, and its probability density function can be fitted by a Gamma function.[12] The amplitudes, ranging from less than a millivolt to over ten millivolts, appear to be bimodally distributed into large and small bumps. The precise forms of both the amplitude and latency distributions appear to vary from preparation to preparation and the factors responsible for this variability have yet to be worked out. Controversy exists as to whether large bumps and small bumps have different latency distributions.

Extensive studies on the statistics of bump occurrence have led to the following two postulates: First, each absorbed photon has a probability of generating a stochastic process whereby either only one bump is produced at some variable time after the photon absorption, or no bump is produced at all. Second, the stochastic processes initiated by absorbed photons are statistically independent of each other. These two postulates have only been tested for very dim lights, where they seem to hold, but they doubtless fail for sufficiently bright lights. Among other things, they imply the theorem that the distribution of intervals between successive bumps evoked by steady light is a negative exponential whose argument is proportional to light intensity.[13] This has been observed experimentally in dim lights.[7, 14] (This theorem, although it seems trivial, is not, and was proved some years after the experimental work on the statistics was done.)

It should be emphasized that the above two postulates are not the only ones consistent with present data. A more precise knowledge of p, the number of bumps per absorbed photon, is at least required before judging if competing statistical hypotheses can be eliminated.

When a receptor is exposed to moderately intense light, the average size of the bumps decreases. This observation has led to the adapting bump model[15,16] which postulates that the receptor potential is the summations of quantum bumps whose average size is determined by the degree of adaptation of the receptor. This model has had considerable success in describing the ommatidial potential as recorded intracellularly at the eccentric cell in Limulus.

Bumps occur in the dark and appear to be statistically independent of light evoked bumps. The origin of these so called dark bumps is obscure. In the lateral eye the relationship between rate of occurrence and temperature suggests that dark bumps result from the spontaneous thermal isomerization of visual pigment molecules.[17] Contrariwise, in the ventral receptor there is evidence that a large proportion of small bumps arise by some process not related to the process producing light evoked bumps.[9] However, there is some experimental evidence supporting the suggestion that many of the small dark bumps in the ventral receptor may result from the bathing solution used, or injury caused by microelectrode impalement.[18,19] The nature of the dark noise in the vertebrate receptor does not appear to be altogether clear but may not arise entirely from spontaneous isomerizations of visual pigment molecules.

To sum up, the occurrence of bumps appears to follow the same laws as the occurrence of shots from the anode of a photomultiplier. Also, as in the photomultiplier, there is considerable gain in the transduction of light. However, bumps differ from photomultiplier shots in two important ways. Bump size is dependent on the past history of illumination and there is wide fluctuation in the time between photon absorption and bump occurrence. In the vertebrate receptor any analogy to a man-made device is difficult to see.

References to discussion comment of S. Yeandle

1. S. Yeandle 1958, *Am. J. Ophthamol.* **46** (3): Part 2: 82.
2. K. Kirshfeld 1966, *In* "The Functional Organization of the Compound Eye" (ed. C. G. Bernhard), p. 291, Pergamon Press, Oxford.
3. J. Scholes 1965, *Cold Spring Harbor Symp. Quant. Biol.* **30**, 517.
4. M. E. Behrens and V. J. Wulff 1965, *J. Gen. Physiol.* **48**, 1081.
5. A. Borsellino, M. G. F. Fuortes and T. G. Smith 1965, *Cold Spring Harbor Symp. Quant. Biol.* **30**, 429.
6. H. Stieve 1965, *Cold Spring Harbor Symp. Quant. Biol.* **30**, 451.
7. M. G. F. Fuortes and S. Yeandle 1964, *J. Gen. Physiol.* **47**, 443.
8. R. Millecchia and A. Mauro 1969, *J. Gen. Physiol,* **54**, 310.
9. S. Yeandle and J. B. Spiegler 1973, *J. Gen. Physiol.* **61**, 552.
10. A. Borsellino and M. G. F. Fuortes 1968, *J. Physiol.* (*Lond.*) **123**, 417.
11. R. Srebro and S. Yeandle 1970, *J. Gen. Physiol.* **56**, 751.
12. R. Srebro and M. Behbehani 1971, *J. Gen. Physiol.* **58**, 267.
13. G. H. Weis and S. Yeandle 1975, *J. Theor. Biol.* **55**, 519.
14. A. R. Adolph 1964, *J. Gen. Physiol.* **48**, 297.

15. W. A. H. Rushton 1961, *In* "Light and Life" (eds. McElroy and Glass), p. 706, Johns Hopkins Univ. Press, Baltimore.
16. F. A. Dodge, B. W. Knight and J. Toyoda 1968, *Science* **160**, 88.
17. R. Srebro and M. Behbehani 1972, *J. Physiol. London* **224**, 349.
18. M. Behbehani and R. Srebro 1974, *J. Gen. Physiol.* **64**, 186.
19. D. S. Bayer 1976, Special Report of Institute for Sensory Research ISR-S-13, Syracuse University, Syracuse, N.Y. 13210.

20

How does Your Research Explain our Inability to See

W. A. H. RUSHTON

Physiological Laboratory, University of Cambridge, Cambridge, U.K. and Institute of Molecular Biophysics, Florida State University, Tallahassee, Florida, U.S.A.

Seventeen years ago I was invited by Professor Piéron to attend the conference on Colour Vision that he had organized at the College de France in Paris. The colour vision experts of the world were there, chiefly Young–Helmholtz men, but the devotees of Hering were also seen, fighting a rearguard action. Although the meeting was dominated by the traditional psycho-physical approach there was a section on electrophysiology, and I was asked to open it. I thought it my part to persuade the psycho-physicists whose interest lay in the sensations produced by light, that the sensationless records of the electrophysiologists are not quite irrelevant to their sensory studies, because they may help—or more likely destroy—the nice simple symmetrical theories psychophysicists invent. Part of what I said on that occasion in 1959 was this:

"If psychophysicists think that Colour Vision is *their* subject, they have a long and distinguished past to justify that view. But the future will not sustain it. Though it is only during the past 20 years that the sciences of biochemistry and electrophysiology have been able to make useful contributions in the field of colour vision, it is certain that this is going to increase, and it will no longer be possible to put forward a theory of colour simply by the economy of the hypotheses and the beautiful symmetry of the equations to which it leads. It will be necessary for all of us to contribute together.

But I have sometimes thought that psychophysicists have not been as enthusiastic as we could wish when some little advance has been made in electrophysiology. This is not surprising, for the electrophysiologists have taken from the psychophysicists their freedom of imagination and the beauty of their elegant concepts, and in return, they have given them a very indigestible diet. Let us heal this divergence of opinion this morning.

First I would say to the psychophysicists 'We must have the kind of basis that biochemistry and electrophysiology give. It is beautiful to have economy of hypothesis, but Nature is not very economical. In the replication of living things she is prodigal, and her forms are clothed in almost infinite diversity. The beauty of the flowers does not drop from the stars, it grows from the earth. If we wish to get symmetrical blooms we must dirty our fingers a little at the roots, which are far from symmetrical or uniform. Let me ask you to descend into the earth and see what in fact are the roots of colour vision. Have some patience'. 'But my friends, the electrophysiologists, do not strain this patience too far. If the psychophysicist is looking for roots don't simply give him mud. Soil that is sterile or even putrid may have its interest for you, but not for him. There must be at least some little growing points that promise to flower later on'."

Turning now to the present symposium there is little need for me to plead with those whose interest in photoreception stems chiefly from visual sensation, that they should take seriously all these contributions at the molecular level. In the first place these contributions are so impressive that they plead for themselves. But in the second place I seem to be the only one left whose interest still centres in that outdated illusion called "visual sensation".

To be sure, the title of our symposium is "Photoreception" and so we may be thankful that the room is not filled with chlorophyl boys and T.V. cameras. But, though every contributor here has studied the performance of some part of an eye, only the first and last titles on our programme mention the *Human Eye*.

The course of my own research interest is perhaps a little curious. No sooner had I pronounced in Paris that psychophysics must yield to objective recordings, than I took up psychophysics myself.

My object has been to wed these two approaches and to show, if I could, some correlation between *what we see* and the condition of the visual pigments in the photoreceptors. But in fact it is not what we see, but what we cannot see that advances our understanding of the mechanism of vision. Let me explain.

From time to time people write to me about the colours they see in surprising conditions—the "colour" of certain musical keys or moods, or the coloured *aura* by which some people are surrounded. The only way one can respond to such information about what they see is "How interesting, Madam, you must possess very remarkable perception". By contrast the other day, an old friend whose sight was failing, complained to me that he saw from time to time curious figures that had no external existence. I focussed the filament of a flash lamp upon the corneo-scleral junction of his eye and he exclaimed "Yes! There they are". The fact that he could *not* distinguish the figures that he had seen from the shadows of his retinal vessels which I had projected onto his retina, gave a satisfactory explanation of what he had been seeing.

My chief interest in visual research is the wedding of objective measurements to subjective appearances. So it leaves me a bit impatient with your restraint, when you publish beautiful investigations about photoreceptor responses and their organization at the objective level, and keep complete silence about their implication for vision. Vision is a lovely lady and you have pushed yourselves into her company, as I certainly hope. But you behave as though she simply was not there. "We've never been introduced" you say "so we behave with propriety when we act as though she were not present". I wonder if you say this because you are afraid that you might misbehave if you *were* introduced. At any rate I can hardly look to you to help on the marriage if you will not so much as notice the bride.

There are two important classes of observations where our inability to see leads to striking conclusions. One is colour matching. Two identical lights of course appear identical; but, as Newton showed as soon as he mixed his spectral lights, two lights that are not too far separated in the spectrum form a mixture whose appearance is indistinguishable in colour from a single intermediate spectral light.

As is well known, Thomas Young was the first to propose a mechanism of 3 resonators, responsive in varying degrees to all visible frequencies of light that explains our inability to distinguish *metameric* matches. In modern dress Young's theory may be stated thus: There are 3 types of cone each with its own variety of photosensitive pigment. Any light will be absorbed in each pigment in accord with its absorption spectrum, the response of the cone being dependent simply upon the total quantum catch. The colour response (for a fixed state of adaptation) will depend only upon the ratio of absorptions in the 3 cones.

If the mechanism were true it must follow, as Clerk Maxwell saw, that every light could be matched by a mixture of 3 primary colours e.g. red + green + blue (with coefficients that sometimes are negative). Maxwell's experiments were the first to show this, and the beautiful confirmation by W. D. Wright was itself confirmed to perfection by W. S. Stiles many years later.

This indicates some of the power derived from being unable to discriminate in a *colour matching* situation. A second and more versatile failure is called "threshold". If the intensity of a uniform field is gradually reduced, we may note that the brightness is diminishing, but (unless there is some fixed comparison field—which brings the situation into the *matching* category) there is no stage in the dwindling brightness that can be distinguished until the field can no longer be seen. This unique level above which the field (or some feature in it) can be seen, and below cannot, is called "threshold". It is the principal measure in psychophysics because it may be applied to so many features of a visual display.

So the wedding that I wished to effect was between the subjective psychophysical measurements of threshold or of matches, and the objective measurements of pigments in rods or cones.

Retinal Densitometer

Figure 1 shows in principle our reflexion densitometer drawn by Dr E. N. Willmer and Fig. 2 a less artistic but more scientific drawing by Dr R. W. Rodieck.

Incident light reflected from the pigment epithelium passes twice through the retina and suffers absorption by the visual pigments at each passage. Thus the emergent light registered by a photomultiplier cell gives information of light loss and hence of the density of pigment that is responsible. A

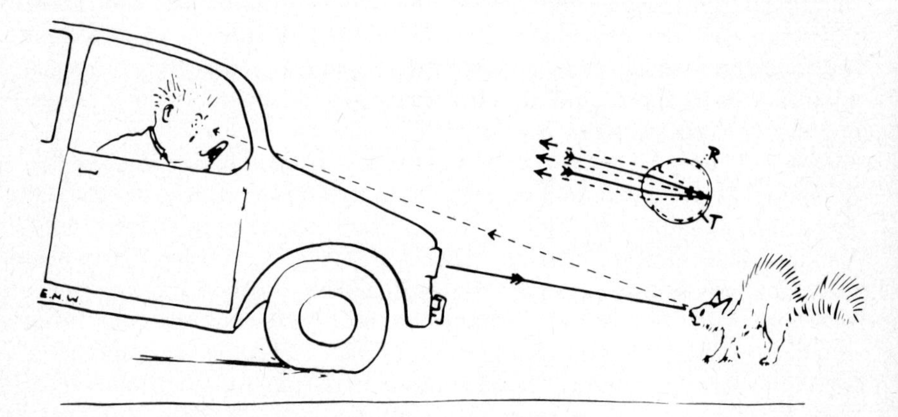

Fig. 1. The inset shows how light entering the cat's eye is reflected at the tapetum, *T*, and emerges after traversing twice the pigmented retina, *R*.

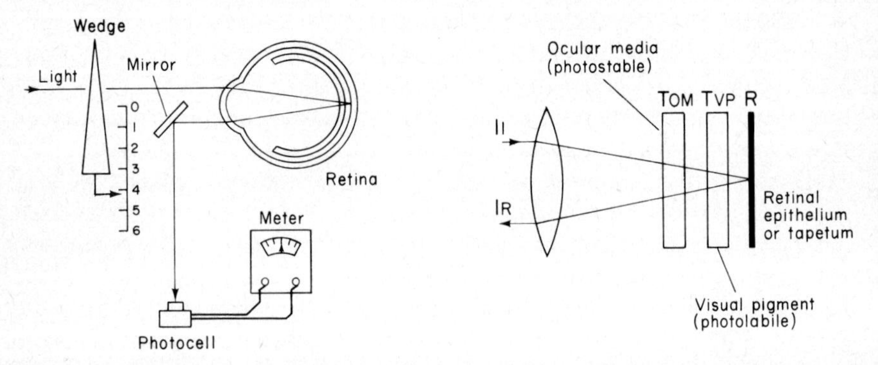

Fig. 2. A photometric wedge in the incident beam is adjusted so that the outgoing beam (measured by the photocell) remains constant. Then any change in retinal pigment density will be equal to the (negative) change in wedge density.

photometric wedge is inserted in the beam to the eye, and is continually adjusted so that the photocell readings remain constant. Then a fall in density of the visual pigment must be compensated by an equal increase in wedge density. And this can be read. The first wedding I shall describe is a relation between the pigment density in the receptors and the visual threshold.

It has long been known that after adaptation to bright light we are nearly blind on entering a dimly lit room. We soon recover somewhat and continue to improve more slowly over the next half hour. It was generally believed that these sensitivity observations were the result of sensitivity loss in the cones and rods that had been partly bleached (i.e. their visual pigment removed by light). But no one knew the quantitative relation between the subjective and the objective.

The relation turns out to be that, for each receptor, the log threshold flash is raised in proportion to the fraction of that receptor's pigment which is in the bleached state at the time.

This is shown for various receptors in the next slides. Figure 3 is taken from

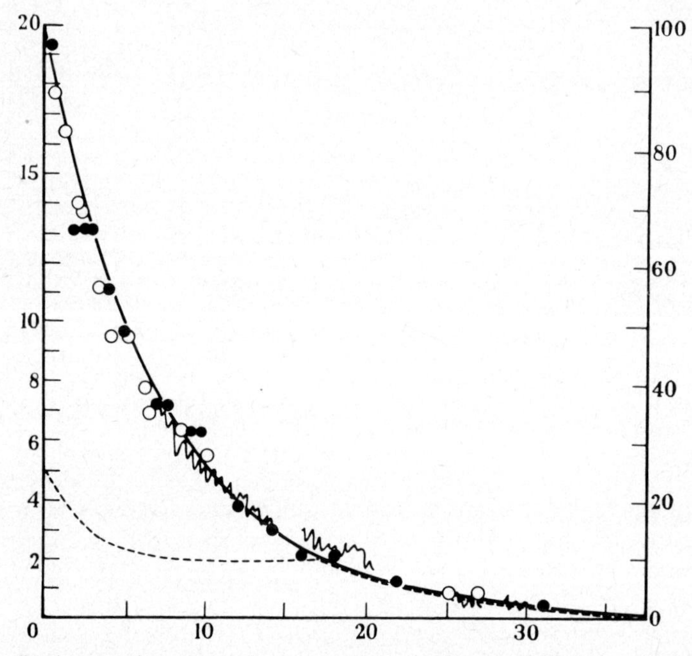

Fig. 3. Dark adaption curve from a rod monochromat. Abscissae: time (min). Irregular line: log threshold traced by subject over a range of 7 log units (scale on left). Dotted curve = cone and rod branches from a normal subject. White and black circles indicate the % rhodopsin in the bleached state (scale on right) for normal and rod monochromat respectively. (From W. A. H. Rushton 1961, *J. Physiol.* **156**, 193, Fig. 1.)

a subject who appears to have only rods in her eyes, so the contamination by cone responses is avoided. The irregular line is the log threshold for seeing the test flash plotted by the subject with an instrument that continually overshot the true threshold in alternate directions.

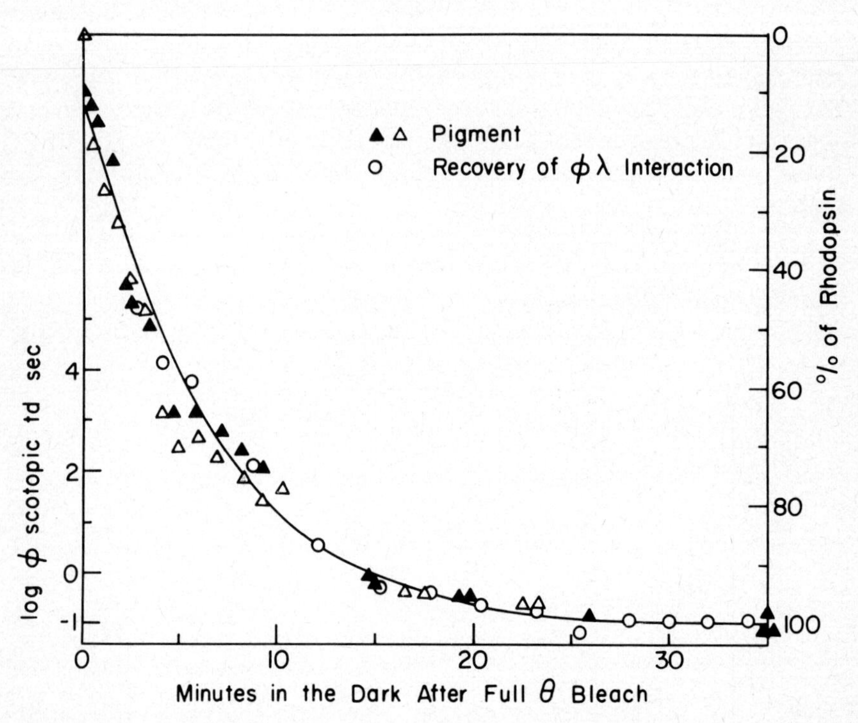

Fig. 4. Dark adaptation measured in a normal eye by Alpern's contrast flash technique. Circles: log threshold. Triangles: % of rhodopsin present (2 runs). (From Alpern, Rushton and Torii 1970, *J. Physiol.* **207**, 499, Fig. 4.)

The pigment density was measured during dark recovery in a second experiment and the pigment curve is seen to coincide with the log threshold curve when suitably scaled and adjusted to coincide at the point of full recovery.

Figure 4 shows the same relation found on the normal eye of Dr Shuko Torii. In this, the contamination by cones was avoided by using Dr Mat Alpern's contrast flash technique, and rod thresholds were followed over a million-fold range, the log thresholds coinciding with the rhodopsin re-generation, measured on Dr Torii following the same bleach by Dr Anne Fulton. Note that thresholds could not be obtained when more than 50% of rhodopsin had been bleached, though flashes much stronger than those shown were tried.

Turning now to cones the same relation was found in measurements on the fovea. Figure 5 shows this for a protanope. The threshold dark adaptation

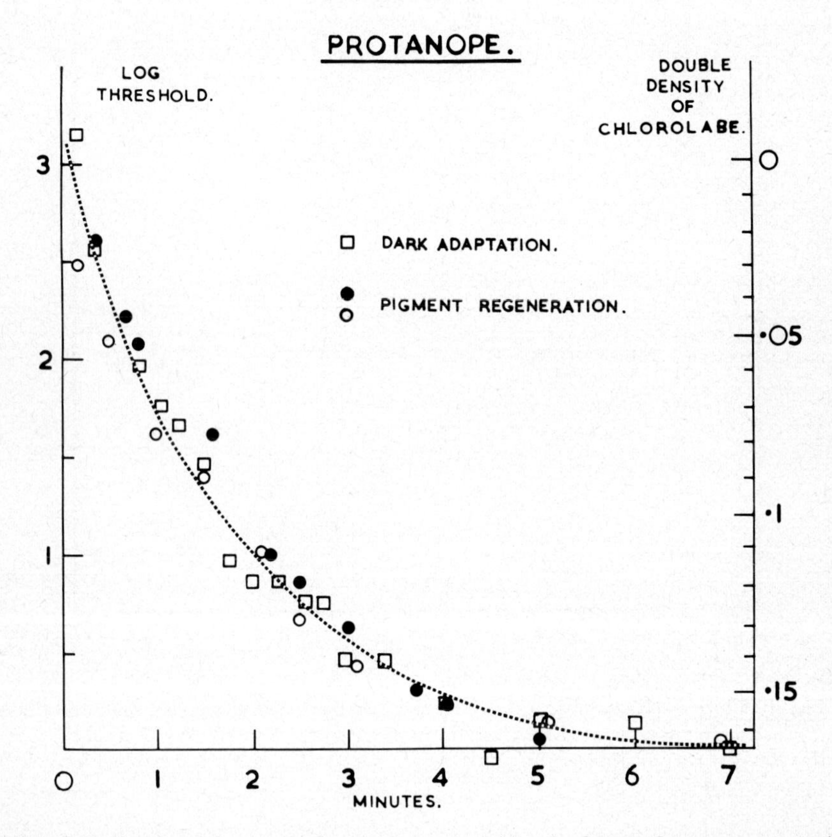

Fig. 5. Dark adaptation curve for cones measured on the fovea of a protanope (red-blind). Squares = log threshold after various minutes in the dark. Circles = fraction of cone pigment, chlorolabe, present (two runs), scale on right. (From Rushton 1963, *J. Physiol.* **168**, 374, Fig. 3.)

curve returns to normal much faster than with rods (as is well known). The cone pigment recovers correspondingly. Figure 6 shows the same with a deuteranope. Figure 7 shows the curve for Dr Howard Baker who has normal vision psychophysically, but appears to have an unusually large amount of red-sensitive cone pigment. We light-adapted him by bleaching with a strong red light (that would not affect the green cone pigment much) and we

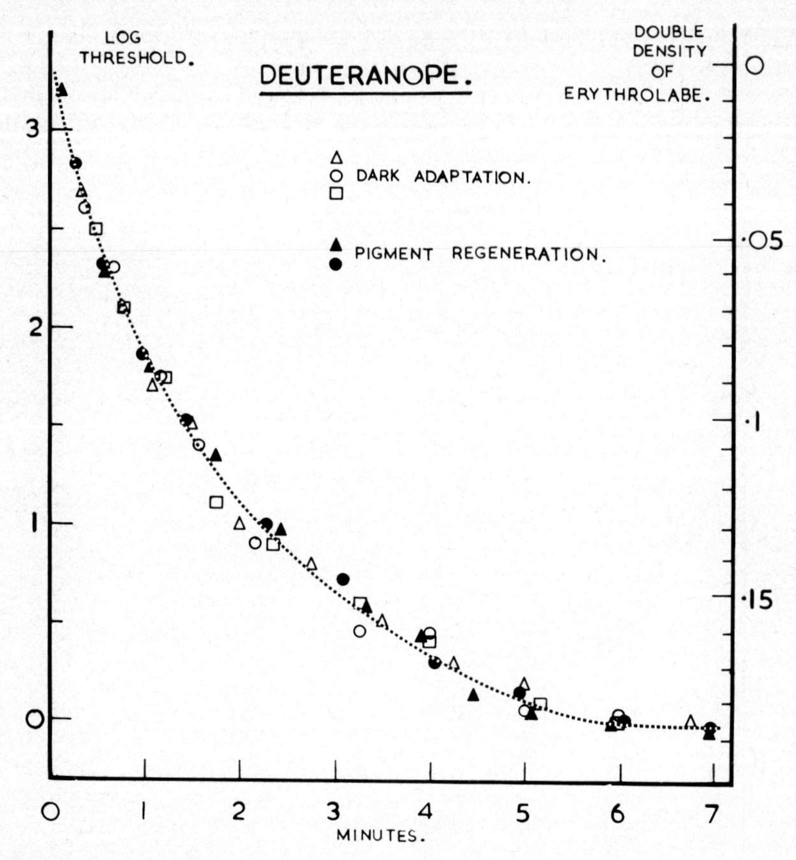

Fig. 6. Dark adaptation curve on the fovea of a deuteranope. Black symbols = fraction of cone pigment, erythrolabe present (scale on right). White symbols = log threshold at various times in the dark (scale on left). (From Rushton 1964, *Amer. J. of Optometry* **41**, 265, Fig. 9.)

measured the dark adaptation on the fovea using a red test flash. The curve drawn near the experimental points is that taken from the deuteranope (Fig. 6).

Dowling many years ago showed the same relation between pigment bleaching and rise in log threshold for the b-wave of the E.R.G. in the rat. But we still do not understand the basis of this logarithmic relation.

Colour Matching. The Red-Green Confusion of Dichromats

The extreme forms of the common red–green anomaly, i.e. those who find red and green identical in colour and can match all colours by a suitable mixture of

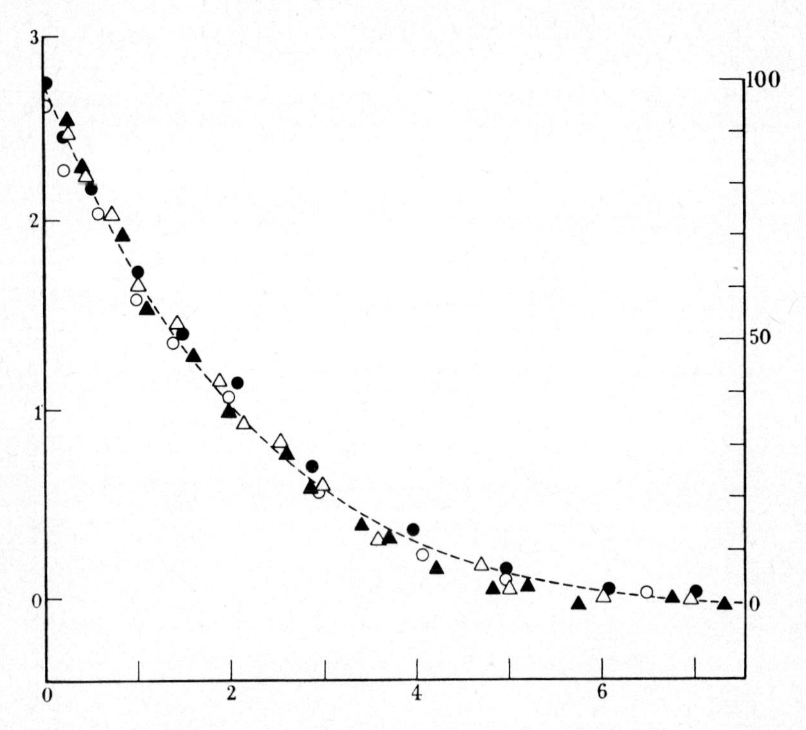

Fig. 7. Dark adaptation curve on a normal fovea. Deep red test flash to measure "red-cone" thresholds (triangles, with scale on left). Circles measure fraction of erythrolabe present (scale on right). (From Baker and Rushton 1965, *J. Physiol.* **176**, 56, Fig. 4.)

red and blue, are called dichromats. They are of two kinds: *protanopes* who see red as so dim a light that the yellow that matches it exactly for them seems to the normal a very dirty yellow indeed. *Deuteranopes* on the other hand need a bright yellow in the match with red. Protanopes have generally been held to lack the red-sensitive cones, but opinion has been divided as to whether deuteranopes lacked green-sensitive cones, or whether they contained both red and green pigments but in some way mixed the two mechanisms—either pigments in the cones or cone signals in their transmission to the brain—so that no red/green discrimination could be made.

A fancy situation may illustrate these two views. An artist has two pots of paint—one red the other green. He intends to paint a picture in the range of colours that can be obtained by mixing these paints on the palette in various proportions. If a prankster decides to reduce this picture to monochromacy, he may proceed in either of two ways. (a) He may remove the red or the green paint pot altogether so that only the other paint is available, or (b) he may mix both paints thoroughly together and leave two pots of mustard colour, so again only one colour is available.

The first is the "loss theory" of deuteranopia due to König, the second the "fusion theory" of Fick and Leber. Densitometry allows us to decide in favour of König.

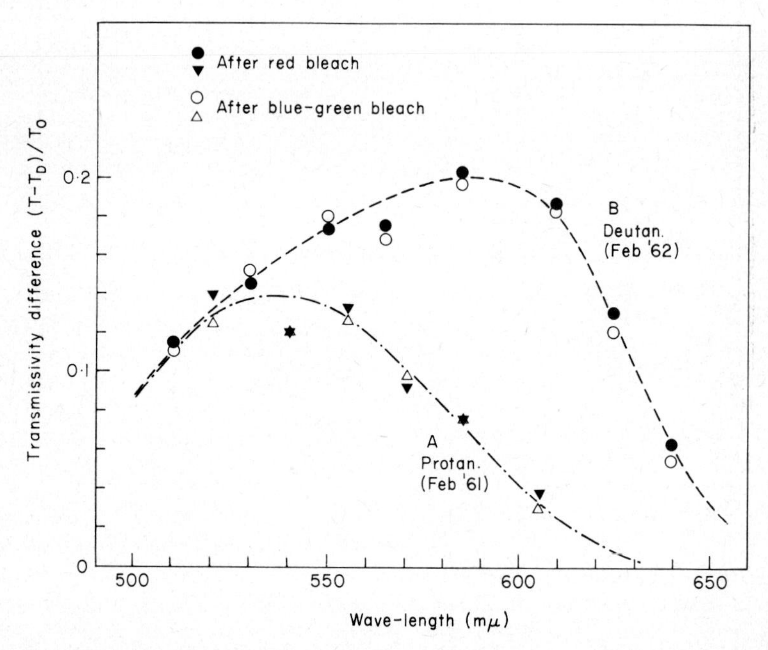

Fig. 8. Difference spectrum, i.e. change in transmissivity to lights of various wave lengths as a result of bleaching foveal pigments. Black symbols = after bleaching with red light, white symbols = after bleaching with blue–green light. Upper curve, a deuteranope, lower protanope (Newton Lecture). (From W. A. H. Rushton 1965, *Nature* **206**, No. 4989, p. 1087, Fig. 1.)

Densitometry gives us information about what pigments are present (mixed or separate) but no information about how receptor signals are processed. Thus densitometry will tell us whether the deuteranope has both the normal red and green pigments as Fick thought, or whether the green was missing as König claimed.

The circles in Fig. 8 plot the difference spectrum of the foveal pigments of a deuteranope. That is, the density change is measured in light of various wave lengths when the foveal pigments are 50 % bleached by either a red light (black circles) or blue–green light (white circles). If Fick were right and both red and green pigments were present (though in some way mixed), clearly red light would bleach chiefly the red-sensitive pigment, blue–green light more the other. But Fig. 8 shows that the difference spectrum (white and black circles) is the same whatever the bleaching light, provided that its intensity is adjusted to give a 50% bleach. This could not occur if these were Fick's red and green pigments. It must occur if there was König's single red-sensitive pigment.

If we are satisfied that deuteranopes have the König loss of green pigment, there will be no great opposition to the idea that protanopes have lost the red pigment, which has been generally accepted since Clerk Maxwell plotted the confusion loci of his father-in-law on the Maxwell triangle, and showed the loci to be straight lines running close to the red corner of the triangle.

In Fig. 8 the black and white triangles show again that the difference spectrum in the protanope is identical whether the 50% bleach is produced by red or by blue–green light.

So protanopes like deuteranopes have only a single pigment in the red–green spectral range. But it is seen that the difference curve for the protanope is by no means the same as that for the deuteranope—in particular that the protanope pigment has very little absorption in the red, which explains the poor sensitivity of these dichromats to red radiation.

Action Spectrum

I have been speaking as though the photolabile pigment we measure was the cone pigment whose quantum catch initiates vision. This fundamental assumption may be refuted or confirmed by a critical experiment.

Over the whole red–green range of the spectrum all lights appear identical in colour to the dichromat, and he can make a perfect match between any pair simply by adjusting the intensity. We have shown that this is because he has but one visual pigment active in this spectral region, so it must follow that lights that appear identical will bleach this pigment at equal rates. How do such matched lights in fact affect the photolabile pigment measured by our densitometer?

The *objective* measurement can be made accurately as follows. If a steady bleaching light is applied, some pigment will be bleached, and a steady state reached when the bleaching rate is exactly balanced by the rate of pigment regeneration. This is proportional to the level of bleached pigment. We choose the 50% level and adjust the bleaching intensity so that this level is maintained. Now the wave length of bleaching light is changed and the intensity changed also so that the equilibrium level is still 50% bleached. This is done for many wave lengths in the red–green range. Since each of these lights has the intensity to bleach in equilibrium to the 50% level, each is bleaching at the same rate, for it is neutralizing the same rate of regeneration. This energy versus wave-length relation, therefore, gives the action spectrum of the photo pigment measured.

The *subjective* half of this experiment is a *matching* situation where each of these bleaching lights in one half-field is matched in brightness by the dichromat against a fixed yellow light which, of course, looks the same colour to him. The results are shown in Figs 9 and 10.

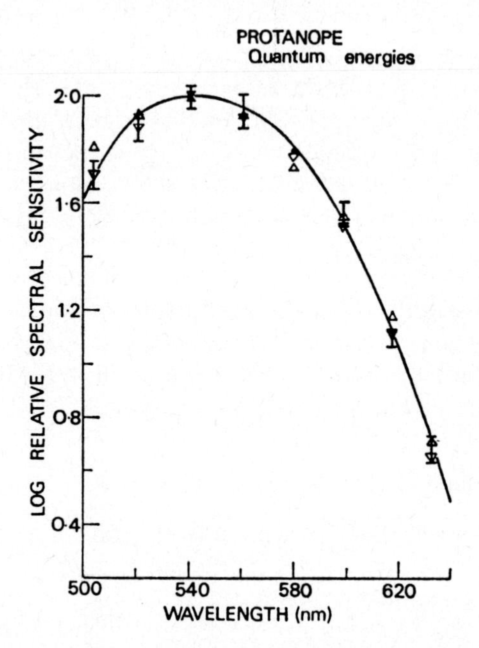

Fig. 9. Action Spectrum of chlorolabe in the protanope. Triangles measure the log light energy of wave length shown which matches a fixed yellow light exactly. Vertical lines show log energy (± 2 S.E.M.) of light of wave length shown, that bleaches chlorolabe half away at equilibrium. (From Mitchell and Rushton 1971, *Vision Res.* **11** 1033, Fig. 1.)

Figure 9 shows the results with the protanope. The little vertical lines give the objective log energy measurements required at each wave length to bleach to 50 % at equilibrium, the length of line being ± 2 S.E.M. (6 measurements). The triangles show the subjective matching log energies required for identity with a fixed yellow half-field. Figure 10 shows the same for the deuteranope.

Since lights of various wave lengths adjusted to bleach our pigment at equal rates appear to the dichromat equally bright, it is very likely that our pigment is the cone pigment upon whose quantum catch the sense of brightness depends.

At this stage you will cry "Enough! You have proved in a few cases, what we have never really doubted, that there is a causal chain between the objective light stimulus, and certain reliable aspects of the sensory response. But what are the links in that chain?"

Here, my friends, I have to come to you, cap in hand saying "Please tell me". I am sure that you are forging the strong links we need for that chain. But at

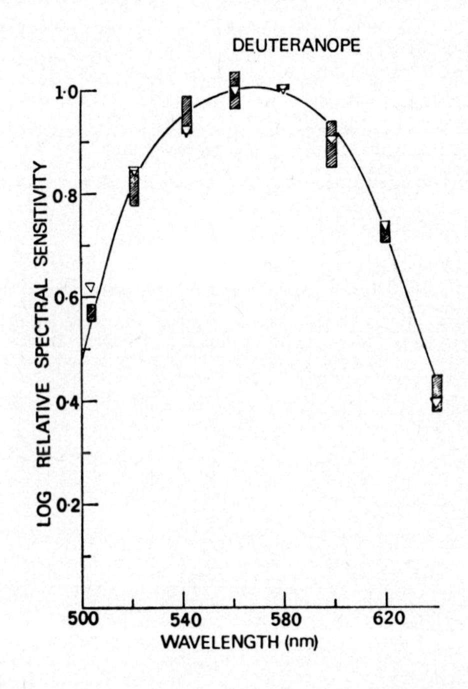

Fig. 10. Action Spectrum of erythrolabe in deuteranope. Measurements plotted as in Fig. 9, but shaded rectangles plot log action spectrum ±S.E.M. (From Mitchell and Rushton as above, Fig. 2.)

present they often do not seem to interlock very well and some of the products look remarkably like scrap iron.

Seventeen years ago in Paris I suggested that you should make the relevance of your researches to vision clearer if you wished the psychophysicists to take you seriously. Today you are so strong that you are indifferent to the opinion of the psychophysicists. But I hope you are not indifferent to the great problem of forging the causal chain between the objective and the subjective. May I ask you to make clear, when you can, how your link may fit in so that in the end we may achieve some understanding of the miracle of vision.

References

1. T. Young 1802, On the theory of light and colours, *Phil. Trans.* **92**, 12–18.
2. J. Clerk Maxwell 1860, On the theory of compound colours, and the relations of the colours of the spectrum, *Phil. Trans.* **150**, 57–84. *Reproduced in* "The Scientific Papers of James Clerk Maxwell", vol. 1, pp. 410–444 (ed. W. D. Niven 1890), Cambridge University Press.
3. W. D. Wright 1929, A re-determination of the trichromatic coefficients of the spectral colours, *Trans. Opt. Soc.* **30**, 141–164.
   W. D. Wright 1929, A re-determination of the trichromatic mixture data, *Medical Research Council, Special Report Series* **39**, 1–38.
4. W. S. Stiles 1955, The basic data of colour matching, *Physical Society Yearbook* 44–65.
   W. S. Stiles and J. M. Burch 1955, N.P.L. investigation of colour matching: interim report, *Optica Acta* **2** 168–181.
   W. S. Stiles and J. M. Burch 1959, N.P.L. colour-matching investigations: final report, *Optica Acta* **6**, 1–26.

Index

Absolute reaction rate theory, 21
Absolute threshold, 1–3, 5–7, 335–345
  light absorbed at, 336
  modifications to experimental determination of, 341
  rod signal and, 169
  quantum efficiency at, 2, 336–338
  variation of, and central effects, 344
Acetylcholine, horizontal cells and, 213
Action spectrum in dichromats, 367–369
Admittance, 79
Albino rats, 103
Ambystoma tigrinum tigrinum (Tiger salamander)
  bipolar and horizontal responses in, 206–211, 219
  gap junctions in, 282, 286
Amino acids, as neurotransmitters, 213
Annular illumination, modelling effects of, 267
Anomalous rectification in horizontal cells, 241–242
Aspartic acid, horizontal cells and, 213, 215
Atomic absorption spectroscopy, 34
ATP, outer segments and, 36–38, 136
Automatic gain control, 4
Axolotl, 181

Bathochromic shift, 25
Bathorhodopsin, 17, 24
Behavioural threshold
  in humans, 336–339
  in turtle, 306, 317
Bicycle-pedal model, 17, 18
Bipolar cells
  centre surround organization of, 209, 210
  cobalt and, 238–240
  conductance changes in, 217, 233
  current voltage curves, 216–218
  in tiger salamander, 206–211
  model for response of, 222, 224
  Off-centre, 219, 236–239, 243, 247
  On-centre, 235–237, 240–244
  photoreceptor coupling and, 187
  Ultrastructure of processes, 279–281, 288
Bleaching
  admittance changes and, 79
  neutron diffraction data and, 75

Bleaching sequence, 18–21
Blocking Molecules, 297
Buffer Experiments, *see* Chelating agents
Bufo Marinus
  rod receptor potential in, 159–165, 310
  single quantum responses in, 306

Calcium
  diffusion time for, in outer segments, 39
  efflux from vesicles, 45–53
  exchange system for, 39
  incorporation in cattle rod outer segments, 35–38
  intracellular injection of, 160, 163
  outer segment binding, 30–33, 131–133
  presynaptic membrane and, 318
  radioisotope methods, 35
  rat rod responses and, 116
  release by light from vesicles, 49–51
  release by single photon, 133
Calcium Hypothesis
  see also Well-Stirred Outer Segment model
  Signal amplification and, 53, 159
  tests of, 99–103, 161–165
6-Carboxyfluorescein (6CF), 107–114
  distribution by vesicle transfer, 111
  pK of, 114
  self-quenching of, 107
  vesicle fusion and, 107–109
Cat, absolute threshold and, 337
CDTA (1,2-Diaminocyclohexane tetra-acetic acid), Vesicle fusion and, 120
Chelating agents
  critique of experiments with, 133–135
  effects on rod responses, 116–121, 132
Chelydra Serpentina
  double cones in, 181
  noise in rods of, 323–326
  rods in, 170, 181, 182, 194
Chlorolabe, action spectrum of, 368
Chromophore, 16–23
  cis-trans isomerization, 16, 17–19
  nature of binding to opsin, 23
  reactive and non-reactive, 23
Ciliary pathway, 183
Cobalt, and horizontal cells, 213
Colour matching, 359, 364–367
  and dichromats, 364–367
Colour vision, 357, 359, 361–369
Concanavalin A, 47
Conductance
  measurements on outer segments, 78–93
  photosensitivity of disc membrane, 79–85
  saturation of disc membrane changes, 89–91
  single channel in cones, 300
  single channel in rod discs, 87
Cones
  annular illumination and, 267
  calcium and, 99
  coupling between, 170, 293–295
  electrical noise and, 291–300
  elementary conductance change in, 300
  green-sensitive, 293, 295
  input to rods, 173
  membrane nonlinearity of, 259
  noise models for, 297–299
  oil droplets and sensitivity in, 314
  in rat retina, 153
  red-sensitive, 195–197, 201, 291–293
    threshold signal in, 316
  spatial sensitivity of, 293–295
  transretinal current and, 258
Cone, ultrastructure of, 276–279, 283–289
Contrast sensitivity, bipolar cells and, 226
Coupling
  Anatomical identity of, 180
  between cones, 177
  between rods, 174–182, 306, 326
  cobalt and rod, 178
  consequences of rod, 187–189
  models for cone, 293–295
  noise reduction and, 187
  single photon response and, 171, 172, 326
  space constant for, 186, 293–295
  as spatial filter mechanism, 191
Current divergence, of photocurrent, 150, 155
Cyprinus Carpio, 232

Dark, adaptation
  absolute threshold and, 3
  human retina and, 361–364
  rod responses and, 117

Dark current, *see also* Photocurrent
  external calcium and, 116
  in rat rods, 102, 141
  sodium ions and, 186
Dark light, determination of, 341, *see also* intrinsic noise
Dark noise, *see* Noise
Dark potential, *see* resting potential
Densitometry
  method of retinal, 360
  X-ray diffraction pattern, 65
Difference spectrum, of rods in *Chelydra*, 172
Disc membrane
  bilayer structure of, 62–69
  bleaching and electron density profile of, 73
  calcium binding to, 35
  electrical measurements on, 77–93
  protein assymetry in, 69–75
  rhodopsin and, 42, 69–75, 77, 89–93
  x-ray diffraction from, 61–75
Dose-response relationships, and horizontal cells, 216

EDTA, (Ethylene diamine tetra acetic acid), and outer segment preparations, 78
Efficiency, *see also* Quantum efficiency
  estimates by random dot patterns, 345–347
  perceptual, 347
  psychophysical and photometric, 336–339
EGTA, (Ethyleneglycol Tetra acetic acid)
  intracellular injection of, 163–165
  outer segment preparation and, 30, 35
  vesicle fusion and, 119
Electron density profiles
  Disc membrane, 62–68
  discrepancies in, 65, 67, 68
Electron microscopy
  of photoreceptor synapses, 276, 278, 282–289
  of rhodopsin-phospholipid vesicles, 48
Electroreceptors
  calcium regulation and, 322
  threshold signal in, 318
Energy analysis in visual pigment photolysis, 16–22
Energy surface
  bicycle-pedal model, 17, 18
  calculated, 18
Enthalpy, 25, 26
Entropy, photolysis and, 25, 26
EPR spectra
  rhodopsin bleaching and, 54–58
  spin labelling and, 54–56
Equilibrium constants, 19, 20
Equilibrium dialysis, calcium binding studied by, 30–33
Equilibrium potential, *see* reversal potential
Erythrolabe, action spectrum of, 369
ESR Spectroscopy, *see* EPR spectra
Exchange carrier systems, in rod outer segments, 39

Feedback
  bipolar cells and, 217
  horizontal-cone, 262, 265–273
  modelling of, 266–7
  morphological evidence for, 281
  synaptic gap resistance and, 266–273
Filter
  rod coupling as, 191
  synaptic pathway as, 202, 333, 334
Flash sensitivity, *see* increment sensitivity
Fluorescence, *see also* 6-carboxyfluorescein
  6CF and change with pH, 114
Fluorescence methods, rat rods and, 107–109
Free energy, 18–22, 25, 26
  summary of photolytic changes, 25
Freeze fracture microscopy, 42, 70
Frequency-of-seeing curve, 326, 327
  false positives and, 340
  turtle cones and, 198

Ganglion cell
  antidromic invasion of, 194
  on-, Off-centre, 195, 196–198
  quantum efficiency determined in cat, 338
  recording in turtle, 194
  signal pathway and, 194–202
Gap junctions, Photoreceptor coupling and, 181, 307

Gecko gecko, 310, 313
Gel electrophoresis, 43–45
Glia, *see* Müller cell
Glow modulator tube, 233
Glutamic acid, Horizontal cells and, 213–215

Horizontal cell
anomalous rectification in, 212, 241, 242
cobalt and, 178, 237–239
coupling between, 229
electrical properties of, 212–216
external and intermediate, 233–235
luminosity-type in turtle, 314
Response inversion of, 241, 253–5
transretinal current and, 234–236
ultrastructure of processes, 276, 278–289
Hydrogen ions
metarhodopsin I and, 79
in rods, 114–116
as transmitter particles, 138

Ideal detector, 336
intrinsic noise and, 340
Increment sensitivity, 311–313
backgrounds and, 310, 321
see also Quantal sensitivity
Increment–threshold curve
for turtle, intracellular determined, 314
for turtle, behaviourally determined, 315
Input resistance
change in bipolar cells, 244
of cone, 300
of horizontal cells, 212–214, 229
of turtle rods, 185
Intensity–response curves
for second-order neurones, 207–210, 221–223
for toad rods, 311
Internal transmitter, *see* Intracellular transmitter
Intracellular transfer, Maximum vesicle fusion in, 109
Intracellular transmitter
see Calcium, Calcium Hypothesis
see Well-Stirred-Outer-segment model
Intrinsic Noise, *see* Noise
Ionic channels, noise and, 298–299
Ionophore
A23187 and rod outer segments, 32, 36–38
A23187 and membrane vesicles, 49
X537A and membrane vesicles, 49–51
evidence for intracellular transmitter and, 99, 132, 161
Isorhodopsin, 18, 23

Junctions
basal, 275, 277–281
close apposition, 309
distal, 279–281
gap, 181, 307
narrow gap type, 279
outer plexiform layer, 275–281
ribbon, 277–279

Lactoperoxidase iodination, 45
Light adaptation
horizontal cells and, 254
photocurrent and, 152
in photoreceptors, 310–313
Limulus, photoreception in, 353–355
Lineweaver–Burke plot, 312
Lipid bilayer, X-ray diffraction and, 62–64, 71–74
Liquid scintillation counter, 35
Lorentzian function, 297, 330
Lumipigments, spectral properties of, 24, 25

Magnesium
competition with calcium, 129–131
rod responses and, 121
Membrane potential, *see* resting potential; *also under* individual cell types
Metal ion buffers, *see* Chelating agents
Metarhodopsin I
hydrogen ion uptake and, 79
tautomeric equilibrium and, 26
Metarhodopsin II, 19, 21, 26
Metarhodopsin III, 92
Michaelis–Menten equation, 312
Micropipette, properties of recording, 334
Müller cell, Photocurrent and, 141, 148–152
Myelin, 72

Necturus, (Mudpuppy), 185
Neutron diffraction, 75
Noise, *see also* Power spectrum, Quantum fluctuations
  in cones, 291–303
  dark, 198, 326–330, 339–341
  extrinsic current and rod, 327–330
  intrinsic, and frequency of seeing curves, 339, 340
  intrinsic, in human vision, 339–344
  intrinsic, and quantum efficiency, 341–343
  light sensitive, 291, 323, 330, 332
  nonlinear corrections and cone, 299
  receptive field in cones and, 293–295
  reduction by photoreceptor coupling, 187, 293–295
  in rods, 326–332
  sources of, 297–299, 333, 343
  synaptic mechanisms and, 188, 318
  television picture quality and, 8–10
  visual sensitivity and, 301–302
Nonlinearity
  of cone membrane, 259
  cone noise and, 299
  of presynaptic membrane, 266

Oil droplets, cone sensitivity and, 314
Opsin
  bleaching sequence and, 16
  linkage of chromophore to, 24, 25
Organomercury compounds, Spin label reagents and, 54
Outer segments, *see* Rod outer segments

Papain, 43
Perfusion technique, for Bufo rods, 160
pH
  estimation of cytoplasmic, 114–116
  invertebrate neurones and, 167
Photochemistry
  in visual transduction, 15–27
  "Era of", 16
  potential energy surfaces in, 16–22
Photocurrent
  estimate for single photon, 186
  localisation of, 145
  model for, 156
  rod network and, 183
  sources and sinks for, 144, 146–156
Photoisomerization, *see also* Chromophore
  noise in cones and, 300
  response produced by single, 324
Photolysis
  equilibrium constants in, 19, 20
  thermodynamics of, 19–22, 25, 26
Photomultiplier
  quantum bumps in *Limulus* and, 355
  signal-to-noise considerations, 2
Photon, *see also* Quantal sensitivity
  rod response to single, 306–309, 323–326
Photon noise *see* Quantum fluctuations, Noise
Photoreceptors
  coupling between, 187–189
  events leading to signalling in, 97–99
  invertebrate, 353–355
  synaptic organization of, 275–289
  threshold signal in, 305–310
PIPES, and rod responses, 118, 128
  pK, 6-carboxyfluorescein and, 115
Poisson distribution *see also* Frequency-of-seeing curve
  photon absorption and, 325
  quantum bumps in *Limulus* and, 353
  toads rods and, 306
Porphyropsin
  in turtle rods, 172, 324
  response to bleach of single, 324
Potential energy surface, *see* Energy surface
Power spectrum
  for cone noise, 297, 298
  of noise in turtle rods, 328–332
Pre-lumirhodopsin, 17
Procion Yellow M4R, 182
Pseudemys Scripta Elegans, 170, 177, 194, 276, 291
  red-sensitivity cone dominated, 314
Psychophysics
  colour vision and, 360–369
  determination of quantum efficiency by, 338
  estimates of retinal noise and, 348
Purkinje Shift, 302

Quantal sensitivity
  illumination area and, 172, 173, 185
  in snapping turtle, 172, 185, 324, 325
  techniques in recording, 170
Quantum bumps, in *Limulus* photoreceptors, 353–355
Quantum catch, 321
Quantum efficiency
  definitions of, 351, 352, 353
  determined for green light, 7
  errors in estimates of, 338
  in high light range, 10
  Human vs. cat, 336–339, 343
  information theory and, 351, 352
  light absorption and, 337
  photography and, 10, 11
  psychophysical estimate of, 343
  variation with background, 4
Quantum fluctuations, *see also* Noise
  behavioural threshold and, 309
  as limitations to visual performance, 336
  performance of eye and, 6–10
  toad rods and, 306–309
Quantum noise, *see* Quantum fluctuations, Noise

Random dot patterns, 345–347
Rapid binding, and noise models, 299
Receptive field
  in bipolar cells, 206
  cone noise and, 293–295
Receptor potential
  in *Bufo* rods, 160–165
  sodium conductance and, 160
Rectification
  in bipolar cell, 240
  in horizontal cells, 241–242
Resistivity, retinal, 146
Resting potential, Dark adaptation and horizontal cell, 256
Retina
  recording techniques of isolated, 142–143, 233
  resistivity of, 146
Retinal, *see also* Chromophore
  geometric isomers of, 23
Reversal potential
  in horizontal cells, 214, 253
  hyperpolarizing bipolars and, 261
Rhodopsin, *see also* Bleaching, chromophore
  as calcium ionophore, 45–53
  cattle, 18
  conformational changes with light, 53–58
  distribution in disc membrane, 68, 69
  equilibrium constant for dissociation, 19
  evidence for clustering, 89–93
  membrane vesicles and, 43–58
  molecular weight, 43, 70, 91
  relation of bleaching to absolute threshold, 3
  response to activation of single molecule of, 87
  thermal reactions of, 92
  as transmembrane protein, 43–45, 70
  water excluding volume of, 80
  X-ray diffraction and, 70–74
Rod disc membrane, *see* Disc membrane
Rod outer segments, (ROS)
  calcium content of cattle, 33–38
  conductance of suspensions of, 78–81
  freeze fracture microscopy in, 70
  metabolism in, 30–40
  passive electrical properties of, 79
  phosphodiesterase activity and, 42
  preparation in TRIS and sucrose, 34
  amplitude–intensity curves, 124–133
  background light and, 310–312
  coupling between, 171, 174–180, 306
  electrical model for, 182
  external calcium and, 117
  extracellular responses of, 103–105, 146–153
  human, 181, 351
  input from cones, 173
  intracellular buffers and, 118–121
  receptive field, 184
  single photon response, 306—309, 323–326
Rose-de-Vries Hypothesis, 335

Scatchard plots, 32
Scene brightness, 7
Schiff's base linkage, 23
SDS-polyacrilamide gels, 43, 44, 45
Sensitivity, *see also* Increment sensitivity
  rod-bipolar transmission, 209

Signal amplification
  calcium hypothesis and, 53
  mechanisms in rods and cones, 99
Signal pathways
  cone to ganglion cell, 194–196, 198, 201
  retinal, 193–202
  rod to ganglion cell, 197, 201
  statistical properties, 198
Signal-to-noise ratio
  diffuse stimuli and, 295
  human vision and, 2, 3
  in rods, 188
  random dot patterns and, 345
Sodium conductance, Light and, 160
Source-sink analysis, *see* Photocurrent
Specific rate constant, in photolytic processes, 21
Spectrophotometry, on single rods and cones, 12, 337
Squid, giant synapse in, 319
Storage Time, 5
Strength–duration curves, 200
Sulfhydryl groups, rhodopsin and, 54–58
Synapse, *see also* Junctions
  bipolar cell response and, 206–211, 217–220
  chemical, 180, 258
  cone pathway and, 199
  feedback to receptors and, 258
  as filter, 202, 332
  horizontal-bipolar, 260
  horizontal-receptor, 251–263
  ribbon, 275, 277
  rod–rod coupling and, 180
  squid giant, transfer function for, 319
  transmission from photoreceptors and, 220

Teleodendrion, 182, 186
Thermodynamics, in bleaching sequence, 20–22
Threshold signal
  central efficiency and, 344–345
  in photoreceptors, 305–310, 317
Tiger salamander, *see Ambystoma Tigrinum Tigrinum*
Transmitter, *see also* Intracellular transmitter
  release from photoreceptors, 220–224
Transretinal current, 231–237, 244–246
Tris, preparation of cattle rods in, 30, 33
Triton flash, 109
Troler effect, 345
Turtle, *see under Pseudemys Scripta Elegans*, *Chelydra Serpentina*

Utilization time, 199

Vesicles
  calcium efflux from membrane, 47–53
  intracellular transfer of 6CF by, 107–114
  in photoreceptor endings, 275
  reconstituted membrane, 43–53
Visible stars, 1
Vision
  colour, 361–369
  electronic, 5, 7–11
  human, Electronic compared to, 1–13
    noise limitations in, 336–355
  quantum limitations to, 1, 2, 339
Visual pigment, *see* Rhodopsin, Porphyropsin
Visual threshold, *see* Threshold
Vitamin A_2, 172
Voltage current curves
  for bipolar cells, 241
  for cone membrane, 267–273
  for horizontal cells, 241, 257
Voltage noise, *see* Noise
Voltage variance, *see* Noise

Weber constant, 313
  and peripheral retina, 321
Weber–Fechner relation, 312
Well-Stirred Outer Segment (WSOS) model, 100–102
  approximations in 101
  critique of, 133–135
  quantitative predictions, 122

X-ray diffraction, 61–75
  see also Disc membrane
  low angle data discrepancies, 62, 65–68
  models for disc membrane and, 62, 69–75

Young–Helmholtz theory, 357